Transmission, Processing, and All-Optical Routing for Ultra-High Capacity Data Center Networking (Second Edition)

Online at: https://doi.org/10.1088/978-0-7503-5849-1

IOP Series in Advances in Optics, Photonics and Optoelectronics

SERIES EDITOR

Professor Rajpal S Sirohi Consultant Scientist

About the Editor

Rajpal S Sirohi is currently working as a faculty member in the Department of Physics, Alabama A&M University, Huntsville, AL, USA. Prior to this, he was a consultant scientist at the Indian Institute of Science, Bangalore, and before that he was Chair Professor in the Department of Physics, Tezpur University, Assam. During 2000–11 he was an academic administrator, being vice-chancellor to a couple of universities and the director of the Indian Institute of Technology, Delhi. He is the recipient of many international and national awards and the author of more than 400 papers. Dr Sirohi is involved with research concerning optical metrology, optical instrumentation, holography, and speckle phenomena.

About the series

Optics, photonics, and optoelectronics are enabling technologies in many branches of science, engineering, medicine, and agriculture. These technologies have reshaped our outlook and our ways of interacting with each other, and have brought people closer together. They help us to understand many phenomena better and provide deeper insight into the functioning of nature. Further, these technologies themselves are evolving at a rapid rate. Their applications encompass very large spatial scales, from nanometers to the astronomical scale, and a very large temporal range, from picoseconds to billions of years. This series on advances in optics, photonics, and optoelectronics aims to cover topics that are of interest to both academia and industry. Some of the topics to be covered by the books in this series include biophotonics and medical imaging, devices, electromagnetics, fiber optics, information storage, instrumentation, light sources, CCD and CMOS imagers, metamaterials, optical metrology, optical networks, photovoltaics, free-form optics and its evaluation, singular optics, cryptography, and sensors.

About IOP ebooks

The authors are encouraged to take advantage of the features made possible by electronic publication to enhance the reader experience through the use of color, animation, and video and by incorporating supplementary files in their work.

Do you have an idea for a book you'd like to explore?

For further information and details of submitting book proposals, see iopscience.org/books or contact Ashley Gasque at Ashley.gasque@iop.org.

A full list of titles published in this series can be found here: https://iopscience.iop.org/bookListInfo/series-on-advances-in-optics-photonics-and-optoelectronics.

Transmission, Processing, and All-Optical Routing for Ultra-High Capacity Data Center Networking (Second Edition)

Le Nguyen Binh
European Research Centre, Huawei Technologies GmbH, Duesseldorf, Germany

IOP Publishing, Bristol, UK

ISBN 978-0-7503-5849-1 (ebook)
ISBN 978-0-7503-5847-7 (print)
ISBN 978-0-7503-5850-7 (myPrint)
ISBN 978-0-7503-5848-4 (mobi)

DOI 10.1088/978-0-7503-5849-1

Version: 20231001

IOP ebooks

British Library Cataloguing-in-Publication Data: A catalogue record for this book is available from the British Library.

Published by IOP Publishing, wholly owned by The Institute of Physics, London

IOP Publishing, No.2 The Distillery, Glassfields, Avon Street, Bristol, BS2 0GR, UK

US Office: IOP Publishing, Inc., 190 North Independence Mall West, Suite 601, Philadelphia, PA 19106, USA

To my parents

To Thi Phuong Nguyen and our son Le Nguyen Lam

Contents

Preface

Early in the 2010s long haul terrestrial and intercontinental telecommunications systems and networks were reaching towards longer transparent distance and increased information capacity with economical implementation of the communications systems. These ultra-high-capacity long haul transmission systems have been installed throughout the global networks. The progress in transmission symbol rates and spectral efficiency in bits/s/Hz is shown in figure P.1. A spectral efficiency reaches 10 bits/s/Hz in 2020. That means that for 100 GHz spacing grid between the carriers one can accommodate 1.0 Tera-bits which is a few million times over the rate of transmission in the early 1980s.

Concurrently the explosion of the Internet is driving the tremendous progress in deployment of data centres (DCs) throughout the world's information networks. Thence the long haul and short haul interconnection between these DCs, so DC networking is possible without going through the traditional telecom network. This has put tremendous pressure on global telecom carriers as they do not own the content of the traffic such as the traffic of YouTube, Facebook, Google etc. However, the infrastructure of these telecom carriers has allowed them to deploy ultra-high capacity mobile networks, thus the access plane of the networks such as 5G/6G/7G (5th, 6th or 7th generation) mobile networks and distributed DC networking. This direction has been strategically developed to challenge the other DC networking communities.

All traditional multi-layer telecommunication networks of global telecom are reconfigured to DC-based networking. The upgrading is to be carried out step by step to ensure the stability of networks and client services. Figure P.1(c) shows a typical rack of servers by which massive data and ultra-high speed transfer of communications travel back and forth. The annual growth of data demand is exponential, as depicted in figure P.1(b). The spectral efficiency, normally specified in bits/s per Hz, is shown in figure P.1(a). Thus modulation and transmission techniques become highly important.

The spectral efficiency of optical transmission over optical fibers in the spectral window C-band and super C-band, have been increased gradually, as shown in figure P.1(a). The attempt to increase both the baud rate and higher order of modulation format has been reported by several research labs and institutes around the world, as illustrated in figure P.1(b). Due to the transmission distance of only a few hundred meters to a few kilometers of fiber, higher baud rate can be transmitted without much distortion or over the distortion limit that can be compensated in the digital signal processor associated with the optical receivers. The annual growth in demand for data transmission and routing is also included, showing further increase in additional fibers or another spectral band, for example the L-band. Therefore, optical routing is even more important in DCs (DC networking—DCN) and their interconnections (DCI).

Therefore, the transmission of ultra-high-speed channels and ultra-high capacity link over short distance has become critical for DCN by both telecom carriers and

(a)

(b)

(c)

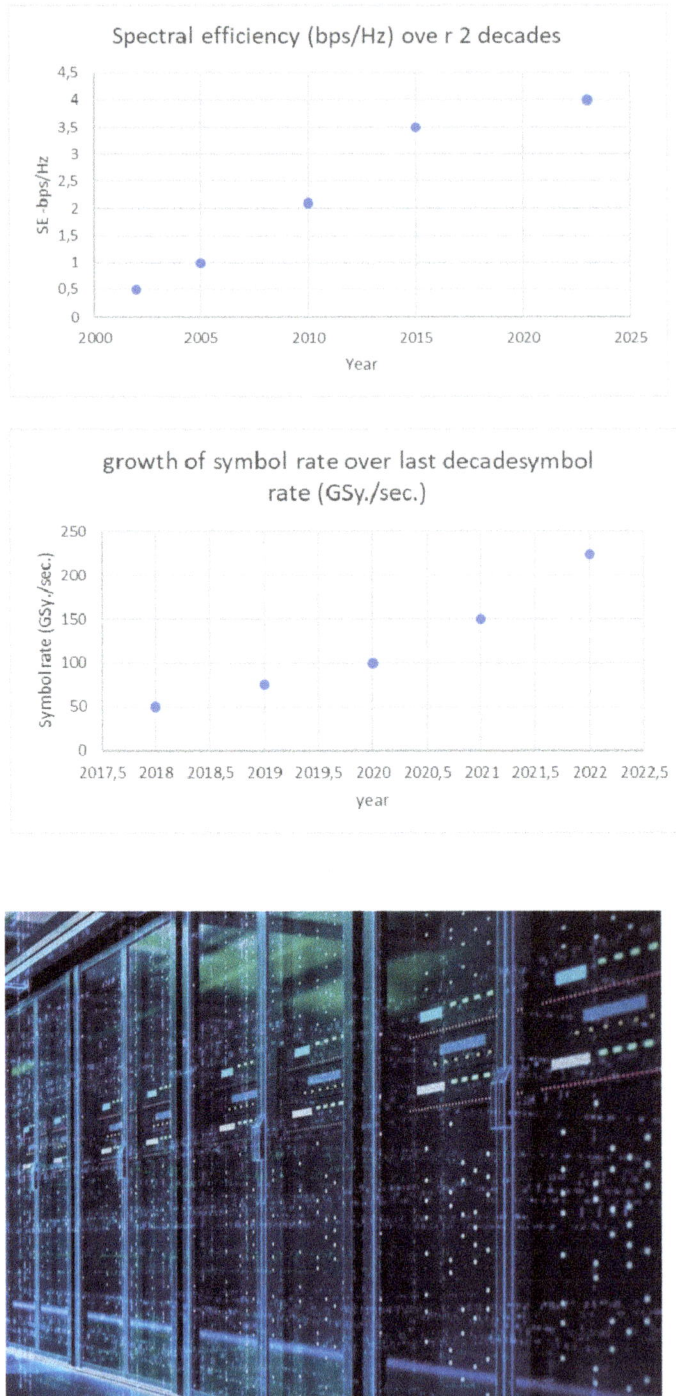

Figure P.1. (a) Growth of spectral efficiency (bps/Hz) in optical fiber transmission; (b) growth of symbol rate (GSy./sec.) over the last decade; (c) racks of a section of a typical DC (data centre) through which optical data communications travel at ultra-high speed to serve data transferring back and forth.

data communications for DCN owners. Further, the spectral regions of silica single mode optical fibers have been extensively used in the O-band (1300 nm window) and C+ (110 channels of 50 GHz spacing), pushing into L-band, even to the 850 nm region with low cost VCSEL (vertical-cavity surface-emitting laser). The spectral extension is to ensure sufficient bands for capacity transport. The global transport capacity would reach Zetta-bits/s. Thus, several possible solutions are offered by research and development communities for such capacity demands.

The possibilities for such ultra-high speed transmission come from the digital processing of coherently detected signals which was not possible in the first-generation coherent systems developed in the 1980s. However, with the baud rate of the signals reaching more than 100 GBaud, digital signal processing (DSP) faced severe problems in the electronic speed of integrated micro-electronics, even at the gate range of less than 7 nm. Thence optical/photonic signal processing (PSP) can assist to overcome these difficulties. The main chapters of this book address the technological principles of these two principal technologies, DSP-based processing of optical systems and PSP systems.

This second edition of the book is to address the following two principal techniques for modern communication transmission of ultra-wideband channels:

1. DSP-based techniques in optical transmission systems with both non-coherent (direct detection) and coherent principles and associate modulation formats.
2. Principles of processing in photonic domain with special emphasis on the association of the neural computing networking and nonlinear optical logic circuits.
3. All optical switching using wavelength selective switches and reconfigurable optical add/drop multiplexer (ROADM) and optical cross connect (OXC) for massive capacity interconnection, especially in DCs.
4. Remarks on bandwidth limited and power-limited scenarios, or the ultra-high capacity and quantum level transmission for absolute security in networking.

The book chapters are to provide an understanding of such technological development for practical engineers, R&D engineers, academic scholars etc, especially DC engineers.

Acknowledgments

I am grateful to Huawei Technologies for the availability of laboratories and facilities to carry out the works in my sojourn years with the company. The chapters of this edition have been initiated while I was a full-time staff member as Technical Director of the European Research Centre in Munich.

I am grateful to Dr Thomas Lee of SHF AG of Berlin Germany and formerly of Nortel Networks for advice and the loan of various optical transmitters and receivers and purchase of 64 Gb s^{-1} BERT for several experiments on Tera-bits/sec, and transmission employing various incoherent and coherent Mary-QAM. I am also indebted to discussions with my colleagues at Keysight Inc. of Deutschland and US Colorado R&D Center, especially Drs Thomas Kirchner and Mike Beyers.

I would like also to sincerely thank my former colleagues at Huawei Technologies, especially Dr Xu Xiaogeng, Bruce Liu, Chen Ming, Dr Mao Bangning, Dr Xie Changsong, and Professor Dr Nebojsa Stojanovic, Dr Jinlong Wei, Dr Song Xiaolu, Dr Sun Xu, Dr Suo Jing, Dr Li Rui and Dr Spiros Mikroulis for fruitful exchanges of ideas and technological information. During my sojourn as a fellow of Mynaric Lasercom GmbH, I acknowledge fruitful discussions with Joachim Horwath, Carsten Fetchman and Axel Hof.

I would like to thank Ashley Gasque, the Senior Commissioning Editor in Optical Sciences of IOP publishing house for her encouragement to the publication of this second edition.

Last and not least, I thank my wife Phuong Nguyen and my son Lam Le Nguyen who have supported me over the years with happiness and putting up with my hours in the preparations of the chapters. My parents, the Le-Nguyen family, have provided me over the years with the education of *learning for life* and their sacrifice in educating their children, and I am indebted to them for their lifetime encouragement. This book is dearly dedicated to my parents.

Le Nguyen Binh, BE (Summa Cum Laude), PhD, Dr Eng., C. Eng.
Principal Expert Consultant to Huawei Technologies, European Research Center, Munich 80992, Deutschland
Fellow, Mynaric Lasercom GmbH, Gilching 82205, Muenchen, Bayern, Deutschland.
Spring 2023.

Author biography

Le Nguyen Binh

Dr. Eng. Prof. Le Nguyen Binh received the degrees of BE (Electronic Eng. Summa cum Laude) and PhD, Dr. Eng. in Electronic Engineering and Integrated Photonics/Optical Communications Eng. from the University of Western Australia, Nedlands, Western Australia. He is a Chartered Engineer. He has published more than 350 journal and conference papers, owns several patents and 15 books in the fields of optical communications, photonic processing, integrated optics and photonics. He was the technical Director of Huawei European Research Center and is currently Fellow of Mynaric Lasercom GmbH of Bayern Deutschland. He is currently a fellow of Mynaric Lasercom GmbH of Gilching, Bayern, Deutschland.

IOP Publishing

Transmission, Processing, and All-Optical Routing for Ultra-High Capacity Data Center Networking (Second Edition)

Le Nguyen Binh

Chapter 1

Historical overview and digital transmission technologies in data centre networking

Undoubtedly optical guided media, the circular optical waveguide and the optical fiber have allowed tremendous progress in transporting huge amounts of information for human societies over the last half century. This introduction chapter briefly outlines the historical development and emergence and merging of the fundamental digital communication technique and optical communications to fully exploit and respond to the challenges of the availability of the ultra-high frequency and ultra-wideband in the optical spectra of optical fiber communication technology. Currently, with the emergence of cloud networking—where the contents are owned by the entertainment and social media delivery corporations such as YouTube, Netflix, Facebook—coupled with communication platforms, networking has evolved into ultra-high capacity both in the core and distributed data centre clouds (DCCs), both in terms of intra- and inter-networking. This has forced the evolution of the traditional telecom carrier networks to adapt DCC networking, taking advantage of their infrastructure. Wireless access networks have evolved into 5G and this has forced the distribution of the cloud into distributed DDCs into local nodes of 5G networking. The extensive infrastructures of telecom network owners give them tremendous leverage over the DC networking (DCN) owners. However, the delivery and processing of signals and channels over these networks are critical for error-free transmission to be achieved.

The processing of these ultra-wideband channels is critical and is reaching the limit in the electronic domain, but offers advantages to the photonic domain. This chapter thus gives introductory remarks and fundamental understanding for the remining chapters of the book. A historical overview and the emergence of the challenges of processing in optical communications is given to summarize the importance of the field.

doi:10.1088/978-0-7503-5849-1ch1

This chapter introduces a historical overview of the optical transport via the optical guided media, from the innovative ideas of 1966 to the beginning of electronic digital signal processing (DSP)-based optical transmission and then to the DSP-based system. We then cover traffic growth via DCN and demand for extreme data rate transmission and interconnection reaching an approximation of Tera-bits/s (Tbps) per wavelength carrier. The basic modulation technique for such signal generation, in both amplitude and phase modulation, as well as frequency division modulation with orthogonality are given. Then, electronic DSP-based optical systems and photonic signal processing-based systems are introduced to provide basic information of the rest of the chapters of this book.

List of abbreviations

ADC	Analog-to-digital converter
ASK	Amplitude shift keying
C-band	Spectral region 1550 nm (40 nm window 1525–1565 nm C+ 1580 nm)
CD	Chromatic dispersion
CDC	Cloud data centre
Co-OFDM	Coherent OFDM
CONN	Convolutional optical neural network
CPFSK	Continuous phase frequency shift keying
DAC	Digital-to-analog converter
DC	Data centre
DCF	Dispersion compensating fiber
DCI	DC interconnection
DCN	DC networking
DD-OFDM	Direct detection OFDM
DDC	Distributed data centres
DFB	Distributed feedBack
DFE	Decision feedback equalizer
DMT	Discrete multi-tone
DPSK	Differential phase shift keying
DSP	Digital signal processing
DWDM	Dense WDM
EDFA	Erbium-doped fiber amplifiers
ENOB	Effective number of bits
FEC	Forward error coding
FFE	Feed forward equalizer/equalization
FFT	Fast Fourier transform
FT	Fourier transform
Gbps	Gbits/s
IFFT	Inverse FFT
IFT	Inverse Fourier transform
IT	Information technology
ITU	International telecommunication union
LO	Local oscillator
Mary-QAM	Multi-level QAM
Mbps	Megabits/s
MCU	Micro-control unit

MLSE	Maximum likelihood sequence estimation
MMF	Multi-mode fiber
MSK	Minimum shift keying
MZDI	Mach–Zehnder delay interferometer
MZM	Mach–Zehnder interferometric modulator
O-band	Spectral region 1300 nm (35 nm window)
O/E	Optical-to-electrical conversion
OA	Optical amplifier
OCNN	Optical computing neural network
OFDM	Orthogonal frequency division multiplexing
OFDR	Optical frequency discrimination receiver
OIF	Optical internet working forum
OLG	Optical logic gate
ONN	Optical neural network
OOK	On–off keying
OPT	Optical FT
ORNN	Optical reservoir neural network
PAM	Pulse amplitude modulation
PMD	Polarization mode dispersion
PSP	Photonic signal processing
QAM	Quadrature amplitude modulation
QPSK	Quadrature phase shift keying
RCNN	Reservoir neural computing network
SFG	Small form-factor pluggable
SMF	Single-mode fiber
SPM	Self-phase modulation
SSMF	Standard SMF
Tbps	Tera-Bits/s
Term	Expansion
TIA	Tans-impedance-amplifier
TO	Thermal optic
TV	Television
WDM	Wavelength division multiplexing
XPM	Cross phase modulation
Zbps	Zetta Bits/s

1.1 Digital optical communications and transmission systems: historical overview

1.1.1 Non-DSP-based optical transmission technologies

Starting from the dielectric waveguide proposed by Kao and Hockham [1, 2] in 1966, the first research phase attracted intensive interest around the early 1970s in the demonstration of fiber optics, and optical communications have progressed tremendously over the last five decades. The first generation lightwave systems were commercially deployed in 1983 and operated in the first wavelength window of 800 nm over multi-mode optical fibers (MMFs) at transmission bit rates of up to 45 Mb s^{-1} [1]. After the introduction of ITU-G652 standard single-mode fibers (SSMFs) in the late 1970s [3], the second generation of lightwave transmission systems became available in the early 1980s [4, 5]. The operating wavelengths were shifted to the second window of 1300 nm, which offers much lower attenuation for silica-based optical fibers compared to the previous 850 nm spectral window region; in particular, the chromatic dispersion (CD) factor is almost zero. These second generation systems could operate at bit rates of up to 1.7 Gb s^{-1} and have a repeaterless transmission distance of about 50 km [6]. Further research and engineering efforts were also pushed for the improvement of the receiver sensitivity by coherent detection techniques and the repeaterless distance has reached 60 km in installed systems with a bit rate of 2.5 Gb s^{-1}. Optical fiber communications then evolved to third generation transmission systems which utilized the lowest attenuation 1550 nm wavelength window and operated up to 2.5 Gb s^{-1} bit rate [7]. These systems were commercially available in 1990 with a repeater spacing of around 60 to 70 km [8]. At this stage, the generation of optical signals was mainly based on direct modulation of the semiconductor laser source or direct detection. Since the invention of erbium-doped fiber amplifiers (EDFAs) in the early 1990s [9–11], lightwave systems have rapidly evolved to wavelength division multiplexing (WDM) and shortly after that, dense WDM (DWDM) optically amplified transmission systems which are capable of transmitting multiple 10 Gb s^{-1} channels. The 10 Gbps systems are dominated by Nortel Networks Inc. and made the decreasing market share of 2.5 Gbps as the main line of Lucent Technologies. This is due to the fact that the loss is no longer a major issue for external optical modulators that normally suffer an insertion loss of at least 3 dB. These modulators allow the preservation of the narrow linewidth of distributed feedback (DFB) lasers. These high speed and high capacity systems extensively exploited external modulation in their optical transmitters. The present optical transmission systems are considered as the fifth generation, having transmission capacity of a few tens of Tb s^{-1}.

Coherent detection, homodyne or heterodyne and intradyne, was the focus of extensive research and developments (R&D) during the 1980s and early 1990s [12–17], and was the main detection technique in the first three generations of lightwave transmission systems. At that time, the main motivation for the development of coherent optical systems was to improve the receiver sensitivity, commonly by 3–6 dB. The repeaterless transmission distance was thus able to be extended to more than 60 km of SSMF (with 0.2 dB km^{-1} attenuation factor). However, coherent optical systems suffer severe performance degradation due to fiber dispersion impairments. In addition, the phase coherence for lightwave carriers of the laser source and the local laser oscillator was very difficult to maintain. In contrast, incoherent detection technique minimize the linewidth obstacles of the laser source as well as the local laser oscillator, and this relaxes the requirement of the phase coherence. Moreover, incoherent detection mitigates the problem of polarization control in the mixing of transmitted lightwaves and the local laser oscillator at multi-THz optical frequency range. The invention of EDFAs, which are capable of producing optical gains of 20 dB and above, has also greatly contributed to the progress of incoherent digital photonic transmission systems to date.

Recent years have experienced a huge increase in demand for broadband communications driven mainly by the rapid growth of multi-media services, peer-to-peer networks, and IP streaming services, in particular IP TV. It is most likely that such tremendous growth will continue in the coming years. This is the main driving force for local and global telecommunications service carriers to develop high-performance and high-capacity next generation optical networks. The overall capacity of WDM or DWDM optical systems can be boosted either by increasing the base transmission bit rate of each optical channel, multiplexing more channels in a DWDM system or preferably, by combining both of these schemes. However, while implementing these schemes, optical transmission systems encounter a number of challenging issues as outlined in the following.

Current extensively deployed 100 Gb s^{-1} transmission systems employ intensity modulation (IM), also known as on–off keying (OOK) and utilize non-return-to-zero (NRZ) pulse shapes via either direct detection or coherent receiving subsystems using four wavelength carriers or over four lanes. The term OOK can also be used interchangeably with amplitude shift keying (ASK)[1]. Moving towards high bit rate transmission such as 100 Gb s^{-1}, the performance of OOK photonic transmission systems is severely degraded due to fiber impairments, including fiber dispersion and fiber nonlinearities. The fiber dispersion is classified into chromatic dispersion (CD) and polarization mode dispersion (PMD), causing an inter-symbol interference (ISI) problem. On the other hand, severe deterioration of the system performance due to fiber nonlinearity results from high-power spectral components at the carrier and signal frequencies of OOK-modulated optical signals. It is also of concern that existing transmission networks comprise millions of kilometers of SSMF which have been installed for approximately two decades. These fibers do not have as advanced properties as state-of-the-art fibers used in recent laboratory 'hero' experiments, and they have degraded after many years of use. Figure 1.1 shows a typical optical transmission system employing single mode fiber as the transmission guided medium, plus the optical modulation, fiber, optical amplification spans, and optical reception subsystems, using direct or coherent techniques.

The total transmission capacity can be enhanced by increasing the number of multiplexed DWDM optical channels. This can be carried out by reducing the frequency spacing between these optical channels, for example from 100 GHz down to 50 GHz, or even 25 GHz and even narrower of 12.5 GHz [18, 19]. The reduction of the channel spacing also results in narrower bandwidths for the optical multiplexers (mux) and demultiplexers (demux). Passing through these narrowband optical filters, signal waveforms are distorted and optical channels suffer the problem of inter-channel cross-talks. The narrowband filtering problems are getting more severe at high data bit rate, e.g. 56 GBaud or/and even 112 GBaud, thus degrading the system performance significantly.

Together with the demand of boosting the total system capacity, another challenge for the service carriers is to find cost-effective solutions for the upgrading process. These cost-effective solutions should require minimum renovation to the

[1] The OOK format simply implies the on-off states of the lightwaves where only the optical intensity is considered. On the other hand, the ASK format is a digital modulation technique representing the signals in the constellation diagram by both the amplitude and phase components.

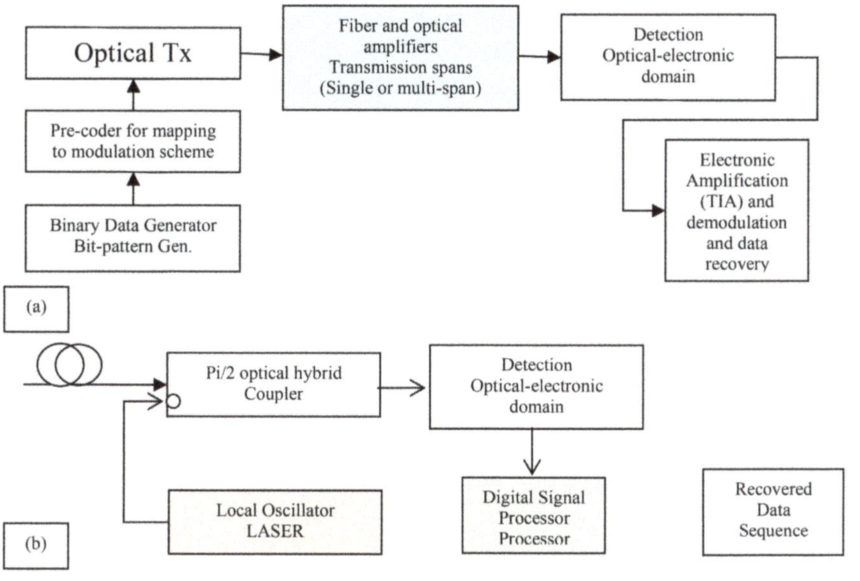

Figure 1.1. Schematic diagram of the modulation. (a) The electronic detection and demodulation of an advanced modulation format optical communications system. (b) Modulation and direct detection if no local oscillator laser is employed. Optical transmission link in single- or multi-cascade spans.

existing photonic and electronic sub-systems, i.e. the upgrading should only take place at the transmitter and receiver ends of an optical transmission link. Another possible cost-effective solution is to extend significantly the uncompensated reach of optical transmission links, i.e. without using dispersion compensation fibers (DCF), thus reducing a considerable number of required inline optical amplifiers, the EDFAs. This network configuration has recently received much interest from both the photonic research community as well as service carriers.

1.1.2 Signal processors DSP-based optical transmission

The bit rate carried by each optical carrier can be increased by improvement in the narrowness of the linewidth of the lasers as light source and as local oscillator with reasonable high power ~40 mW in the 1550 nm. Furthermore, the advancement of DSPs in the microelectronic technology together with the ultra-high sampling rate, via analog-to-digital converter (ADC) and digital-to-analog converter (DAC), of the received optical signals in continuous form sampled at very high bit rate and digital processors, permit the compensation of fiber impairments and phase recovery of system clocking as well as the determination of the frequency offset of the local oscillator (LO) laser and that of the optical carrier, hence overcoming the severe problem of homodyne reception systems faced by the first generation of coherent transmission systems in the early 1980s.

Over the last decades, extensive R&D has demnstrated coherent reception incorporating DSP that can push the bit rates per wavelength channel to 100 Gb s^{-1} and then 800G and beyond, employing 25 G/100 GBaud polarization multiplexing and QPSK (2 bits/symbol) and/or M-ary-QAM to aggregate to 200 Gbps or 400 Gbps/800

Gbps. Furthermore, the channels can be pulse shaped by using DACs to pack the channels into super-channels to generate Tbps per channel with sub-carriers. The advances of ultra-high sampling rate ADCs and DACs have reached 64 GSa/s to 200 GSa/s which allow the DSP to recover the clock, hence the sampling rate and time, combating the linear and nonlinear impairments due to CD and PMD and self-phase modulation (SPM), cross-phase modulation (XPM) and other effects. The transmission for 100 Gbps would reach 3500 km in a field trial and 1750 km for 200 Gbps over optically amplified and non-DCF fiber span transmission distance. However, the power consumption is extremely high, more than 15 W for 200 GSa s^{-1} sampling rate. It is noted that the power consumption of a DSP is increased as a cubic exponential with respect to the increasing factor of the sampling rate.

In addition to coherent reception, the use of DSP is also applied to IM/DD transmission of signals, which can take the modulation formats such PAM-n [20] or DMT (discrete multi-tone) [21–23] or OFDM (orthogonal frequency division multiplexing). DMT is indeed OFDM for intensity modulation and direct detection (IMDD) transmission.

1.1.3 Increasing transmission capacity over short-reach distance

Capacity needs to be increased, and in early 2010 multi-core fibers (MCFs) were demonstrated in which a common cladding and several single-mode cores are embedded. Further optical amplifiers were developed via cladding pumping for structuring optically amplified MCF spans for medium haul transmission links.

However, when the baud rate reaches 112 GBaud and beyond, the power consumption and electronic speed of the DSP create severe difficulties. Thus, by processing signals in the photonic domain, photonic signal processing (PSP) offer significant advantages as compared with DSP in the electronic domain. This is one of the main topics of this book. Naturally, PSP is a processing technique in analog domain without memory or at most the memory length is of ONE.

The transmission distance and bit rate are evolving dependent on the networking demands. Cloud and DC networking are evolving at a very fast pace. Furthermore, the 5G network is developing and was expected to be extensively deployed from 2020 onwards, demanding extreme capacity of accessing devices and the local cloud. In addition, the intra-DC interconnections require even higher rates and all-optical interconnection and switching. The transmission technologies are thus evolving to meet these demands, i.e. ultra-high rate with shorter distance with minimum latency. The next section addresses these technological evolutions.

1.2 Digital optical transmission in evolving networking

1.2.1 Traffic growth

It is astounding that the growth of telecommunication traffic never stops. We have observed this since 1966, the year of the invention of the idea of guiding light over a dielectric waveguide, with a bit rate of few Mbits s^{-1} over only a few kms of MMFs. To date the distance has shortened to 2–80 km for intra-DC transport to metro access and the bit rate can reach 800 Gbps or even beyond. Demand on the data transport over the internet has exploded due to the resources, big or giant data pipes etc.

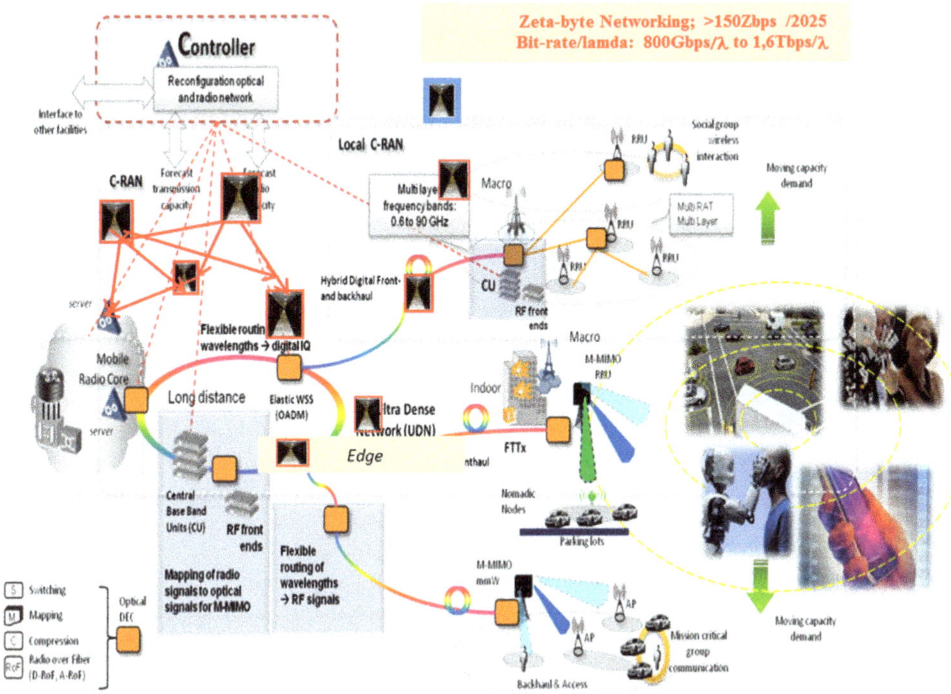

Figure 1.2. DC distribution in a mixed traditional telecom and DC networking in expected 5G network scenarios.

Data storage and delivery on demand through DCs or clouds is required for mixed data communication of traditional telecom and cloud networking for 5G and Internet of Things (IoT). The total capacity of information to be transport can reach Zeta-bits per second (figure 1.2).

The growth in communications of humans and machine-to-machine and cloud computing drives the growth of information pipelines to become larger and larger, as shown in figure 1.3. The expected evolution of the transport network can be shown in the schematic of figure 1.4, where the demands in large bandwidth can reach to multi-Tbps per optical carrier. So how can the telecom carriers and data networking corporations evolve to meet the common networking requirements in the near future for the benefit of global communities?

A market overview of the transport technology and total bit rate per transmission link is illustrated in figure 1.5 regarding the 100G and the emerging 400G market. Furthermore, the 800G and Tera-bps are also emerging to supply the demands of traffic transport, as observable in figure 1.4. This figure gives the timeline of standards developed for 100G and 400G by international bodies. 400G standard has already been issued in 2017. This has put a high importance of this rate and beyond.

These data rates are increased in order to supply capacity for the evolving DCN with accommodation for growth, especially for 5G mobile services. The

Figure 1.3. The explosive traffic growth (in growth rate) in cloud computing, mobility communications and Internet Protocol (IP) video drives the needs of larger bandwidth pipelines and machine-to-machine 5G.

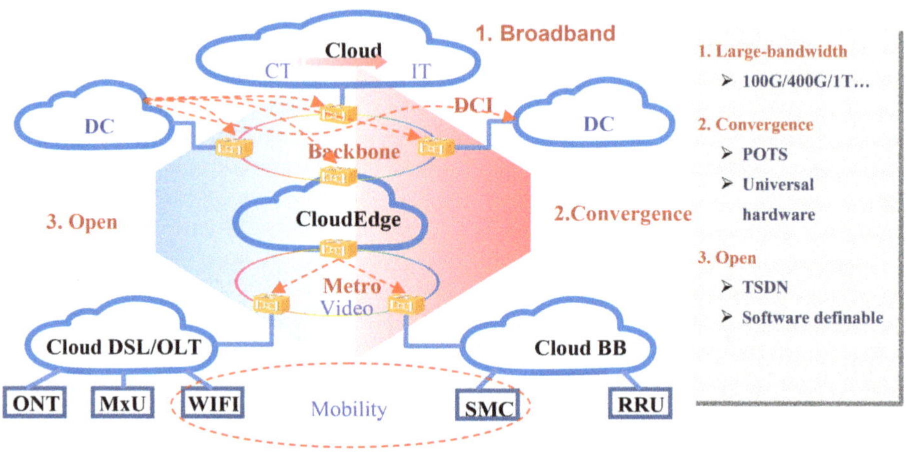

Figure 1.4. Expected future transport network under software control architecture from broadband to distribution of clouds and DCN. Labels: RRU = radio remote unit; DC = data centre; B = broadband; DCI = DC interconnect; WIFI = wireless fiber; DSL = digital service line; OLT = optical line transport; ONT = optical transport unit; MxU = multiple cross unit; SMC = service mobile control; TDSN = time division data network.

evolutionary structures of DC and emergence of modified DC distribution structure can be seen in figures 1.6 and 1.7. Figure 1.7 illustrates generic overall optical networks in which testing using traffic generators for various formats by Anritsu Network Master Series, is used[2]. This gives an overall view of optical networking from access networks to broadcasting to cloud/DC to core terrestrial and global

[2] Anritsu product Network master Series, Anritsu Coporation, Feb-2022.

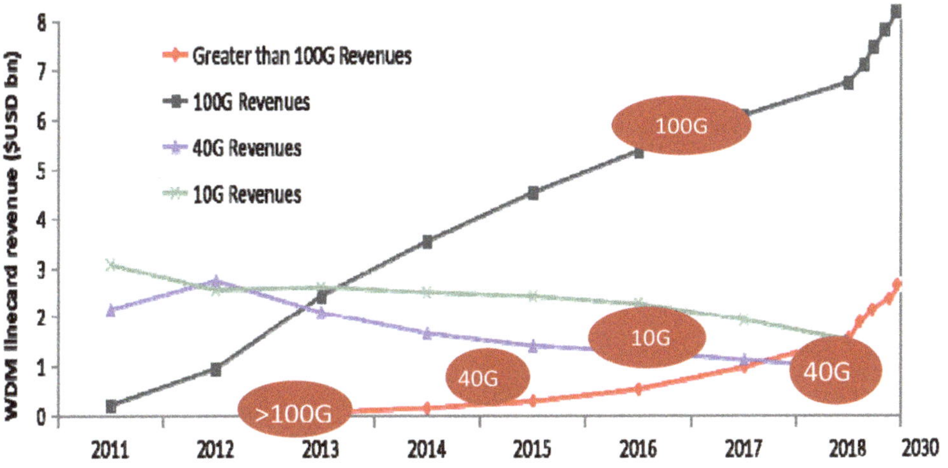

Figure 1.5. Market overview: 100G is surging and 400G is emerging with expected exponential growth in the next decade 2018 to 2030. It is expected in 2023 that the tbit rate can reach 800 Gbps and then 1.6 Tbps per wavelength carrier with baud rate reaching 256 GBaud plus overhead for FEC (forward error coding).

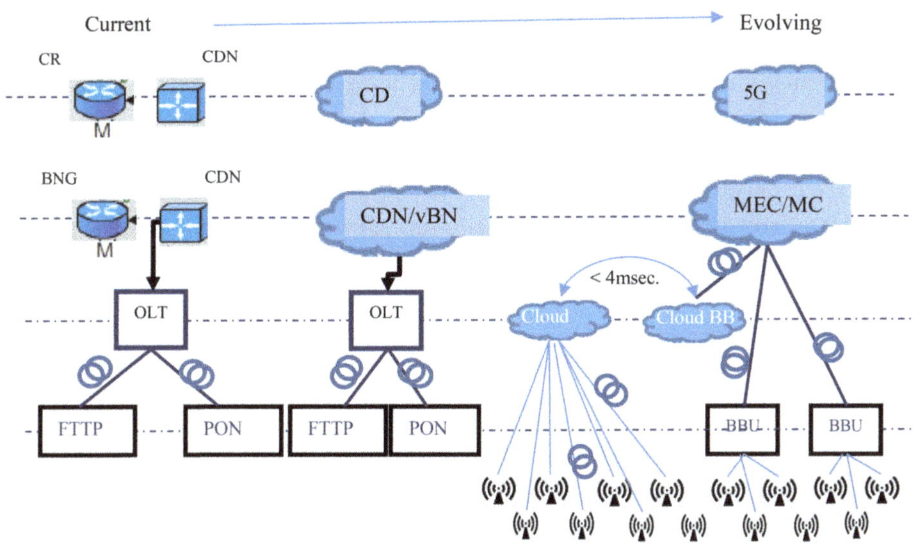

Figure 1.6. Evolution of current transport to cloudified networks for 5G. Labels: CDN = content delivery networks; FBB = fixed broadband; CDN = current delivery network; vBNG = variable business network group; OLT = optical line terminal; MCE = mobile cloud engine; BBU = base-band unit; FTTP = fiber to the premise; PON = passive optical networks. All optical paths are fiber-based cabling.

networks and metro networks. A metro network can be defined as being as large as a network of a major city in which the population can reach 10–20 million people. Thence the core network can be a continental one that interconnects several metro networks.

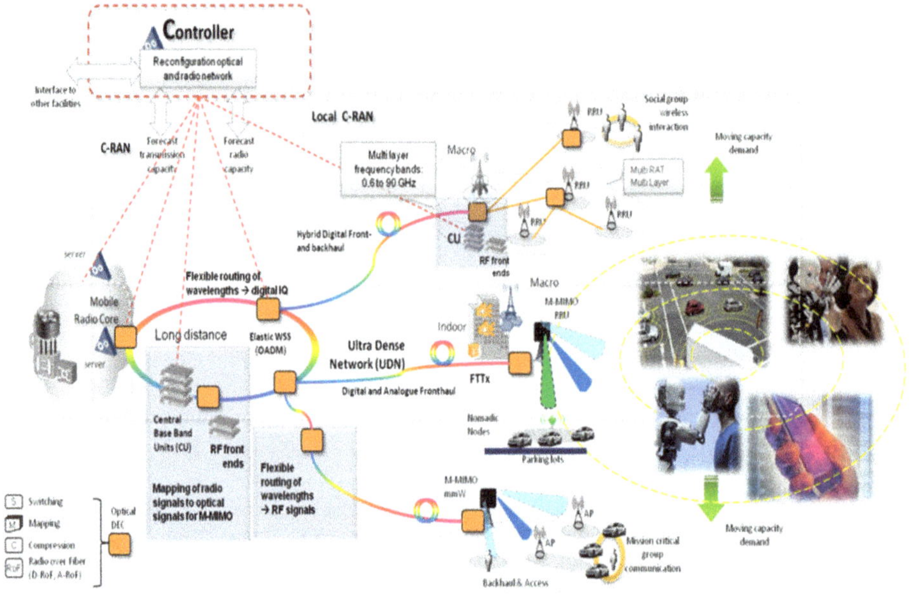

Figure 1.7. Optical networks and traffics testing, Ethernet rates of 100G, 200G, 400G.

1.2.2 Transmission technologies for intra- and inter-DC, cloud networking

DCs have evolved significantly in recent years, adopting technologies such as *virtualization* to optimize resource utilization and increase IT flexibility. As an enterprise IT needs to continue to evolve toward on-demand services, many organizations are moving toward cloud-based services and infrastructure. A focus has also been placed on initiatives to reduce the enormous energy consumption of DCs by incorporating more efficient technologies and practices in DC management. DCs built to these standards have been termed as 'green DCs'. The interconnections between servers and interconnection between DCs are critical for communications and serving demands. Thus the technological understandings of these transmission links are the main subjects of this book.

1.2.3 DCs and cloud computing—a historical overview

A DC is a facility used to house computer systems and associated components, such as telecommunications, storage and server systems. It generally includes redundant or backup power supplies, redundant data communications connections, environmental controls (e.g. air conditioning, water cooling and fire suppression) and various security devices. A large DC is an industrial-scale operation using as much electricity as a town. Some DCs can consume an aggregate power which is equivalent to that generated from a medium scale nuclear power station or about a few hundred megawatts per hour (figure 1.8).

DCs have their roots in the huge computer rooms of the 1940s, one of the earliest examples of which is shown in figure 1.9. Early computer systems, complex to

Figure 1.8. An operation engineer overseeing a network operations control room of a DC. CERN DC. Image source: Hugo van Meijeren.

operate and maintain, required a special environment in which to operate, the computer center. Many cables were necessary to connect all the components, and methods to accommodate and organize these were devised, such as standard racks to mount equipment, raised floors, and cable trays (installed overhead or under the elevated floor). A single mainframe required a great deal of power, and had to be cooled to avoid overheating.

During the boom years of the microcomputer industry, and especially during the 1980s with the operating speed reaching several GHz and sampling rates of few hundreds GSa s^{-1}, users started to deploy computing systems everywhere, in many cases with little or no care about operating requirements. However, as information technology (IT) operations started to grow in complexity, organizations grew aware of the need to control IT resources. The advent of Unix from the early 1970s led to the subsequent proliferation of freely available Linux-compatible PC operating systems during the 1990s. These were called ‘servers’, as time sharing operating systems like Unix rely heavily on the client–server model to facilitate sharing unique resources between multiple users. The availability of low-cost networking equipment, coupled with new standards for network structured cabling, made it possible to use a hierarchical design that put the servers in a specific room inside the company. The use of the term ‘data center’ (DC), as applied to specially designed computer rooms, had already gained popular recognition.

The boom of DCs came during the dot-com bubble of 1997–2001. At this time the transmission of 10 Gbps and 40 Gbps using $LiNbO_3$ modulator in transmitter and self-homodyne detection without any DSP-based processing in the receiving system was achieved. Companies needed fast Internet connectivity and non-stop operation to deploy systems and to establish a presence on the Internet. Installing such

(a)

(b)

Figure 1.9. (a) A control room in a computer center of a university circa 1962. Now more or less a cloud DC with massive server sections and interconnections via ultra-high-speed links; (b) part of a modern DC of server racks. Memory per bit space is at least more than one trillion times smaller compared with 1970s size. Cloud sections of the DC include hybrid cloud, public and private clouds.

equipment was not viable for many smaller companies. Many companies started building very large facilities, called Internet DCs (IDCs), which provided commercial clients with a range of solutions for systems deployment and operation. New technologies and practices were designed to handle the scale and the operational requirements of such large-scale operations. These practices eventually migrated toward the private DCs, and were adopted largely because of their practical results. DCs for cloud computing are called cloud DCs (CDCs). But nowadays, the division of these terms has almost disappeared and they are being integrated into the term 'data center' (DC).

With the increased employment of cloud computing, business and government organizations began to scrutinize DCs to a higher degree in areas such as security, availability, environmental impact and adherence to standards. Standards documents from accredited professional groups, such as the Telecommunications

Industry Association (TIA), specify the requirements for DC design. Well-known operational metrics for DC availability can serve to evaluate the commercial impact of a disruption. Development continues in operational practice, and also in environmentally-friendly DC design. DCs typically cost a lot to build and to maintain and have high power consumption.

1.2.3.1 Requirements for modern DCs

IT operations are a crucial aspect of most organizational operations around the world. One of the main concerns is business continuity; companies rely on their information systems to run their operations. If a system becomes unavailable, company operations may be impaired or stopped completely. It is necessary to provide a reliable infrastructure for IT operations, in order to minimize any chance of disruption. Information security is also a concern. For this reason, a DC has to offer a secure environment which minimizes the chances of a security hacking. A DC must therefore keep high standards for assuring the integrity and functionality of its hosted computer environment. This is accomplished through redundancy of mechanical cooling and power systems (including emergency backup power generators) serving the DC along with fiber optic cables (figure 1.10).

The Telecommunications Industry Association's Telecommunications Infrastructure Standard for DCs specifies the minimum requirements for telecommunications infrastructure of DCs and computer rooms including single-tenant enterprise DCs and multi-tenant Internet hosting DCs. The topology proposed in this document is intended to be applicable to any size DC.

The DC equipment may be used to: (i) operate and manage a carrier's telecommunication network; (ii) provide DC based applications directly to the carrier's customers; (iii) provide hosted applications for a third party to provide services to their customers; and (iv) provide a combination of these and similar DC applications.

Figure 1.10. Racks of telecommunications equipment in part of a DC in back plane.

Standardization and modularity can yield savings and deficiencies in the design and construction of telecommunications DCs. Modularity is important for scalability and thus easier low-cost growth, even when planning forecasts are less than optimal. For these reasons, telecommunications DCs should be planned in repetitive building blocks of equipment, and associated power and support (conditioning) equipment when practical. The use of dedicated centralized systems requires more accurate forecasts of future needs to prevent expensive over construction, or perhaps worse—under construction that fails to meet future needs.

The 'lights-out' DC is a DC that, ideally, has all but eliminated the need for direct access by personnel, except under extraordinary circumstances. All of the devices are accessed and managed by remote systems, with automation software control operations for energy savings, hence reducing operational expenditure costs and providing the ability to locate the site further from population centers.

There is a trend to modernize DCs in order to take advantage of the performance and energy efficiency increases of newer IT equipment and capabilities, such as cloud computing. This process is also known as DC transformation.

Organizations are experiencing rapid IT growth but their DCs are aging. The industry research company International Data Corporation (IDC) puts the average age of a DC at nine years old. The growth in data (>200 Zeta-bytes by 2025) is one factor driving the need for DCs to modernize. The typical projects within a DC transformation initiative include standardization or consolidation, virtualization, automation and security.

1.2.3.2 Virtualization

There is a trend to use IT network functions virtualization (NFV) technologies to replace or consolidate multiple DC equipment, such as servers. Virtualization helps to lower capital and operational expenses, and reduce energy consumption. Virtualization technologies are also used to create virtual desktops, which can then be hosted in DCs and rented out on a subscription basis.

1.2.3.3 Automation

DC automation involves automating tasks such as provisioning, configuration, patching, release management and compliance. As enterprises suffer from few skilled IT workers, automating tasks make DC operations more efficient. Software defined networking (SDN) in the optical domain or in the software environment is critical for automatic control centers to reconfigure the DC networking and linking to other DCs.

1.2.3.4 Security

In modern DCs, the security of data on virtual systems is integrated with existing security of physical infrastructures. The security of a modern DC must consider physical security, network security, and data and user security. Physical and software security are both critical. Quantum key distribution (QKD) is now reaching a state of absolute secure.

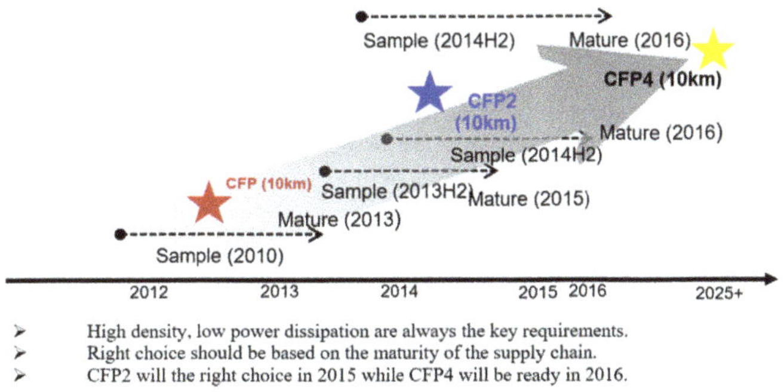

Figure 1.11. 100G client-side module evolution with the physical size from CFP to CFP2 and then CFP4 with the availability of samples for testing and maturity. CFP = compact form factor pluggable.

1.2.3.5 Carrier neutrality

Currently, many DCs are run by Internet service providers solely for the purpose of hosting their own and third party servers. However, traditionally DCs were either built for the sole use of one large company, or as carrier hotels or network-neutral DCs. These facilities enable interconnection of carriers and partners, and act as regional fiber hubs serving local business in addition to hosting content servers.

1.2.4 100G, 400G and beyond transmission technologies for cloud networking

In brief, the 100G module consists of four lanes (wavelength) and each lane carries 25 Gbps. The inter-link between servers in a DC requires the size and power consumption to be as small as possible. The CFP[3] to CFP2 and then CFP4 preferable for the size, as illustrated by figure 1.11, for 100G module from client access link or in DC inter-sever connections. Some technical details are described in chapter 4. Figure 1.12 gives the timelines of the standardization for 400G and 100G. The standard 100G has already been issued and was developed quickly for several deployments in DC networking and transmission[4],[5]. This is to ensure that multi-vendor modules can be used, thus lowering the cost with these plug-and-play approaches. 400G rate is also based on the modulated channels on four lanes but with 56 GBaud and PAM4 (2 bits/symbol) to give 100G per lane. So, four lanes would aggregate to 400G.

1.3 Photonic signal processing

1.3.1 Motivations and background

Transmission of data channels have been re-structured as follows:

---

[3] CFP module = C form-factor pluggable (*CFP*) module; C = centum (100)

[4] IEEE Std 802.3ba Media Access Control Parameters, Physical Layers, and Management Parameters for 40 Gb/s and 100 Gb/s Operation, Approved by IEEE and ANSI on 17 June 2010, http://standards.ieee.org/findstds/standard/802.3ba-2010.html

[5] IEEE P802.3bs 400 GbE Task Force, http://ieee802.org/3/bs/public/14_05/index.shtml

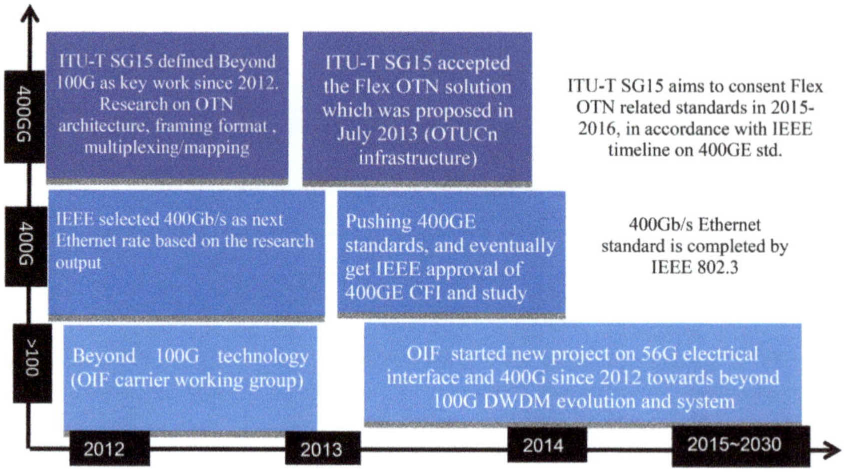

Figure 1.12. Standard timelines for 200G and 400G. OIF = optical Interchange forum; IEEE= Institute of Electrical and Electronic Engineers.

- Long haul and high bit rate under coherent technique. Low-cost coherent transmission modules for short haul ultra-high bit rate interconnections.
- For short haul DC transmission DCN and DCI inter-channel transmission using coherent and IM/DD with bit rate reaching 800 Gbps and 250 Gbps, respectively. The basic baud rate can now reach 112 GBaud and even higher
- DSP with sampling rate reaching 200 Gbps is needed. This creates a lot of strain on the electronics [24].

Thus, the processing of signals in the photonic domain will avoid this constraint and would offer much higher rate due to the high-speed sampling in the optical domain. For a 1.0 ps sampling time it requires a waveguide delay length of only about 120 μm. Thus, for 10 ps pulse period, sampling 10 times of 1.0 Tera-samples per second presents no difficulty[6].

Optical neural network computing has also been a modern topic of interest since 1990. But its use and application in transmission has not been investigated. Neural computing allows the distribution of field and intensity and their weighting to hidden layers of neurons so that several effective algorithms of computing can be processed. This is a very powerful computing technique.

We propose here for the first time the high probability of using optical neural computing (ONC) for processing of signals in the optical domain, PSP. ONC takes in the optical samples from the optical sampler, and then distributes them to other neurons with variable weighting. Optical neurons are designed using multi-mode

[6] Note: for 1.0 ps delay. Under SOI (Si on Insulator) integrated photonic waveguide, a length required is 120 um as the effective ref index is 2.55. Thus for 10 optical samples or slicing of the optical signals a total length of the samplers would be in order of 1.2 mm long

interference (MMI) devices, incorporating an optical phase shifter to vary the output field strength of the MMI outputs. Thus, MMI acts like optical neurons [25, 26].

Recently published articles on optical logic circuits employing nonlinearity of phases in SOA (semiconductor optical amplifier) have given us a new look at PSP [27], because such logic would assist in the optical decision making for recovery of 'high' or 'low' states, thus in data transmission systems.

The availability of optical logic gates (OLGs) determines the final decision that can be obtained by a summation of all optic branches inserting into the input nodes or making comparison logically to output an amplified and saturated optical signal. The OLG is structured with an SOA embedded in the branches of a Mach–Zehnder interferometer (MZI), the nonlinearity in phase of the SOA offers many nonlinear properties which are important for optical decision or optical logical functions (refer to chapter 5).

Therefore, our innovative aspects are thus based on an optical sampling structure so that (i) the optical signal can be sliced into a number of optical sampled slices; (ii) which are then fed into the ONC with a distribution whose weighting coefficients can be tuned by optical phase shifters, the optical neurons as termed for the first time. Performing the neural computing algorithms in the optical domain; coefficients or weighting are tunable via thermal tuning of phase shifters on branches of inputs to optical neurons (see chapter 5); (iii) thence a summation of all outputs of the hidden intermediate neuron layers to a decision neuron whose output is fed into the OLG so that a 'high' or 'low' optical state is obtained. It is then straightforward from this point to electrical recovery signals.

In formulating the photonic processors, we base our various structures on the optical sub-modules as described above, such as detection of deferentially coded optical sequence and FEC (forward error coding) (see Patent 2 idea attached in appendix part (d)). Several other processing algorithms can be developed from this basic/fundamental principle.

Our proposal, therefore, gives a detailed account of various novel structures of PSP but demonstrates only one basic PSP structure using the three basic optical sub-modules, as briefly described above and in the following.

It is believed that such PSP proposed here will lead to several patents and lead the global R&D on photonic processors, especially for telecommunication systems for data centres DCN and DCI.

1.3.2 Generic innovative models of photonic signal processors

This section outlines the generic models for PSPs so as to give the distinctions of our models in dealing with optical channels as follows:

PSP model in time domain without optical feedback, but electrical control feedback whose electrical signals derived from the output of the processing system.

- PSP model in time domain with optical feedback;
- PSP model in spectral domain;

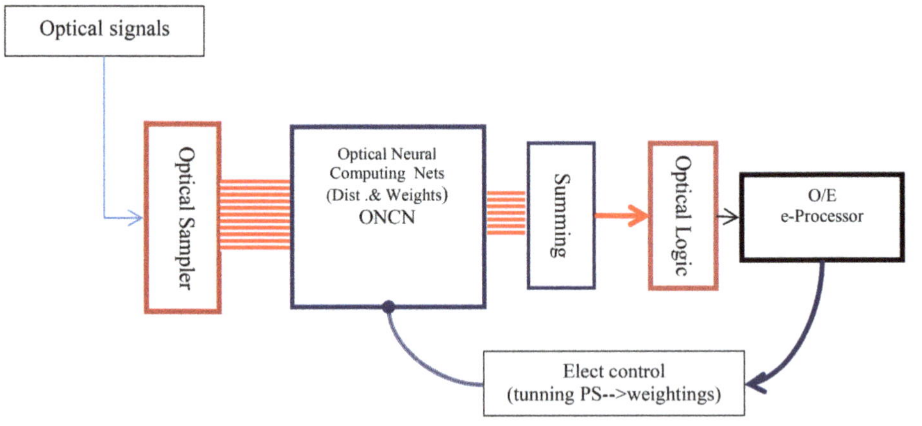

Figure 1.13. Generic model of time-domain PSP with electrical feedback to control the weightings of ONCN.

In principle, we can see the operational points. In the case of temporal domain PSP:

- In time-domain PSP the optical signals are converted into samples and from serial-to-parallel (S/P) conversions by the use of optical samplers (see figure 1.13).
- The optical samples are then processed in an optical computing neural network (OCNN) by distribution into intermediate layers (the hidden layers) whose paths are weighted by tuning the coupling coefficients of the optical couplers (e.g. multi-mode interferometric coupler).
- The outputs of the OCNN are then summed to give a resulting optical output which is then fed into an optical logic circuit whose output gives a 'high' or 'low' amplified optical signal which is then converted into recovered electrical signal.
- The electronic processor for this PSP can be much simpler and possibly a simple and low speed micro-controller unit (MCU) to control and tune the optical section via thermo-optic (TO) effects.

In the case of optical feedback time-domain PSP (see figure 1.14), the optical output of the optical logic circuit is tapped and then further processed via a second OCNN whose delay time can be within one symbol period or longer as desired. The outputs of the OCNN-1 and OCNN-2 are compared (e.g. via an optical NAND), thus this is more or less an operation of differential processing. Such differential processing can be highly useful for differentially coded sequence as in differential coding or forward error coded (FEC) sequence.

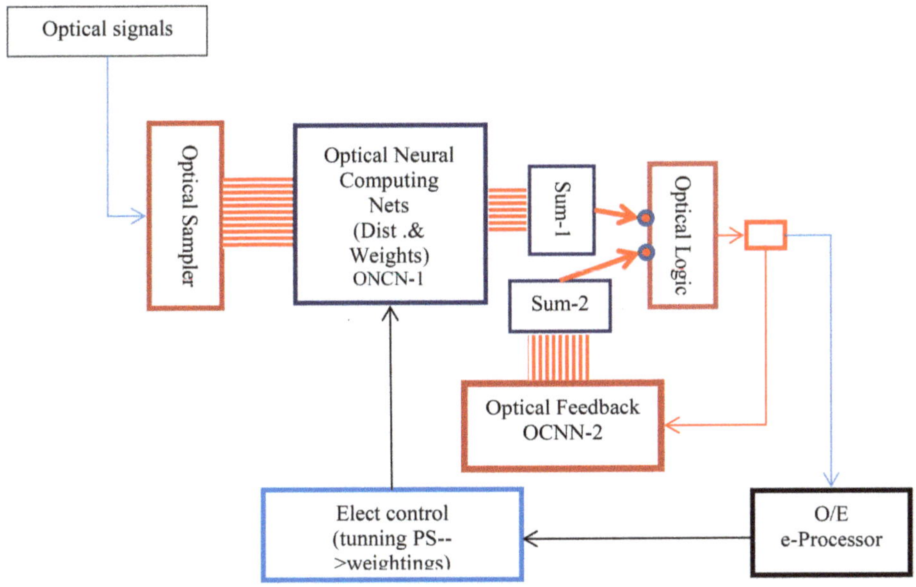

Figure 1.14. Generic model of time-domain PSP with optical feedback to OCNN-2 to compare with OCNN-1 by one-bit (or longer) delay in the optical logic circuit plus e-feedback control of the weightings of ONCN-1.

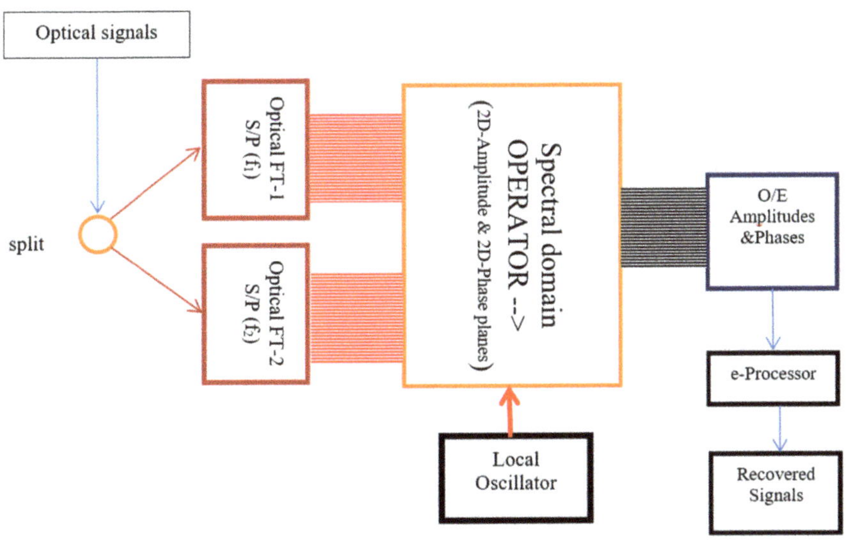

Figure 1.15. Generic model of spectral domain PSP. Labels: S/P = serial-to-parallel; FT = Fourier transform; O/E = optical-to-electrical converter (photodetector bank); 2D = two dimensional; e– = electrical/electronic.

In the case of spectral-domain PSP (SPSP), we have the following. The input signal sequence is split into two and then Fourier transformed into two spectral regions of fundamental frequencies f_1 and f_2, in parallel form (see figure 1.15).

These spectral FT optical sections are fed into an optical operator which can be a bispectrum processor whose outputs are two 2D planes of amplitude and phase via a coherent detection sub-system. OCNN in spectral domain can be used in the SPSP with a multiplication in optical domain to represent the equivalence of time-domain auto- or cross-correlation operations if the processing is more complex.

In summary, we can see that both time domain and spectral domain PSPs can be used in receiving sub-systems or pre-processing optical transmitter or in optical switching or routing. The applications of these generic models are given in this proposal. The demonstration of a PSP is to use a simple model of time-domain PSP without optical feedback model.

We can compare the performance and processing parameters of a PSP in contrast to a typical e-DSP-based optical receiver as tabulated in table 1.1.

1.4 Modulation techniques for ultra-broadband

1.4.1 Modulation formats and optical signals generation

Modulation is a process of facilitating the transfer of information over a medium. In optical communications the process of converting information so that it can be successfully sent through the optical fiber is called optical modulation.

There are three basic types of digital modulation techniques. These are ASK, frequency shift keying (FSK) and phase shift keying (PSK) in which the parameter is the carrier whose amplitude, frequency or phase are varied to represent the information which is to be sent. Digital modulation is a process of mapping such that the digital data of ‘1’ and ‘0’ or symbols of ‘1’ and ‘0’ that convert it into some aspects of the carrier, such as the amplitude, the phase or both amplitude and phase, and then transmits the modulated carrier which is lightwaves in this book context. The modulated and transmitted lightwave carrier is then remapped at the reception systems back to a near copy of the information data.

1.4.1.1 Binary level

Modulation is a process that facilitates the transport of information over the medium, in this book our medium is the optical guided fiber and associate photonic components. In digital communications there are three basic types of digital modulation techniques: ASK, PSK and FSK. All these techniques vary a parameter of a sinusoidal carrier to represent the information ‘1’ and ‘0’.

In ASK the amplitude of the lightwave carrier, normally generated by a narrow linewidth laser source, is changed in response to the digital data and everything else is kept fixed. That is, bit ‘1’ is transmitted by the lightwave carrier of a particular amplitude. To transmit ‘0’ the amplitude is changed keeping the frequency unchanged, as shown in figure 1.16. NRZ or RZ can be assigned depending on the occupation of the state ‘1’ during the time length of a bit period. For RZ, normally only half of the bit period is occupied by the digital data.

In addition to NRZ and RZ formats, in optical communications the carrier can be suppressed under these formats so as to achieve non-return-to-zero carrier suppression (NRZ-CS) and return-to-zero carrier suppression (RZ-CS). This is normally generated by biasing the optical modulator in such a way that the carriers passing through the two parallel paths of an interferometric modulator are π-phase

Table 1.1. Comparing PSP and e-DSP-based optical receiver.

Parameters	e-DSP Optical receiver	PSP	SPSP	Remarks
Sampling	Limited by e-sampler <200 GSa/s	Limited by delay optical waveguide >0.11 ps and optical phase modulation of optical switching in sampler	Limited by optical delay in Opt FT (MZDI).	PSP sampling rate ~10–20 DSP
Noises	ENOB < 4.5 for 100–200 GSa/s	Not limited by ENOB but optical noises if opt. OA is used	Same as PSP	OA = optical amplifier; PSP noise performance much more superior
Processing speed	Limited by electronic processor (FPGA Or ASIC) and memory length	Much faster than e-DSP but limited by memory—max memory length is ONE. Longer if OA used but oscillation problem	Not limited by memory length and possible processing of long sequence by using Wiener–Minchin theorem and processing in frequency domain of recovered elect sequence.	E-DSP memory length is long and >1-Million-bit length
Accuracy	Limited by number of bit resolution Max 8-bit at 200 GSa/s.	Limited by optical samples by samplers and optical losses.	Limited by opt. FT or MZDI, hence number of optical filters.	PSP affected by optical responses of optical samplers. e-DSP not much due to logical/ digital levels.
Complexity	E-DSp possible processing complex signals but processing time is long due to sequential process, especially in correlation	Simpler with addition in opt domain for multiplication in time domain and optical multiplication for correlation.	Addition for correlation operation in spectral domain.	PSP much simpler and low noise—but limited by response time as PSP operation is in analog process.
Digital/analog processing	Digital	Analog	Analog	Limited by response time if analog operation.
Power consumption	High (>20 W for 200 GSa/s)	<100 mW for 100 GBaud rate.	<100 mW	High for e-DSP
Cost	High	Low if not limited by integrated Photonic integrated circuit technology and memory	Same as PSP	PSP superior advantage if limited factors are not critical factor.

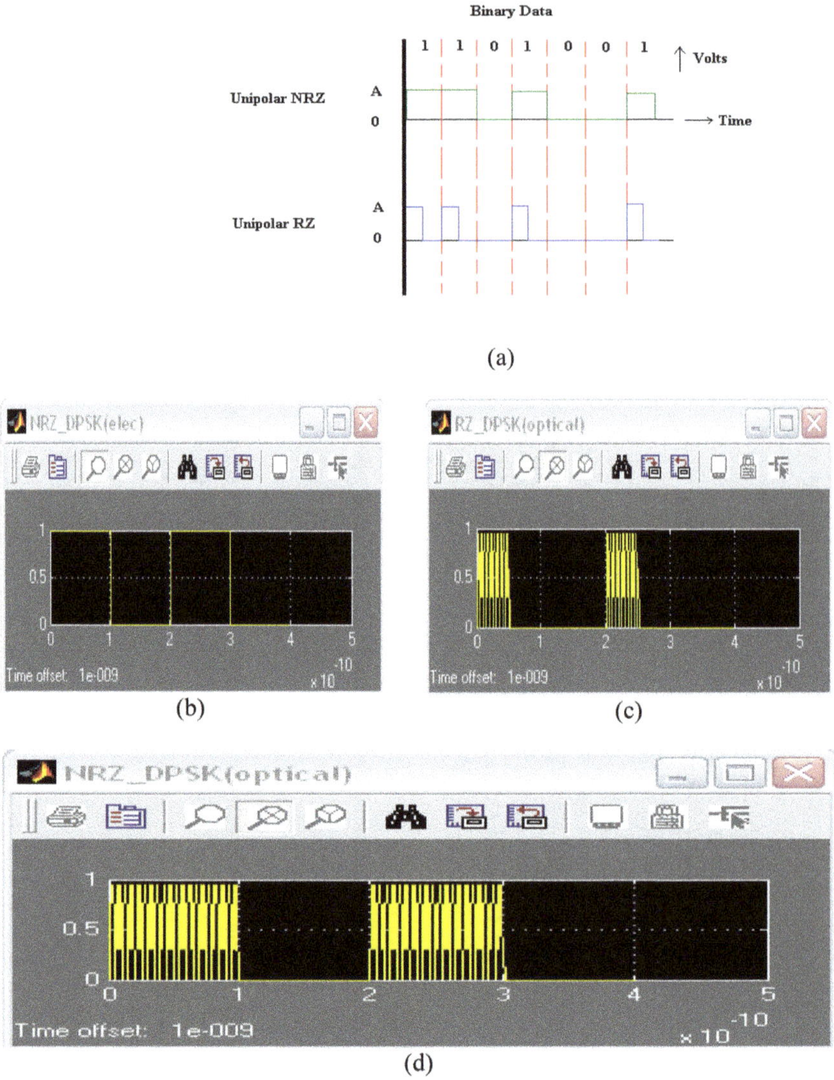

Figure 1.16. (a) Schematics of NRZ and RZ pulse amplitude modulated formats for a sequence of {1 0 1 0 1 0 1 0 1 0 1 0}. (b) Generated ASK signals with carrier (not to scale and high density area)-data and carrier modulated, (c) NRZ, and (d) NZ formats.

shift difference with each other. Thus, the carrier at the center frequency is suppressed but the sidebands of the modulated signals remain.

In PSK the phase of the lightwave carrier is changed to represent the information. The phase in this context is the shift of angle at the start at which the sinusoidal carrier starts. To transmit a '0' the phase would be shifted by π and a '1' with no change of phase. The phase angle can be changed and take a value of a set of phases corresponding to the mapping of the symbols, as shown in figure 1.17.

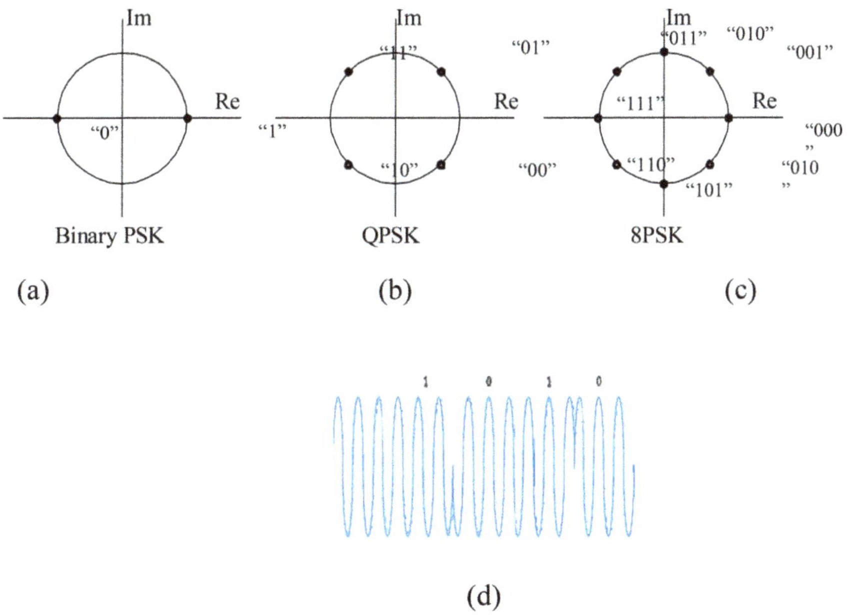

Figure 1.17. Signal-space constellation of discrete PM (a) binary PSK (b) quadrature binary PSK (c) 8-PSK. (d) Phase of the carrier under modulation with π phase shift of the BPSK at the edge of the pulse period.

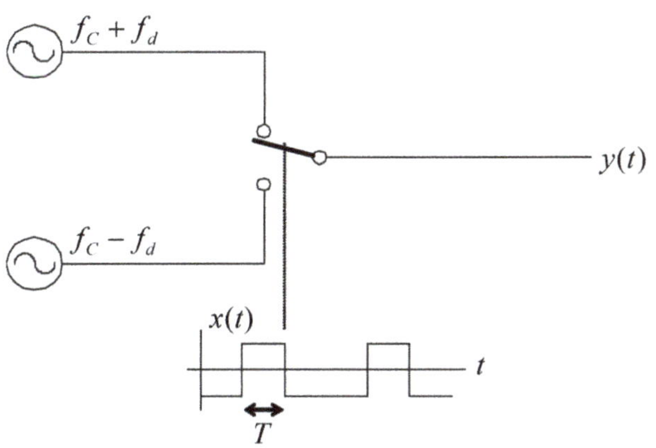

Figure 1.18. Schematic diagram of an FSK transmitter. f_d is the deviation frequency from the carrier frequency f_c.

In FSK the frequency of the carrier represents the digital information. One particular frequency is assigned to a '1' and another frequency is assigned to the '0', as shown in figure 1.18. An FSK can be considered as continuous phase modulation, for example the continuous phase modulation MSK (minimum phase shift keying) whose frequency separation f_d is selected such that the signals carried by these frequencies are orthogonal.

ASK can be combined with PSK to create a hybrid modulation scheme such as quadrature amplitude modulation (QAM) where both the phase and amplitude of the carrier are changed at the same time. The carriers are expected to follow a similar pattern to that of the DPSK in figure 1.17(d) but with different frequencies of the carrier under the envelope of the bit '0' and '1'. For MSK signals the carrier frequency is chirped up or down depending on the '0' or '1'. That is, the phase of the carrier is continuously varied during the bit period, and the carrier frequencies of the bits are such that there is an orthogonality of the carriers and the signal envelope.

1.4.1.2 Binary and multi-level

Additional degrees of freedom for detection can be used to effectively enhance the capacity due to effective equivalence of the multi-level and symbol rate, thence the detection of the received optical signals. A widely used and matured detection scheme for optical signals is direct detection, in which the optical power $P = [E]^2$, the square of a complex optical field amplitude. The photodetector (PD) would not be able to distinguish between a '0' or 'π' phase shift of the carrier lightwave embedded within the pulse. The carrier phase can only be possibly extracted if and only if there is photonic processing to extract the phase at the front end of the receiver. Thus a + or− field complexity would be seen as identical in the PD.

This ambiguity of the phase detection process would allow one to shape the optical spectra of optical signals to induce a modulation format more resilient to the distortion effects accumulated during the transmission process.

Formats making use of the tri-level can be illustrated in figure 1.19 and could be termed as pseudo-multi-level or tri-level or poly-binary signals. These tri-level signals can be represented in terms of the *phase* or *frequency* of the lightwave carrier.

The use of more than two symbols to encode a single bit of information is to increase the information bits carried by a symbol. However, the transmission is still at the symbol rate B_S. Under optical transmission, the −1 and +1 can be coded in terms of the variation of phase of the carrier as there is no negative intensity representation unless the field of the lightwaves is used. The tri-level uses $\{+|E|, -|E|, \text{and } 0\}$ (in field amplitude) or in optical domain the phase of the carrier can be coded as $\{+\pi/2, -\pi/2, \text{and } 0\}$ and its equivalent phase representation can be

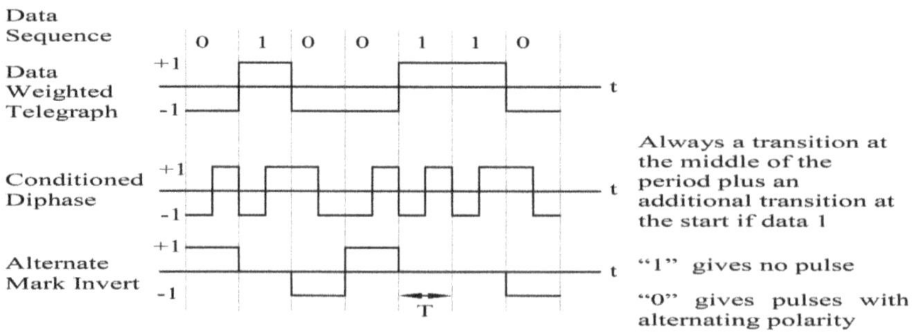

Figure 1.19. Illustration of pseudo-multi-level or poly-binary base-band signals. Binary data sequence, weighted signals, diphase RZ and alternate mark inversion formats.

mapped to $\{0, |E|^2\}$ at the optical receiver, for example the duo-binary format that would be described in a later chapter. A phase difference of π and 0 between the three levels would minimize the pulse dispersion as they are propagating along the fiber due to the relative phase difference of π hence destructive interference of any pulse spreading due to dispersion.

This tri-level must not be mixed up with the truly multi-level signaling in which log2M bits are encoded on N-symbols, and then transmitted at a reduced rate $B_s/\log_2 N$. Both multi-level amplitude or amplitude-phase shift keying and differential quadrature phase shift keying (ODQPSK) are multi-level optical modulation techniques.

1.4.1.3 In-phase and quadrature phase and amplitude modulation

Another form of modulation that would enhance the capacity of the transmission is the use of the orthogonal channels in which the information can be coded into the in-phase and quadrature as shown in its polar or Cartesian coordinates, as shown in figure 1.20.

QPSK is the most commonly used in the differential and non-differential phase modulation in which I and Q components are used extensively due to its bit error rate (BER) and its corresponding energy per bit being very similar to that of PSK and doubling the capacity of that of PSK. QPSK is an extension of the binary PSK signals but with the phase change of only π/2 instead of π. Mathematically the signal $s(t)$ can be expressed as

$$s(t) = A_c p_s(t)\cos\left(2\pi f_c t + i\frac{2\pi}{M}\right); \quad i = 1, \ldots, M \tag{1.1}$$

where $p_s(t)$ is the pulse shaping of the data, M is the quantized level or the total number of phase states of the modulation and I is the phase modulation index. The QPSK can be combined with ASK to generate quadrature amplitude modulation (QAM) where the phase and amplitude can be used to map a symbol of data information into one of the points on the signal space.

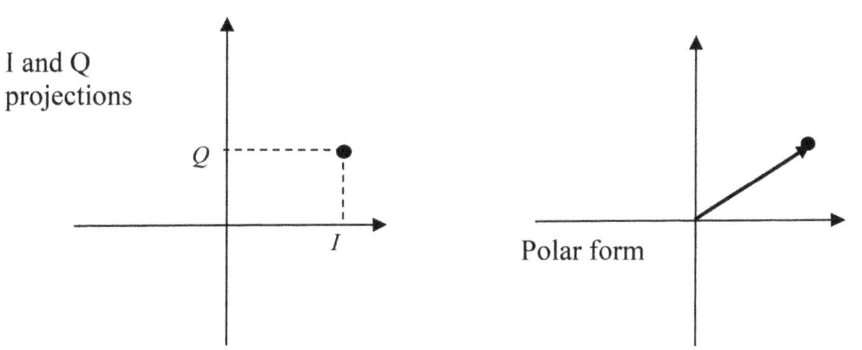

Figure 1.20. Signal vectors plotted in signal space (a) Cartesian coordinate and (b) polar form.

1.4.1.4 External optical modulation

External modulation is the essential technique for modulating the lightwaves so that its linewidth preserves its narrowness and only the sidebands of the modulation scheme dominate the spectral property of the generated passband characteristics. Figures 1.21 and 1.22 show the typical structure of optical transmitters for generation of NRZ and RZ optical signals. The laser is always switched on and its lghtwaves are modulated via the electro-optic (E/O) modulator using the principles of interferometric constructive and destructive interference to represent the ON and OFF states of the lightwaves. The RZ can then similarly be generated but with an additional optical modulator that would generate periodic optical pulses whose width is half that of the bit period. The phase and frequency modulation can also be generated using these E/O modulators by biasing conditions and control of the amplitudes of the electrical pulses. These optical transmitters are described in [28]. We note here that the fiber that connects between the two modulators of figure 1.20 must be of polarization maintaining (PM) type. Otherwise there would be polarization fluctuation and hence reduction of the coupling of the lightwave power one to the other.

The laser source would normally be a narrow linewidth laser that is turned on at all times to preserve its narrowness characteristics. The lightwaves are generated and

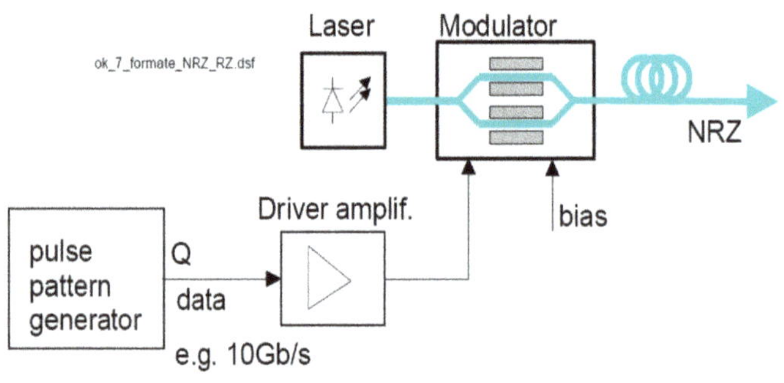

Figure 1.21. Generation of optical signals of format NRZ using an external modulator.

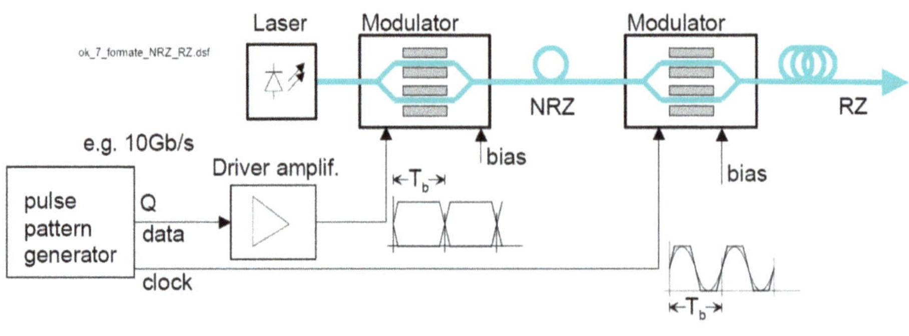

Figure 1.22. Generation of optical signals of format RZ.

coupled to the optical modulator via the pigtails of both devices. The modulator would be driven by data pulse sequence output of a bit pattern generator condition to appropriate driving level required by the V_π and the phase variation of the carrier if phase modulation is needed. When a modulation format is necessary then an electronic pre-coder is required to code the serial sequence to appropriate coding. The pre-coder can be a differential coding, multi-level coding or inverse fast Fourier transform (IFFT) to generate orthogonal data sub-channels in the case of the OFDM.

The pulses of all modulation formats can take the form of NRZ or RZ. For RZ format an additional optical modulator is required to generate or condition the '1' NRZ to RZ, as shown in figure 1.22. The second modulator can exchange its position with that of the other modulator without affecting the generation of the modulation formats. This second modulator is usually called the *pulse carver*. It is noted that if the RZ modulator is biased such that the phase difference at the biasing condition is π phase difference, then we would have a carrier suppression of the carriers at the central location of the generated spectra but the sidebands of the optical signals. The bandwidth of the modulator determines the rise time and fall time of edges of the pulse sequence shown in figures 1.21 and 1.22. The details of the optical transmitters for different modulation formats are given in [29].

1.4.1.5 Advanced modulation formats

The above-described problems facing contemporary optical fiber communications can be effectively overcome by utilizing spectrally efficient transmission schemes via the implementation of advanced modulation formats. A number of modulation formats have recently been reported as alternatives for the OOK format, including RZ pulses in OOK/ASK systems [19–21], DPSK[1] [19–23][2] and more recently, minimum-shift keying (MSK) [30–34]. These formats are adopted into photonic communications from the knowledge of wire-line and wireless communications.

DPSK has received much attention in the last few years, particularly when it is combined with RZ pulses. The main advantages of RZ DPSK are: (i) a 3-dB improvement in the receiver sensitivity over OOK format by using an optical balanced receiver (see chapter 4, [28]), and (ii) high resilience to fiber nonlinearities [6, 19, 28, 35] such as intra-channel self-phase modulation (SPM) and inter-channel cross-phase modulation (XPM). Several experimental demonstrations of DPSK long-haul DWDM transmission systems for 10 Gb s^{-1}, 40 Gb s^{-1} and higher bit rates have been reported recently[1] [22, 36–39]. However, there are few practical experiments addressing the performance of cost-effective 40 Gb s^{-1} DPSK—10 Gb s^{-1} OOK hybrid systems for gradually upgrading the existing installed SSMF transmission infrastructure [40]. In addition, the performance of 40 Gb s^{-1} DPSK for use in this hybrid transmission scheme has not been thoroughly studied. Therefore, one of the main contributions of this research is to prove the feasibility of overlaying 40 Gb s^{-1} DPSK channels on the existing 10 Gb s^{-1} network infrastructure for implementing the hybrid systems.

The MSK format offers a spectrally efficient modulation scheme compared to the DPSK counterpart at the same bit rate. As a subset of continuous phase frequency shift keying (CPFSK), MSK possesses spectrally efficient attributes of the CPFSK family. The frequency deviation of MSK is equal to a quarter of the bit rate and this frequency deviation is also the minimum spacing to maintain the orthogonality between two shifted frequencies of FSK. On the other hand, MSK can also be considered as a particular case of offset differential quadrature phase shift keying (ODQPSK) [41–44], which enables MSK to be represented by I and Q components on the signal constellation. The advantageous characteristics of optical MSK format can be summarized as follows: (i) Compact spectrum, which is of particular interest for spectrally efficient and high-speed transmission systems. This also provides robustness to tight optical filtering; (ii) high suppression of spectral side lobes in the optical power spectrum compared to DPSK. The roll-off factor follows f^{-4} rather than f^{-2} as in the case of DPSK. This also reduces effects of inter-channel cross-talks; (iii) no high-power spectral spikes in the power spectrum, thus reducing fiber nonlinear effects compared to OOK; (iv) as a subset of CPFSK, MSK can be detected either incoherently based on the phase or the frequency of the lightwave carrier or coherently based on the popular I–Q detection structure; and (v) Constant envelope property, which eases the measure of the average optical power.

Several studies have been conducted recently on the generation and direct detection of externally modulated optical MSK signals[3], [4], [5] [24]. However, there is little research investigating the performance of externally modulated MSK format for digital photonic transmission systems, particularly at high bit rates such as 40 to 100 Gb s^{-1} over four lanes[3] [45, 46]. Furthermore, if MSK can be combined with a multi-level modulation scheme, the transmission baud rate would be reduced in addition to the spectral efficiency of the MSK formats. This is of great interest for long-haul and metropolitan optical networks and thus, provides the main motivation for proposing the dual-level MSK modulation format in this research. In addition, the potential of optical dual-level MSK format transmission has yet to be explored. Therefore, another main contribution of this research is to provide comprehensive studies on the performance of MSK and dual-level MSK modulation formats for long-haul and metropolitan optical transmission systems.

1.4.1.6 Orthogonality spectral modulation: OFDM and DMT

Orthogonal frequency division multiplexing (OFDM) is an attractive modulation format that recently received lots of attention in practical and R&D fiber optic communities [47]. The main advantage of optical OFDM is that it can cope with a virtually unlimited amount of inter-symbol interference (ISI). This is possible as the whole band is divided by several sub-carriers and modulation formats can be assigned to each carrier of much narrower bandwidth. Thus, the dispersion becomes negligible. In high-speed optical transmission systems, ISI caused, for instance by chromatic dispersion (CD) and polarization mode dispersion (PMD), is serious in long- and short-haul systems whose bit rate is higher than 56–100 GBaud, especially at these ultra-high rate and multi-level modulations in the optical domain. In this sub-section some of the basics of an optical OFDM system are given. Furthermore, high-speed transmission experiments up to 122 Gbps are briefly presented.

In principle, the basic concept behind OFDM is the division of a high bit rate data stream into several low bit rate orthogonal sub-carrier streams. In general, the sub-carriers are generated in the electronic digital domain, therefore these systems typically consist of many sub-carriers (typically more than 64, and should be in multiple numbers of). In these systems, channel estimation is realized by periodically inserting training symbols. In fiber optic transmission systems, the OFDM systems where the sub-carriers are generated in the optical domain are also created. These systems are sometimes referred to as coherent WDM systems [47]. Coherent WDM systems typically have few sub-carriers and do not use training symbols, but rely on blind channel estimation instead.

In optical OFDM systems, the front end of the transmitter consists of an optical modulator, where the OFDM signal is unconverted to the frequency of an optical carrier. The front end at the receiver consists of either a coherent (Co-) or a direct detection (DD-) scheme. DD-OFDM is realized by sending the optical carrier along with the OFDM band so that direct detection with a single photodiode can be used at the receiver. In a Co-OFDM system, the optical carrier is suppressed at the transmitter and a local oscillator (LO) and optical hybrid is required, such as that shown in figure 1.21. The superior performance of Co-OFDM with respect to optical signal-to-noise ratio (OSNR) requirements, PMD tolerance and spectral efficiency makes it an excellent candidate for long-haul transmission systems, whereas DD-OFDM, which requires fewer components at the receiver than Co-OFDM, is more suitable for cost-effective short reach applications. The spectra of both these two OFD schemes are shown in figure 1.23. The DSP processing of DD-OFDM is much simpler as only the real part of the OFDM signals are required to be generated. The alternative name for DD_OFDM is DMT (discrete multi-tone). For DMT the total capacity can be increased using the bandwidth extending to 6 dB rather than only 3 dB. For the sub-carriers in the 3–6 dB the sub-carrier can be modulated at lower order of modulation formats so that the sensitivity of the receiver can be increased to compensate for the lower sensitivity by the 3 dB lower band, as shown in figure 1.24. OFDM and DMT are described in detail in chapter 3.

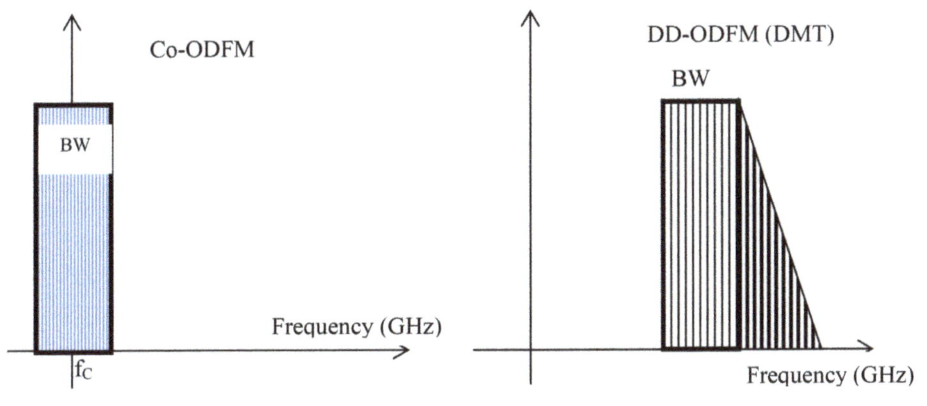

Figure 1.23. Spectra of optical OFDM-Co-OFDM and DD_OFDM. Sharp roll off at spectrum boundary → compact channels multiplexing.

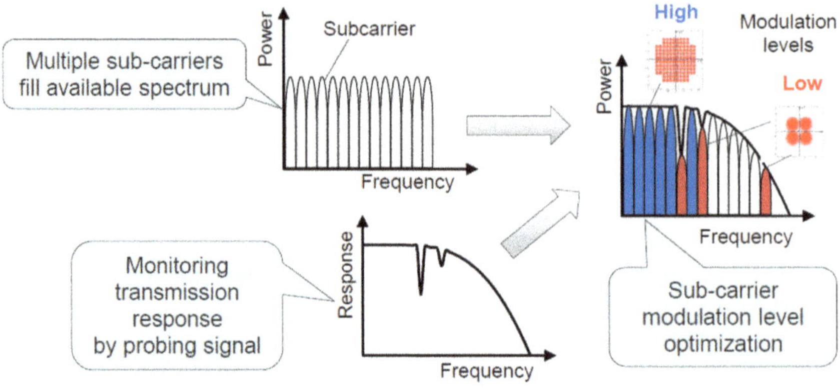

Figure 1.24. Principles of DD_OFDM (DMT) modulation scheme using carriers and modulation formats in the band 0–6 dB band 'high' and 'low'.

1.5 Opto-electronic reception and processing

The reception sub-system in an optical communication link is an important part. Its sensitivity is critical for determining the transmission speed and reach distance. The O/E is the opto-electronic device (PD) which converts the optical energy into electrical signals. However, if a local oscillator laser is together injected into such a PD then the optical fields of the transmitted signals and the local laser are mixed in the PD to give a product in the RF spectral region, the DC component due to the residual power of the local oscillator and that of the signals which are positioned at DC (direct current). These DC components are normally filtered and only the RF component remains detected, whose phase and amplitude reflect those of the signals multiplied by the amplitude of the LO. Thus, it can be considered as an amplification of the signal field by the amplitude of the LO with its phase. Therefore, phase dependent modulation is effectively detected by this coherent mixing and detection. This section gives a brief introduction to these reception sub-systems.

1.5.1 Incoherent optical receivers

The modulation formats studied in this research, optical DPSK and MSK-based formats, can be demodulated incoherently by using an optical balanced receiver which employs a Mach–Zehnder Delay Interferometer (MZDI). In the case of optical DPSK format, MZDI is used to detect differentially coded phase information between every two consecutive symbols [6, 28, 29, 48]. This detection is carried out in the photonic domain as it is beyond the speed of the electrical domain, especially at very high bit rates of 40 Gb s^{-1} or above. The MZDI balanced receiver is also used for incoherent detection of optical MSK signals[3], [5] [24] by also detecting the differential phase of MSK-modulated optical pulses. However, using the MZDI-based detection scheme, it is found that optical MSK provides a slight improvement for the CD tolerance over DPSK and OOK counterparts[5], [6].

As a subset of the CPFSK family, MSK-modulated lightwaves can also be incoherently detected based on the principles of optical frequency discrimination. Thus, an optical frequency discrimination receiver (OFDR) employing dual narrowband optical filters and an optical delay line (ODL) is proposed in this research. This receiver scheme effectively mitigates CD-induced ISI effects and enables breakthrough CD tolerances for optical MSK transmission, as reported in my first-authored papers[6] [25]. In addition, the feasibility of this novel receiver is based on recent advances in the design of optical filters, in particular the micro-ring resonator filters. Such optical filters have very narrow bandwidths, e.g. less than 2 GHz (3 dB bandwidth), and they have been realized commercially by Little Optics [49, 50]. This research thus provides a comprehensive study on this OFDR scheme, from the operational principles to the analysis of the receiver design, and onto the performance of OFDR-based MSK optical transmission systems.

1.5.2 DSP—coherent optical receivers

Coherent detection and transmission techniques were extensively exploited in the mid-1980s to extend the repeaterless distance a further 20–40 km of SSMF with an expected improvement of the receiver sensitivity of 10–20 dB depending on the modulation format and receiver structure using phase or polarization diversity.

In general a coherent receiver would operate on the beating of the received optical signals and that of the field of a local laser oscillator. The beating optical signals are then detected by the PD with the phase of the carrier preserved that permits the detection of the phase of the carriers. Hence the phase modulation and continuous phase or frequency modulation signals can be processed in the electronic domain. With the advancement of digital electronic processor, the processing of the received signals either in the IF or base band of the heterodyne and homodyne detection, respectively, can be processed to determine the phase of the modulated and transmitted signals. Coherent receivers for different modulation formats are described at appropriate sections of the chapters of the book.

The three types of coherent receivers, namely homodyne, heterodyne and intradyne detection techniques, are possible and dependent on the frequency difference of zero, intermediate frequency greater or smaller than the passband of the signals between the local oscillator laser and that of the signal carrier. With modern advanced optically amplified fiber communications, broadband ASE noises always exist and under coherent detection the beating between local laser source and ASE dominate the electronic noise of the receiver at the front end. These noise considerations are described in [29].

The advanced aspects of coherent reception systems are currently pushed to a generation by incorporating the DSP sub-systems and naturally the ultra-high sampling rate ADC and DAC permitting the bit rate per channel to reach 100 Gb s^{-1} and higher by polarization division multiplexing (PDM) QPSK or M-QAM modulation scheme with M of 16 or 32 or even higher. The DSP in association with advanced algorithms has overcome a number of difficulties faced by 'analog' coherent reception techniques extensively reported in the 1980s, such as frequency matching of the local oscillator and that of the signal carrier; the clock recovery and generation of sampling for re-timing; the compensation of dispersion due to CD and PMD impairments and nonlinear distortions. These DSP techniques are focused in this second edition with four chapters and another chapter on optical hardware and digitalization circuitry for such processing systems.

1.5.3 e-DSP electronic equalization

Electronic equalizers have recently become one of the most potential solutions for future high-performance optical transmission systems. The Si–Ge technological development has enhanced the electron mobility and hence produced shorter rise and fall time of the pulse propagation, thus increasing the processing speed. Sampling rate can now reach several giga-samples/sec (GSa s^{-1}). that enables the processing of 10 Gbps bit rate of the data channel without any difficulty. Hence it is very probable that the electronic processing and equalization can be implemented in real systems in the very near future (figure 1.25).

The channel is single-mode optical fibers dispersive in the negative or positive factors and with some residual dispersion. Thus, the distortion of the pulse is purely phase distortion prior to the detection by the PD which follows the square law rule for direct detection case. On the other hand, for coherent detection the beating between the local oscillator and the signal in the PD would lead to the preservation of the phase and one could be considered a pure phase distortion. Again, in the case of direct detection after the square law detection the phase distortion is then transferred to the amplitude distortion. In order to conduct the equalization process, it is important for us to know the impulse and step responses of the fiber channel $h_F(t)$ and $s_F(t)$. We will then give the fundamental aspects of equalization using feed forward equalization, decision-feedback equalization with maximum mean square error (MMSE) or maximum likelihood sequence estimation of Viterbi algorithm (MLSE).

1.5.3.1 Feed forward equalizer

FFE is a linear equalizer which has been the most widely studied and a transversal filter structure would offer a linear processing of the signal prior to decision. The structure of an FFE transversal filter consists of a cascade delay of the input sample and at each delay the signal is tapped and multiplied with coefficients. These tapped signals, whose delay tap time is the bit period, are then summed to give the output

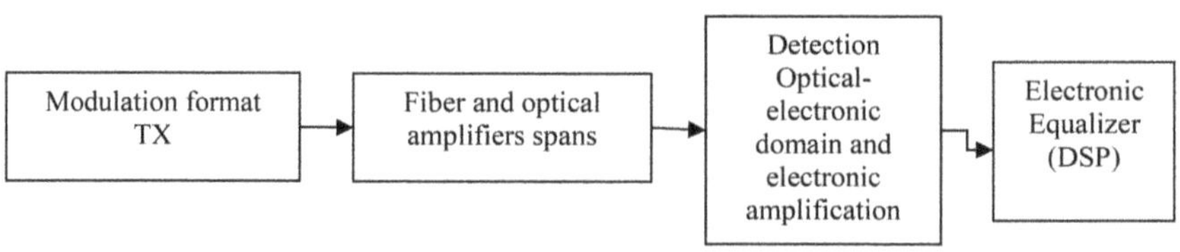

Figure 1.25. Schematic diagram of the location of the electronic equalizer at the receiver of an advanced optical communications system. DSP = digital signal processor.

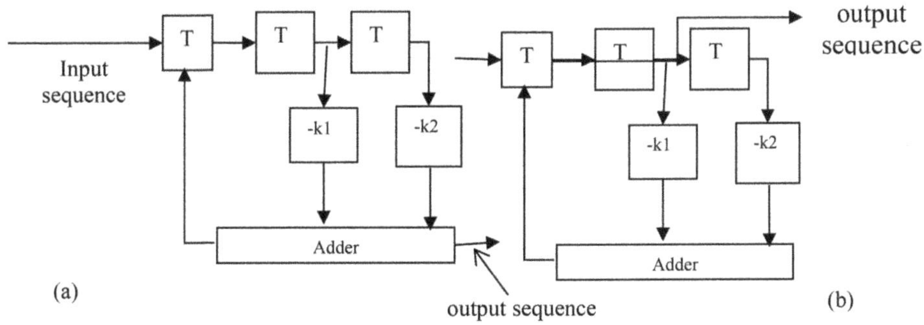

Figure 1.26. Linear (a) feed forward equalization (transversal equalization) and (b) feedback equalization scheme. $z^{-1} = T$ represents delay unit of sampling in frequency domain or discrete domain.

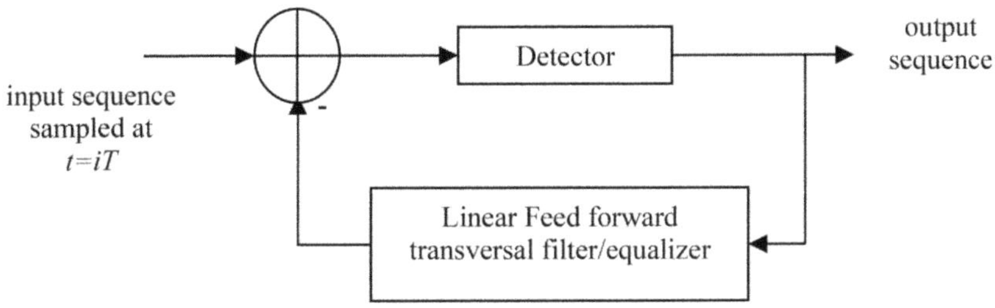

Figure 1.27. Schematic diagram of receiver using nonlinear equalization by decision directed cancellation of ISI.

sample. The coefficients of the transversal filter must take values that match with the channel so the convolution of the channel impulse response and that of the filter result in a unity so as to achieve complete equalized pulse sequence at the output. Figures 1.26(a) and (b) show the linear equalization scheme that uses either feed forward or feedback equalization, the difference between these two schemes is the tapped signals either at the output of the transversal filter or at the output of the feedback that minimize the input sequence.

On the other hand, a decision-feedback equalizer (DFE) differs from that of the linear equalizer with a decision detector that would determine the signal amplitude required for feedback to the difference error at the input as shown in figure 1.26.

1.5.3.2 Decision feedback equalization

The FFE method is based on the use of a linear filter with adjustable coefficients. The equalization method that exploits the use of previous detected symbols to suppress the ISI in the present symbol being detected, is termed as decision-feedback equalization.

The decision-feedback equalizer (DFE) is illustrated in figure 1.27. It consists of m coefficients and m delay taps. Each tap spacing (delay in time domain) equals the bit duration T or sampling time which may be much smaller. From figure 1.26(b) we can see that the received signal sequence goes through the forward filter first. After making decisions on previously detected symbols, the feedback filter provides information from the previously detected symbols for present estimating.

1.5.3.3 Minimum mean square error equalization

Consider next the general case where the linear equalizer is adjusted to minimize the mean square error due to both ISI and noise. This is called the MMSE (minimum mean square error equalization).

1.5.3.3.1 Placement of equalizers

Linear and nonlinear equalization are possible as they are well known in the field of signal processing. The principles of equalization with equalizers placed at the transmitter, or the receiver or sharing between the transmitter and receiver are critical for practical networks.

1.5.3.3.2 MlSE electronic equalizers

Amongst the electronic equalization techniques, maximum likelihood sequence estimation (MLSE), which can be implemented effectively with the Viterbi algorithm has attracted considerable research interest [51–55]. However, most of the studies on MLSE equalizers have focused on either ASK or DPSK formats [43, 44, 47, 56]. Apart from a co-authored paper reported recently [26], there has not been any study on the performance of MLSE equalizers for optical MSK transmission, especially when OFDR is used as the detection scheme. The performance of OFDR-based MSK optical transmission systems is significantly enhanced with the incorporation of post-detection MLSE electronic equalizers as the ISI problem caused by either fiber dispersion impairments or tight optical filtering effects are effectively mitigated. Therefore, this decision process considered as an equalizer is included as a case study in a chapter of this book to comprehensively investigate the performance of MLSE equalizers for OFDR-based MSK optical transmission systems.

1.6 100G PDM-QPSK 25GE performance

A set-up of 100 Gbps PDM-QPSK transmission of a commercial product in compact form is conducted as shown in figure 1.28 using a traffic generator (Anritsu 1040A) and the optical transceiver of PDM-QPSK formats. The receiver sensitivity is given in figures 1.29 and 1.30 which shows that a sensitivity of −42 dBm can be obtained or equivalently five photons per bit. This is the limit of coherent detection. The LO laser is independent with respect to the laser used as the source at the transmitter.

1.7 Data pulse shaping and DWDM

1.7.1 Pulse shaping

Figures 1.31(a)–(f) show sequential (a) time domain pulse shapes with zero crossings at sampling points so that the samples at the sampling points can offer maximum pulse amplitude without much inter-symbol interference (ISI); (b) correspondingly the frequency shape of the shaping of pulses; (c) eye diagram of pulse shaped

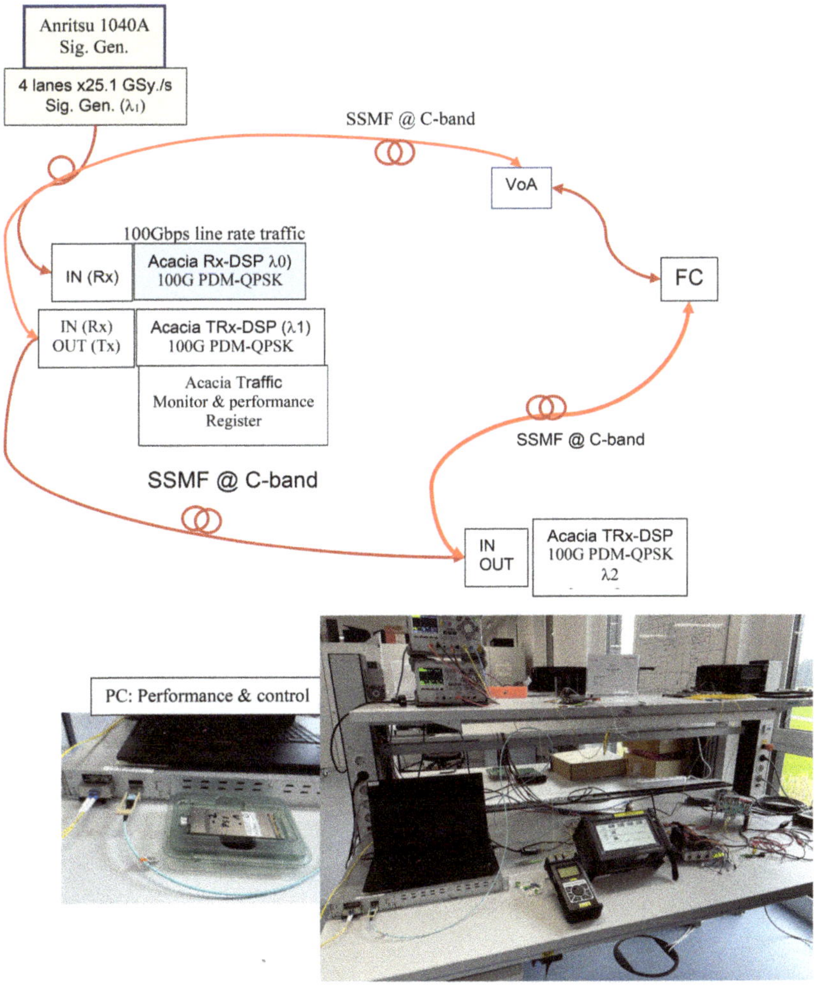

Figure 1.28. Schematic and photo of the 100G optical transceivers (Acacia 100G PDM-QPS module) with two wavelength and bidirectional transmission in a B2B (back-to-back) incorporating a VOA (variable optical attenuator). SSMF = standard single-mode fiber or Corning SMF-28, FC = fiber coupler, VoA = variable optical attenuator, Rx = receiver, Tx = transmitter, OTRX = optical transceiver, DSP = digital signal processor.

sequences; (d) optical spectrum of packed channels of 5 and 10 modulated carriers. (e) and (f) are the eye diagram and frequency spectrum of one modulated carriers and optical spectrum of unshaped 10 modulated channels of 100G.

The pulse shaping of $\sin(x)/x$, or root raise cosine, thus, allows the packing of adjacent modulated channels of the considered channel as much as possible without much ISI between one another. Recently, a commercial manufacturer has provided this feature as an algorithm at the DSP of the transmitter section of the OTRx. We thus take this advantage into considerations for packing and increase the total

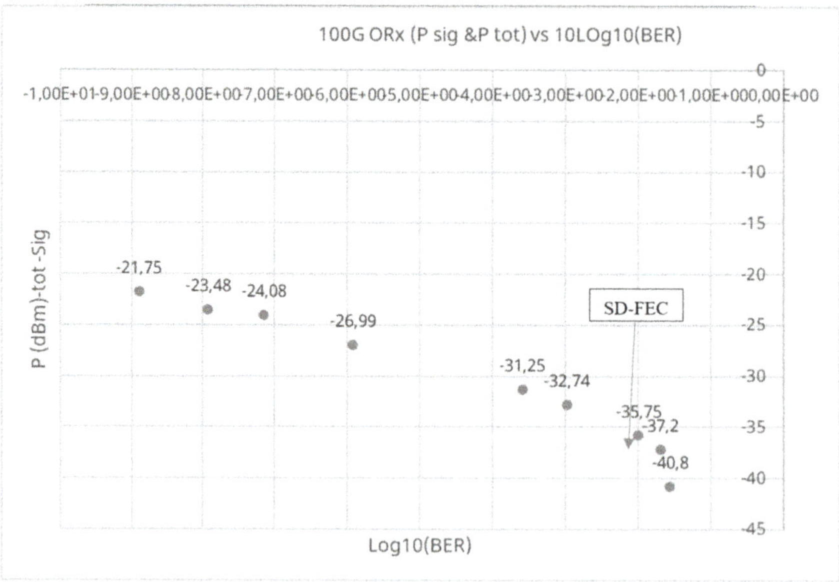

Figure 1.29. Plots of BER versus signal average power (dBm) without any optical amplifier as a pre-amp optical reception sub-system. Note: SD-FEC 2e–1 achieved at −42 dBm for 100 Gbps (PDM-QPSK @25 GSy/s) or about five photons/bits as expected from SoA & CA WP1100 as reported. Legend: P tot—total optical received power as registered by TRx. including noises. P_{sig} = extracted signal power. Note: No optical pre-amplifier is used. 10log10(BER) = −2 at −37.2 dBm.

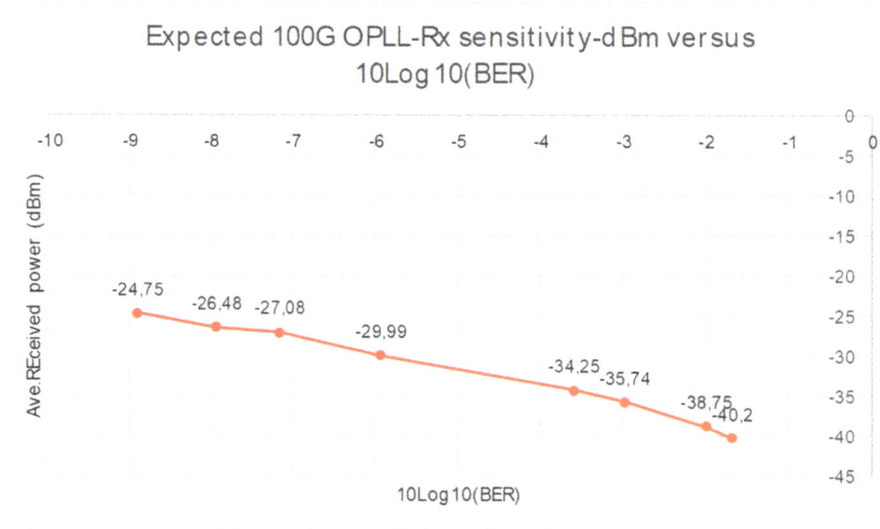

Figure 1.30. Plots of BER versus signal average power (dBm) without any optical amplifier as a pre-amp optical reception sub-system, but incorporating OPLL in DSP processing. Note: expected BER based on optical transceiver 100G of Acacia PDM-QPSK transceiver. 25.2 GSy/s. 100G PDM-QPSK transceiver. 10log10(BER) = −2 dB at −38.75 dBm receiver sensitivity.

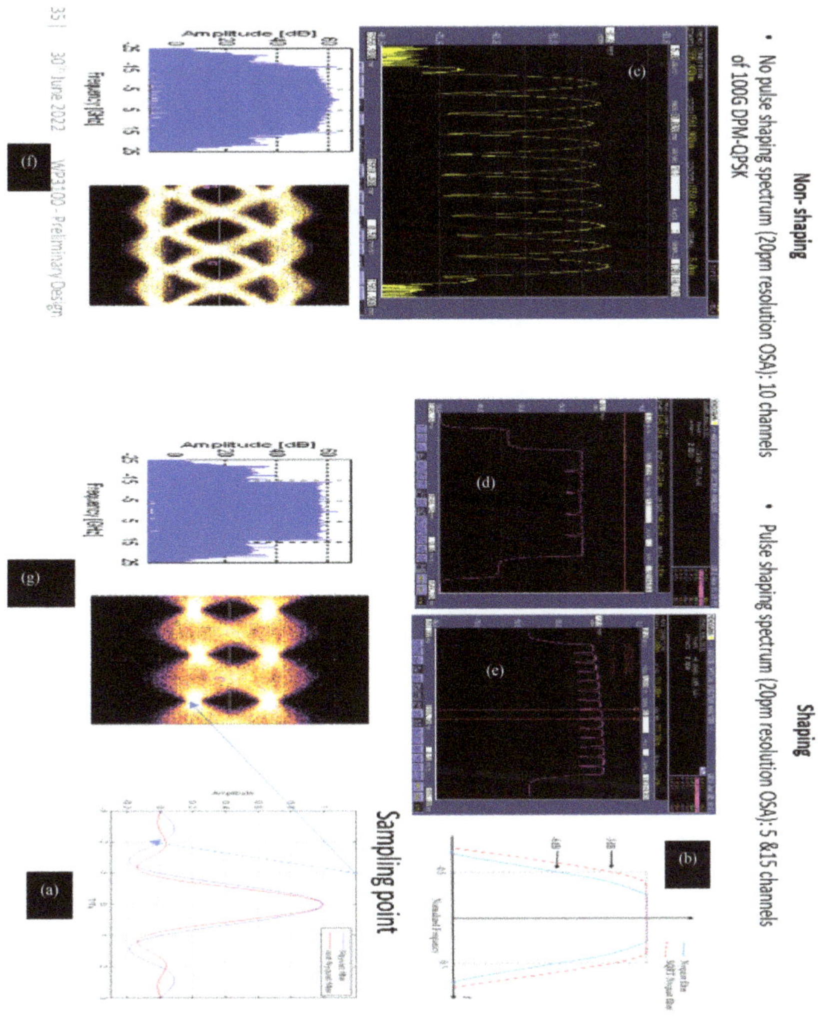

Figure 1.31. Pulse shaping (RRC/RC—root raise cosine)—a commercial TRx. (a) Time domain pulse shaping —note zero crossing at sampling point leading to no error contribution of samples of eye opening of (d). (b) Frequency domain of pulse shaping of (a); (c) optical spectra of 5 and 10 channels under pulse shaping at 100G QPSK25; (e) spectrum of 10 channels under no pulse shaping; and (f) frequency spectrum and eye diagram of non-shaping pulse sequence [57].

transmission capacity. The demonstrator would investigate this important feature for a Tbps space transmission system.

1.7.2 OSNR and BER

It is expected that the BER of 2e−3 under SD-FEC or soft defined FEC is employed as specified by the manufacturer of OTRx. The symbol rates and required OSNR are given in figures 1.32(a) and (b) under certain symbol rates and overhead of different

(a)

Mode	Bits/Symbol	Symbol rate GB/s
100Gbps OTU4 non-diff 27% SDFEC DP-QPSK	2	34.71754971
200Gbps OTU4 non-diff 27% SDFEC DP-QPSK	2	69.43509943
200Gbps OTU4 non-diff 27% SDFEC DP-P-16QAM	3	46.29006629
200Gbps OTU4 non-diff 27% SDFEC DP-16QAM	4	34.71754971
200Gbps OTU4 non-diff 27% SDFEC DP-32QAM	5	27.77403977
300Gbps OTU4 non-diff 27% SDFEC DP-P-16QAM	3	69.43509943
300Gbps OTU4 non-diff 27% SDFEC Fractional QAM BICHM56	3.4375	60.59790496
300Gbps OTU4 non-diff 27% SDFEC DP-16QAM	4	52.07632457
400Gbps OTU4 non-diff 27% SDFEC DP-16QAM	4	69.43509943

: Symbol rates for different modes

(b)

	50Gbps	100Gbps	150Gbps	200Gbps	250Gbps	300Gbps	350Gbps	400Gbps	450Gbps	500Gbps	550Gbps
1 bits/sym	6.8	9.8									
1.5 bits/sym		9.8	11.6								
2 bits/sym		9.8	11.6	13.1							
2.5 bits/sym			13.0	14.4	15.5						
3 bits/sym			13.7	15.1	16.1	17.3					
3.5 bits/sym				15.9	17.0	18.0	19.0				
4 bits/sym				16.4	17.4	18.4	19.4	20.4			
4.5 bits/sym				17.7	18.8	19.9	20.8	21.6	22.6		
5 bits/sym					20.0	21.0	22.0	22.7	23.7	24.8	
5.5 bits/sym					21.4	22.4	23.5	24.4	25.2	26.2	27.7
6 bits/sym						23.6	24.7	25.8	26.7	27.7	29.0

Table 20: Typical OSNR Sensitivities with 27% FEC OH

Note: Typical OSNR sensitivity is an average of large sample size at room temperature, not including wavelength dependence, input power dependence, temperature dependence and aging.

Figure 1.32. (a) TRx (100G–400G) symbol rate/SD-FEC; and (b) OSNR for different No. bits/symbol—typical commercial TRx.

FEC schemes, respectively, for PDM-QPSK. OSNR for 2 and 4 bits/symbol or QPSK and 16QAM equivalent to QPSK25 and QPSK50 as discussed in the previous section, the OSNR is 9.8 dB and 13.1 dB, respectively, as expected. The difference of about 3 dB is expected due to the doubling of the bandwidth of the QPSK50 as compared to QPSK25, thence the doubling in noise power because the total noise is integrated over the entire bandwidth of the sub-system, especially the electronic transimpedance amplifier that follows the balanced detector pair.

1.8 Organization of the chapters

The principal motivations of this book are to describe the processing of signals under the contemporary scenario of transmission technologies with ultra-high speed but short-reach, as well as processing in the photonic domain taking advantage of the neural computing network. The employment of digital communications techniques in modern optical communications is now used under the scenario of cloud networking at extreme sped, thus reaching the sampling rate limits of the electronic sampler in ADC or DAC. Short reach to medium transmission link is so much more important.

Furthermore, these techniques are essential for modules, integrated photonics and processing in either electronic or photonic domains for ultra-high capacity global DC inter-networking and especially in the intra-DC networking and interconnecting between DCs as well as ultra-high capacity access for distributed DC networking in 5G generation networks.

Chapter 2 gives a summary of the principles of optical transmission, especially the modulation formats and receiving techniques to achieved ultra-wideband signals with band-limited devices. The fundamental principles of digital communications employing either coherent or incoherent reception and associated modulation formats are described with a focus on the technological development and limitations in the optical domain. The enabling technologies and research results and demonstration in laboratory experimental platforms for the development of high performance and high capacity next generation optical transmission systems impose significant challenges on the engineering of optical transmission systems in the near future and techniques for next generation data communications links.

The presentation of this book follows the progression of the digital communications and integration of modulation techniques in optical communications under self-coherent, direct detection and intensity modulation, and coherent reception techniques, see also chapter 4.

The essential properties of optical fibers as the transmission medium for long haul, metro and access optical networks, is introduced and given in the appendix. Both geometrical and profile structures as well as the propagation of a modulation envelope under the influence of interference due to the different propagation velocities due to guided mode confinement of the fiber core and wavelength-dependent refractive index are summarily treated to give insights into the propagation of lightwaves and distortion on the modulated signals so that the equalization and compensation in optical and DSP electronic domains can be developed in later chapters of the book.

The chapters are organized as follows:

In chapter 2, on modern DC networking (DCN):

- Evolution of DCN and traditional telecom networks;
- Internet content transport for high capacity demand society;
- Data transmission networks;
- Basic rates, capacity and server clusters;
- Inter data-center transport;
- Core and 'metro' transport;
- Latency and switching.

Chapter 3 on optical transmission systems at ultra-high speed:

- Data optical communications at 400G and higher;
- Tera-bits s^{-1} transport in DCN (DC networking) and DCI (inter-DC connect);
- Electronic processing in digital domain for ultra-speed transport;
- Modulation formats and capacity.

Chapter **4** on optical switching for DC networking:

- Electronic and optical switching for data communications;
- Photonic switching: technique and technology;
- Photonic switching devices and solutions for DCN and DCI.

Chapter 5 on photonic signal processing (PSP):

- Why PSP for data communications: speed and power consumptions in DCN and DCI and impact of electronic processing and switching;.
- PSP techniques;
- Optical neural computing and PSP;
- Multi-channel processing in optical domain.

Chapter 6 on tera-bits s^{-1} transport systems and DCs:

- >100 GBaud optical transmission technology;
- Electronic digital processing technology;
- Coherent and incoherent optical transmission technique;
- >250 Gbits s^{-1} IM/DD (intensity modulation and direct detection) with special modulation technique;
- >800 Gbits s^{-1} coherent transport technology.

Chapter 7 on band-limited and bandwidth-limited scenarios in transmission.

Appendices provide some background on a number of specific technical descriptions related to optical transmission and measurement techniques in data center networking.

- Appendix A: optical fibers including standard SMF (SSMF) and enhanced SSMF and large effective area fibers.
- Appendix B: Balanced coherent reception techniques and noise considerations.
- Appendix C: Power budget.
- Appendix D: Inband OSNR under optically amplified and ROADM-based transmission.

Furthermore, chapter 6 addresses several physical aspects of all optical routing employing wavelength selective switches and thence ROADM (reconfigurable optical add/drop mux) in which MEMS (micro-electro-mechanical switch) to steering the beams in micro-optic and grating structures. Thence this chapter gives detailed operations and structures of all optical routing DCN and DCI and the structures of DC internal connection and interconnection of DCI for networking.

Chapter 7 then gives a simple view on the transmission at the two extreme limits, the power-limited limit and the bandwidth-limited scenarios. For multi-Tbps transmission the systems reach the bandwidth limit at which a number of signaling dimensions have already been exploited including the higher order complex modulation formats, the sampling rate, the multiplexing of wavelength carriers, the diversity of polarizations and phases. The power-limited transmission is when the power of the signals is extremely low, or at the quantum level. The transmission is the continuous variable quantum key distribution (CV-QKD) technique or the energy per bit is well below the energy of one photon. The application of this transmission is for secure transport of secret keys for safe transmission, especially under banking, secret document and images transmission both inter- and internal DC networking.

On the other limit, the power-limited case is when the signal average power level is well below the noise level and reaches the quantum level of a fraction of the photon energy in a bit or key. This is considered to be the most secure states and keys for optical transmission. The continuous variable quantum key distribution.

Four appendices are given to (i) basics of single-mode fiber properties for transport of ultra-high speed channel signals, including the extension of the spectrum band to super C-band; (ii) receiver sensitivity under coherent transmission and reception of polarization division multiplexing quadrature amplitude modulation (PDM-mQAM), especially the noise model; (iii) and (iv) techniques for measurement of OSNR under PDM-mQAM modulation.

References

[1] Kao C and Hockham G 1966 Dielectric-fibre surface waveguides for optical frequencies *Proc. IEE* **113** 1151–8

[2] Kaminow I P and Li T 2002 *Optical Fiber Communications* IVB (Amsterdam: Elsevier Science)

[3] Lin C, Kogelnik H and Cohen L G 1980 Optical pulse equalization and low dispersion transmission in single-mode fibers in the 1.3–1.7 mm spectral region *Opt. Lett.* **5** 476–8

[4] Gnauck A H, Korotky S K, Kasper B L, Campbell J C, Talman J R, Veselka J J and McCormick A R 1986 Information bandwidth limited transmission at 8 Gb/s over 68.3 km of single mode optical fiber *Proc. of OFC'86, Paper PDP6 (Atlanta, GA)*

[5] Kogelnik H 1985 High-speed lightwave transmission in optical fibers *Science* **228** 1043–8

[6] Agrawal G P 2002 *Fiber-Optic Communication Systems* 3rd edn (New York: Wiley)

[7] Chraplyvy A R, Gnauck A H, Tkach R W and Derosier R M 1993 8×10 Gb/s transmission through 280 km of dispersion-managed fiber *IEEE Photon. Technol. Lett.* **5** 1233–5

[8] Kogelnik H 2000 High-capacity optical communications: personal recollections *IEEE J. Sel. Top. Quant. Electron.* **6** 1279–86

[9] Giles C R and Desurvire E 1991 Propagation of signal and noise in concatenated erbium-doped fiber amplifiers *IEEE J. Lightwave Technol.* **9** 147–54

[10] Becker P C, Olsson N A and Simpson J R 1999 *Erbium-Doped Fiber Amplifiers, Fundamentals and Technology* (San Diego, CA: Academic)

[11] Farries M C, Morkel P R, Laming R I, Birks T A, Payne D N and Tarbox E J 1989 Operation of erbium-doped fiber amplifiers and lasers pumped with frequency-doubled Nd: YAG lasers *IEEE J. Lightwave Technol.* **7** 1473–7

[12] Okoshi T 1982 Heterodyne and coherent optical fiber communications: recent progress *IEEE Trans. Microwave Theory Tech.* **82** 1138–49

[13] Okoshi T 1987 Recent advances in coherent optical fiber communication systems *IEEE J. Lightwave Technol.* **5** 44–52

[14] Salz J 1986 Modulation and detection for coherent lightwave communications *IEEE Commun. Mag.* **24** 38–49

[15] Okoshi T 1986 Ultimate performance of heterodyne/coherent optical fiber communications *IEEE J. Lightwave Technol.* **4** 1556–62

[16] Henry P S 1990 *Coherent Lightwave Communications* (New York: IEEE Press)

[17] Elrefaie A F, Wagner R E, Atlas D A and Daut A D 1988 Chromatic dispersion limitation in coherent lightwave systems *IEEE J. Lightwave Technol.* **6** 704–10
[18] Charlet G *et al* 2005 Comparison of system performance at 50, 62.5 and 100 GHz channel spacing over transoceanic distances at 40 Gbit/s channel rate using RZ-DPSK *Electron. Lett.* **41** 145–6
[19] Cho P S, Grigoryan V S, Godin Y A, Salamon A and Achiam Y 2003 Transmission of 25-Gb/s RZ-DQPSK signals with 25-GHz channel spacing over 1000 km of SMF-28 fiber *IEEE Photonics Technol. Lett.* **15** 473–5
[20] Chagnon M, Osman M, Poulin M, Latrasse C, Gagné J-F, Painchaud Y, Paquet C, Lessard S and Plant D 2014 Experimental study of 112 Gb/s short reach transmission employing PAM formats and SiP intensity modulator at 1.3 μm *Opt. Express* **22** 2108–36
[21] Olmedo M I, Zuo T, Jensen J B, Zhong Q, Xu X, Popov S and Monroy I T 2014 Multiband carrierless amplitude phase modulation for high capacity optical data links *J. Lightwave Technol.* **32** 798–804
[22] Kai Y, Nishihara M, Tanaka T, Takahara T, Li L, Tao Z, Liu B, Rasmussen J and Drenski T 2013 Experimental comparison of pulse amplitude modulation (PAM) and discrete multi-tone (DMT) for short-reach 400-Gbps data communication *Proc. Eur. Conf. Optical Communications Conf. (ECOC)* paper Th.1.F.3
[23] Tanaka T, Nishihara M, Takahara T, Yan W, Li L, Tao Z, Matsuda M, Takabayashi K and Rasmussen J 2014 Experimental demonstration of 448-Gbps+ DMT transmission over 30-km SMF *Proc. Optical Fiber Communication Conf. (OFC)* paper M2I.5
[24] Schares L *et al* 2006 Terabus: terabit/second-class card-level optical interconnect technologies *IEEE J. Sel. Top. Quant. Elect.* **12** 1032–44
[25] Shen Y *et al* 2017 Deep learning with coherent nanophotonic circuits *Nat. Photon.* **11** 441–6
[26] Paquot Y 2017 Optoelectronic reservoir computing *Sci. Rep.* **2** 287
[27] Fjelde T *et al* 2000 Demonstration of 40 Gbits/s all-optical logic XOR in integrated SOA-based interferometric wavelength converter *Elect. Lett.* **36** 1863–4 (ii) See also IEEE Younis R M *et al* 2000 Fully integrated AND and OR optical logic gates *Photon. Technol. Lett.* **26** 1900–3
[28] Gnauck A H and Winzer P J 2005 Optical phase-shift-keyed transmission *IEEE J. Lightwave Technol.* **23** 115–30
[29] Binh L N 2008 *Digital Optical Communications* (Boca Raston, FL: CRC Press, Taylor and Francis)
[30] Mo J, Yi D, Wen Y, Takahashi S, Wang Y and Lu C 2005 Optical minimum-shift keying modulator for high spectral efficiency WDM systems *Proc. ECOC'05* **4** 781–2
[31] Huynh T L, Sivahumaran T, Binh L N and Pang K K 2007 Narrowband frequency discrimination receiver for high dispersion tolerance optical MSK systems *Proc. Coin-Acoft'07, Paper TuA1–3 (Melbourne, Australia)*
[32] Huynh T L, Sivahumaran T, Binh L N and Pang K K 2007 Sensitivity improvement with offset filtering in optical MSK narrowband frequency discrimination receiver *Proc. Coin-Acoft'07, Paper TuA1–5 (Melbourne)*
[33] Sivahumaran T, Huynh T L, Pang K K and Binh L N 2007 Non-linear equalizers in narrowband filter receiver achieving 950 ps/nm residual dispersion tolerance for 40 Gb/s optical MSK transmission systems *Proc. OFC'07, Paper OThK3 (Anaheim, CA)*
[34] Sakamoto T, Kawanishi T and Izutsu M 2005 Optical minimum-shift keying with external modulation scheme *Opt. Express* **13** 7741–7

[35] Lazaro J A, Idler W, Dischler R and Klekamp A 2004 BER depending tolerances of DPSK balanced receiver at 43 Gb/s *Proc. IEEE/LEOS Workshop on Advanced Modulation Formats 2004* pp 15–6
[36] Kim H and Gnauck A H 2003 Experimental investigation of the performance limitation of DPSK systems due to nonlinear phase noise *IEEE Photon. Technol. Lett.* **15** 320–2
[37] Bhandare S, Sandel D, Abas A F, Milivojevic B, Hidayat A, Noe R, Guy M and Lapointe M 2004 2/spl times/40 Gbit/s RZ-DQPSK transmission with tunable chromatic dispersion compensation in 263 km fibre link *Electron. Lett.* **40** 821–2
[38] Mizuochi T, Ishida K, Kobayashi T, Abe J, Kinjo K, Motoshima K and Kasahara K 2003 A comparative study of DPSK and OOK WDM transmission over transoceanic distances and their performance degradations due to nonlinear phase noise *IEEE J. Lightwave Technol.* **21** 1933–43
[39] Binh L N and Huynh T L 2007 Phase-modulated hybrid 40 Gb/s and 10 Gb/s DPSK DWDM long-haul optical transmission *Proc. OFC'07, Paper JWA94 (Anaheim, CA)*
[40] Ito T, Sekiya K and Ono T 2002 Study of 10 G/40 G hybrid ultra long haul transmission systems with reconfigurable OADM's for efficient wavelength usage *Proc. ECOC'02, Paper 1.1.4 (Copenhagen, Denmark)*
[41] Iwashita K and Takachio N 1989 Experimental evaluation of chromatic dispersion distortion in optical CPFSK transmission systems *IEEE J. Lightwave Technol.* **7** 1484–7
[42] Proakis J G 2001 *Digital Communications* 4th edn (New York: McGraw-Hill)
[43] Proakis J G and Salehi M 2002 *Communication Systems Engineering* 2nd edn (Englewood Cliffs, NJ: Prentice-Hall) pp 522–4
[44] Pang K K 2005 *Digital Transmission* (Melbourne: Mi-Tec Media Pty. Ltd)
[45] Iwashita K and Matsumoto T 1987 Modulation and detection characteristics of optical continuous phase FSK transmission system *IEEE J. Lightwave Technol.* **5** 452–60
[46] Mo J, Wen Y J and Wang Y 2007 Performance evaluation of externally modulated optical minimum shift keyed data *Opt. Eng.* **46** 035001
[47] Shieh W and Djordjevic I 2009 *OFDM for Optical Communications* 1st edn (New York: Academic)
[48] Huynh T L, Binh L N and Pang K K 2006 Optical MSK long-haul transmission systems *Proc. SPIE APOC'06, Paper 6353-86, Thu9a*
[49] Elrefaie A F and Wagner R E 1991 Chromatic dispersion limitations for FSK and DPSK systems with direct detection receivers *IEEE Photon. Technol. Lett.* **3** 71–3
[50] Little B E 2003 Advances in microring resonator *Integrated Photonics Research Conf. 2003, Invited Paper*
[51] Alic N, Papen G C, Saperstein R E, Milstein L B and Fainman Y 2005 Signal statistics and maximum likelihood sequence estimation in intensity modulated fiber optic links containing a single optical preamplifier *Opt. Express* **13** 4568–79
[52] Brent L *et al* 2006 Advanced ring resonator based PLCs *LEOS 2006 – 19th Annual Meeting of the IEEE Lasers and Electro-Optics Society* pp 751–2
[53] Curri V, Gaudino R, Napoli A and Poggiolini P 2004 Electronic equalization for advanced modulation formats in dispersion-limited systems *IEEE Photon. Technol. Lett.* **16** 2556–8
[54] Agazzi O E, Hueda M R, Carrer H S and Crivelli D E 2005 Maximum-likelihood sequence estimation in dispersive optical channels *IEEE J. Lightwave Technol.* **23** 749–62
[55] Napoli A 2006 Limits of maximum-likelihood sequence estimation in chromatic dispersion limited systems *Proc. OFC'06, Paper JThB36 (Anaheim, CA)*

[56] Qi J, Mao B, Gonzalez N, Binh L N and Stojanovic N 2013 Generation of 28 GBaud and 32 GBaud PDM-Nyquist-QPSK by a DAC with 11.3 GHz analog bandwidth *Proc. OFC2013 (San Francisco, USA)*

[57] Binh L N 2013 *Digital Processing: Optical Transmission and Coherent Receiving Techniques (Optics and Photonics)* (Boca Raton, FL: CRC Press)

IOP Publishing

Transmission, Processing, and All-Optical Routing for Ultra-High Capacity Data Center Networking (Second Edition)

Le Nguyen Binh

Chapter 2

Data centre networking

This chapter briefly describes the general aspects of data center networking (DCN), including some historical evolution of the networking of DCs, both inter- and intra-structures. The transmission technological development of pulse modulation formats and quadrature amplitude modulation (QAM), or frequency-domain coherent orthogonal frequency division multiplexing (Co-OFDM) and direct detection OFDM (DD-OFDM), for increasing the total capacity rate with band limited components. Interconnection technology is also briefly described.

2.1 Evolution of DCN and traditional telecom networks

A DC is a pool of resources (computational, storage and networking) interconnected using a communication network [1, 2]. Data Center Network (DCN) holds a pivotal role in a data center, as it interconnects all of the data center resources together. DCNs need to be scalable and efficient to connect tens or even hundreds of thousands of servers to handle the growing demands of cloud computing [3, 4]. Presently, DCs are constrained by the interconnection network [5].

2.1.1 Types of data center networks

2.1.1.1 Three-tier DCN

The legacy three-tier DCN architecture shown in figure 2.1 follows a multi-rooted tree-based network topology composed of three layers of network switches, namely access, aggregate, and core layers [6]. The servers in the lowest layers are connected directly to one of the edge layer switches. The aggregate layer switches interconnect multiple access layer switches together. All of the aggregate layer switches are connected to each other by core layer switches. Core layer switches are also responsible for connecting the DC to the Internet. The three-tier DCN is the

doi:10.1088/978-0-7503-5849-1ch2

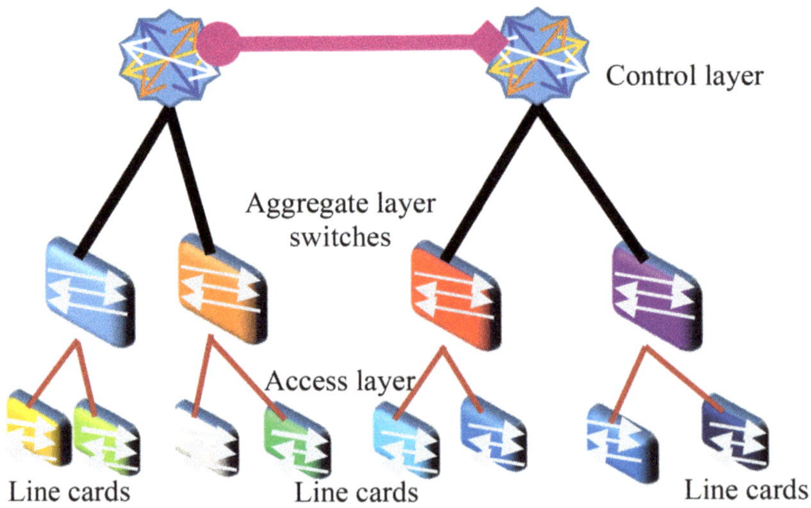

Figure 2.1. Three-tier structure of a DCN: core, aggregation and access layers. All connection paths are optical fibers of single-mode type, the standard single-mode fiber (SSMF)[1].

common network architecture used in DCs [6]. However, three-tier architecture is unable to handle the growing demand of cloud computing [7]. The higher layers of the three-tier DCN are highly oversubscribed [3]. Moreover, scalability is another major issue in three-tier DCN. Major problems faced by the three-tier architecture include, scalability, fault tolerance, energy efficiency, and cross-sectional bandwidth. The three-tier architecture uses enterprise-level network devices at the higher layers of topology that are very expensive and power hungry [5].

Fat tree DCN

Fat tree DCN architecture handles the over-subscription and cross-section bandwidth problem facing the legacy of three-tier DCN architecture. Fat tree DCN employs commodity network switches-based architecture using Clos topology [3]. The network elements in fat tree topology also follow hierarchical organization of network switches in access, aggregate, and core layers. However, the number of network switches is much larger than the three-tier DCN. The architecture is composed of k pods, where each pod contains, $(k/2)^2$ servers, $k/2$ access layer switches, and $k/2$ aggregate layer switches in the topology. The core layers contain $(k/2)^2$ core switches where each of the core switches is connected to one aggregate layer switch in each of the pods. The fat tree topology offers 1:1 over-subscription ratio and full bisection bandwidth.[3] The fat tree architecture uses a customized addressing scheme and routing algorithm (figure 2.2). The scalability is one of the major issues in fat tree DCN architecture and the maximum number of pods is equal to the number of ports in each switch [7], as shown in figure 2.3.

[1] See appendix A of the book for properties of single mode fibers

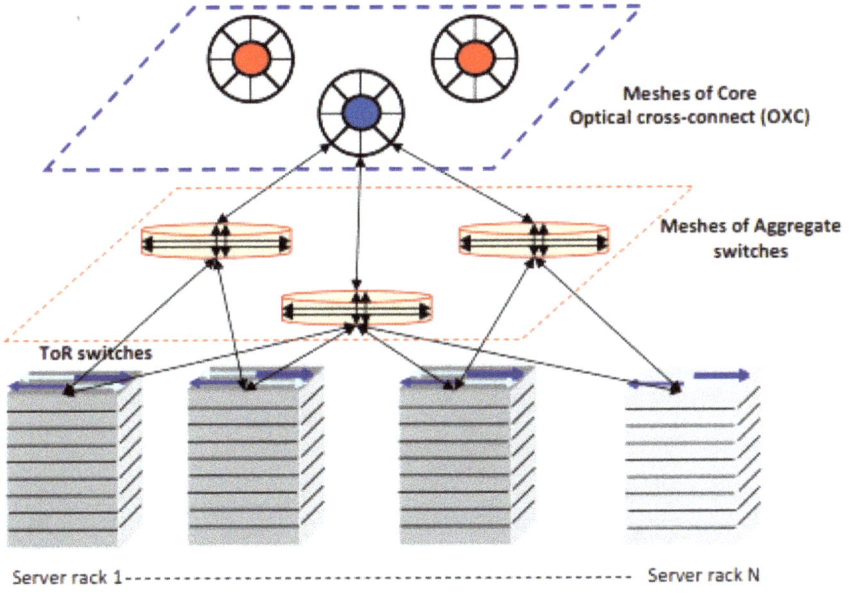

Figure 2.2. Multi-core aggregated switches three-tier DC with interconnection networking with clustered servers racks. ToR = top of rack.

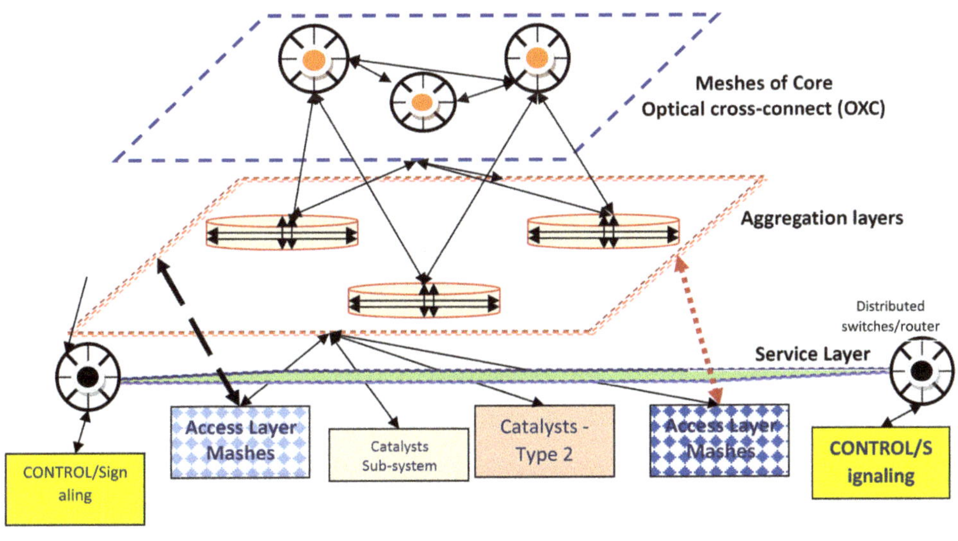

Figure 2.3. Detailed access layers and access interfaces devices with distinct service layers in lower and upper levels before core layers. All lines are optical by standard SMF.

DCell

DCell is a server-centric hybrid DCN architecture where one server is directly connected to many other servers [4]. A server in the DCell architecture is equipped with multiple network interface cards (NICs). The DCell follows a recursively built

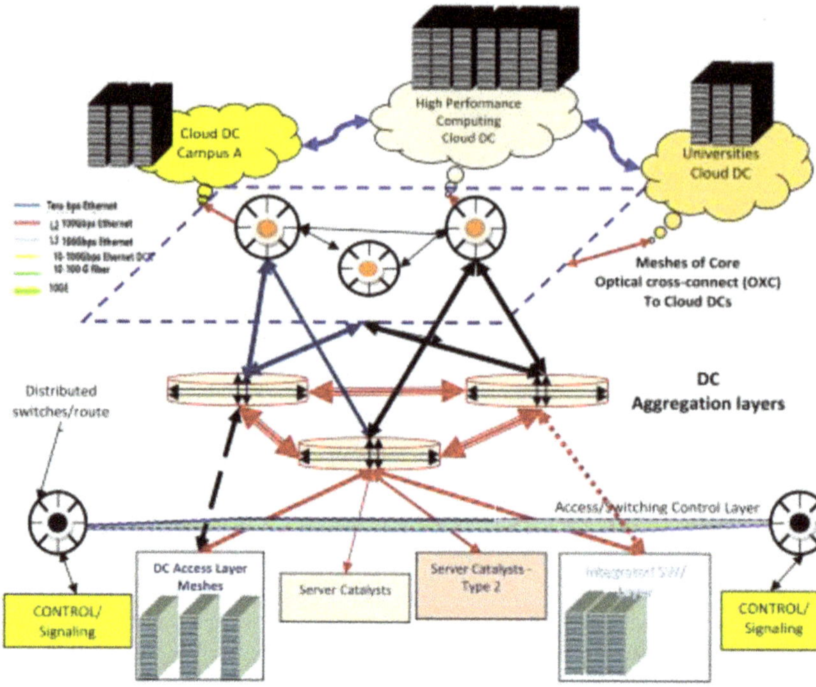

Figure 2.4. Arrangement structures of DC with stack aggregation and access layer in enterprise DC networks, mostly for university academic campuses.

hierarchy of cells. A $cell_0$ is the basic unit and building block of DCell topology arranged in multiple levels, where a higher level cell contains multiple lower layer cells. The $cell_0$ is the building block of DCell topology, which contains n servers and one commodity network switch. The network switch is only used to connect the server within a $cell_0$. A $cell_1$ contains $k = n + 1$ $cell_0$ cells, and similarly a $cell_2$ contains $(k \times n + 1)$ $Dcell_1$. The DCell is a highly scalable architecture where a four-level DCell with only six servers in $cell_0$ can accommodate around 3.26 million servers (figure 2.4). Besides very high scalability, the DCell architecture depicts a very high structural robustness [8]. However, cross-section bandwidth and network latency is a major issue in DCell DCN architecture [1].

Others

Other well-known DCNs can include: BCube [9], Camcube [10], FiConn [11], Jelly fish [12], and Scafida [13]. A qualitative discussion of different DCNs along with benefits and drawbacks associated with each one are referred to in figure 2.5.

2.1.2 Challenges

Scalability is one of the foremost challenges to the DCNs. With the advent of cloud paradigm, DCs are required to scale up to hundreds of thousands of nodes. Besides offering immense scalability, the DCNs are also required to deliver high cross-

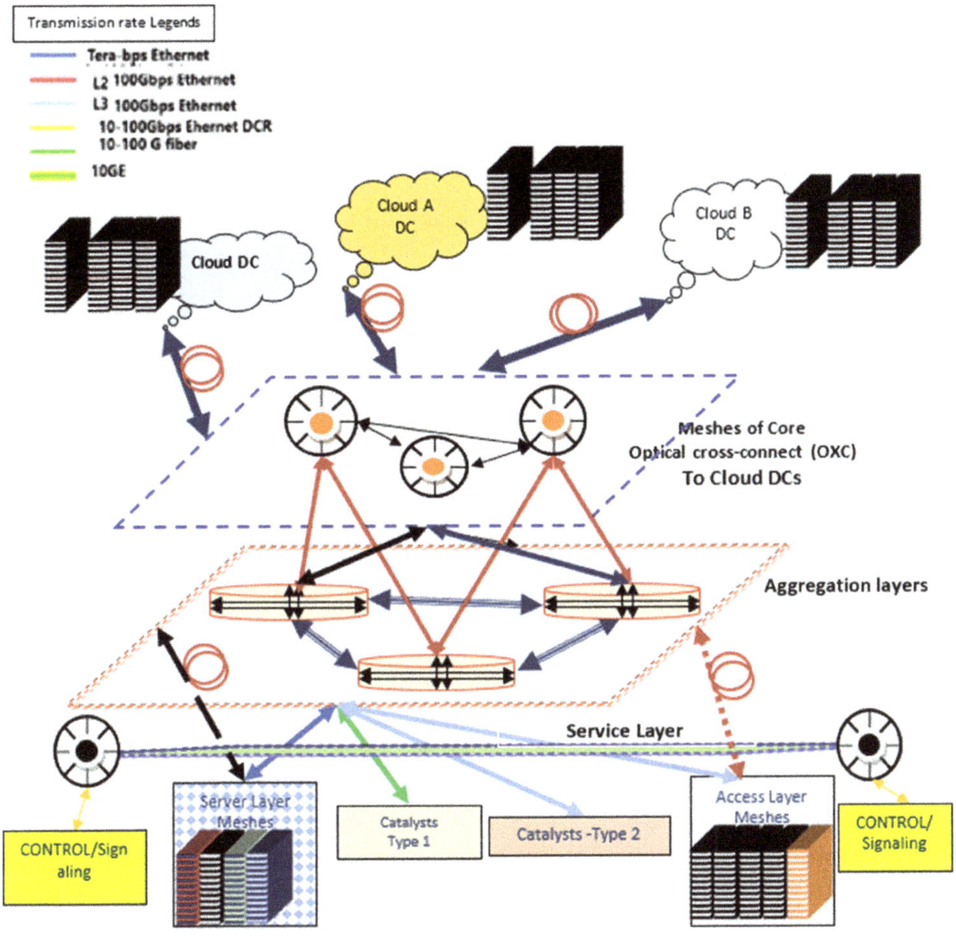

Figure 2.5. Cisco model of transmission interconnection rates in DCN and layers [6] and figure 2.7.

section bandwidth. Current DCN architectures, such as three-tier DCN offer poor cross-section bandwidth and possess very high over-subscription ratio near the roots [3]. Fat tree DCN architecture delivers 1:1 over-subscription ratio and high cross-section bandwidth, but it suffers from low scalability limited to k as the total number of ports in a switch. DCell offers high degree of scalability, but delivering very poor performance under heavy network load and one-to-many traffic patterns (figure 2.6).

2.1.3 Performance of DCNs

A quantitative analysis of the three-tier, fat tree, and DCell architectures for performance comparison (based on throughput and latency) is performed for different network traffic patterns [1]. The fat tree DCN delivers high throughput and low latency as compared to three-tier and DCell. DCell suffers from very low throughput under high network load and one-to-many traffic patterns. One of the

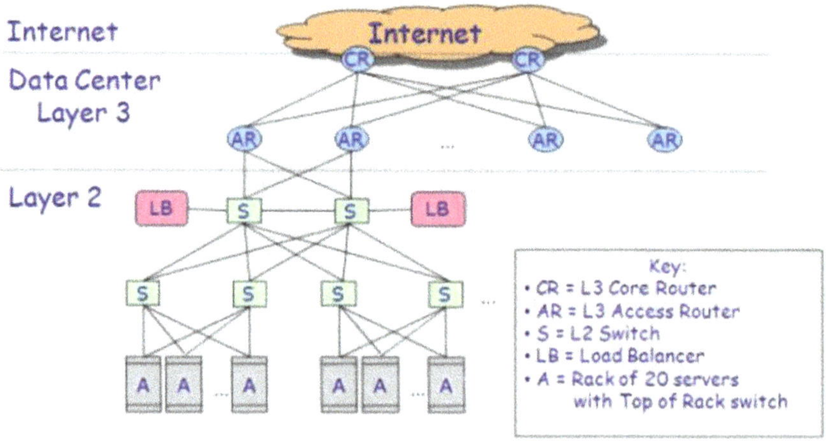

Figure 2.6. Multi layer network with internet networking layer, inter-data center networking (DCI) with access routers, and servers and intra-data-center networking (DCN).

major reasons for DCell's low throughput is very high over-subscription ratio on the links that interconnect the highest level cells.

Structural robustness and connectivity of DCNs

The DCell exhibits very high robustness against random and targeted attacks and retains most of its node in the giant cluster after even 10% of targeted failure [9], multiple failures whether targeted or random, as compared to the fat tree and three-tier DCNs [14]. One of the major reasons for high robustness and connectivity of the DCell is its multiple connectivity to other nodes that is not found in fat tree or three-tier architectures

Energy efficiency of DCNs

The concerns about the energy needs and environmental impacts of DCrs are intensifying. Energy efficiency is one of the major challenges of today's Information and communications technology (ICT) sector. The networking portion of a DC is accounted to consume around 15% of overall cyber energy usage. Around 15.6 million MWh of energy is utilized solely by the network's infrastructure within DCs world-wide. The energy consumption by the network infrastructure within a DC is expected to increase to around 50% in DCs. IEEE 802.3az standard has been finalized in 2011 that makes use of adaptive link rate technique for energy efficiency [15]. Moreover, fat tree and DCell architectures use commodity network equipment that is inherently energy efficient. Workload consolidation is also used for energy efficiency by consolidating the workload on a few devices to power-off or sleep for idle devices [16].

2.1.4 Communication interconnection speed

The transmission rate must be increased per fiber transmission line for interconnection to be as high as possible so as to minimize the total number of fibers between

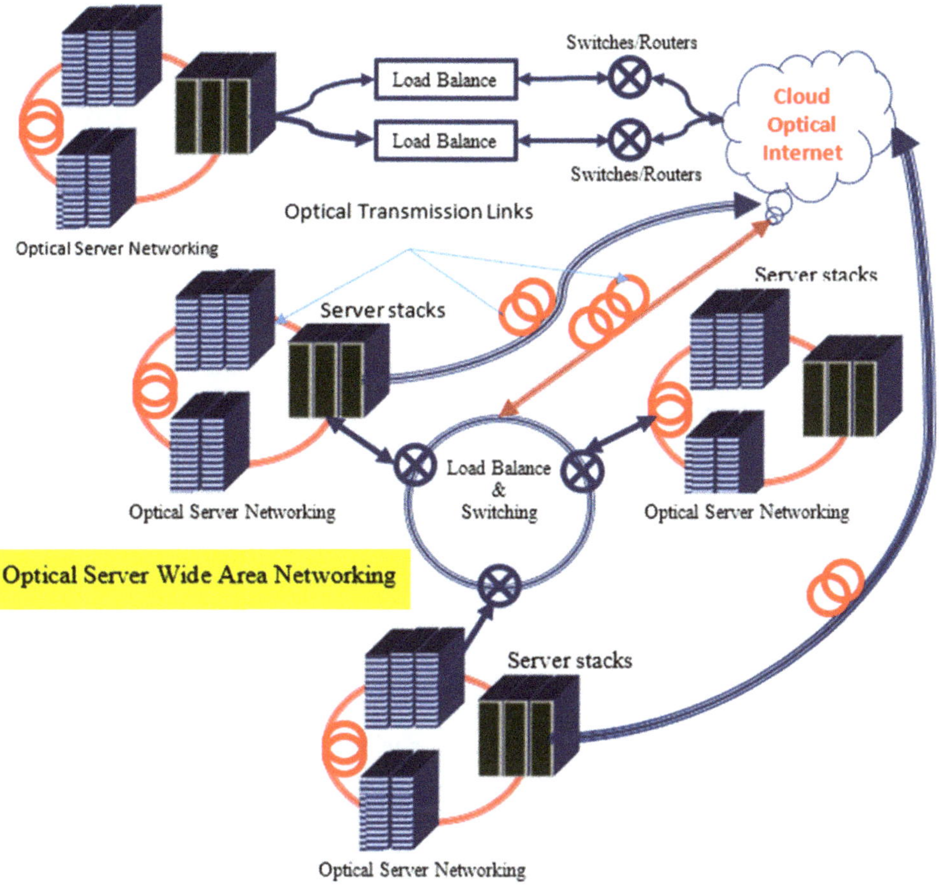

Figure 2.7. DC in contemporary global Internet in distribution mode of DCs. Labels: LAN = local area network; WAN = wide area network; IP = internet protocol.

servers to core switches and hence the complexity and bulkness of the cabling in DC, as shown in figure 2.7. Currently 400G has been standardized and higher aggregate speed over four wavelength channels or four lanes is under planning. It is one of the main aims of this book to describe and provide an understanding of the transmission technologies for such DC environments.

2.2 Telecom carriers and challenges from data center networking

Almost all carriers around the world, such as Korean, Telstra (Australia), Deutsche Telecom (Germany), Vodafone, Telefonica etc, have been taking urgent action in flattening their traditional telecommunication networks to meet the demands of simplicity, minimum latency and high speed capacity transport. These are the principal superior properties expected from DCN communities such as Google, Facebook, Microsoft etc. Both theoretic studies and demonstration of 5G works have been carried out to test some 5G features on live networks. The next sections of this chapter also demonstrate deployment tests of 5G in commercial environments [17]. We thus can expect that the flattening of traditional telecom networks can match those of DCN with the edge cloud distributed DCs of small or medium capacity as compared with concentric DCs.

A generic scenario of such 5G networking is depicted in figure 2.7, which gives a view of 5G transport networks in a densely populated city, long-haul interconnect transport of telecom carrier networks and DCs and user-concentrating environments, as well as machine-to-machine Internet of Things.

2.2.1 Evolution to 5G transport DC-based network

We expect the following evolution of the transport networks:

- Massive evolution of transport network from IP-based architecture of 4G to cloudified architecture for 5G driven by changes in ultra-high capacity and dynamic demand wireless and core network architecture of evolutionary traditional telecom and cloud DC networking.
- Aggregation of DC interconnect to metro core (MC) networks. The low latency and wideband services of 5G drives the down shifting of the backbone node to metro node. Alternatively, the whole terrestrial or intercontinental networks can be viewed as a metro transport network in which low cost low latency and ultra-high capacity must be met. These are the new challenges to be met by all but not only the DCN or telecom carriers or effectively all are 5G service providers (5GSP).
- As far as the radio access network (RAN) is to be densely distributed, metro D-RAN layers are formed and flattened, on which the cloud data units and centralized units can be aggregated.
- Cloud-RAN can enhance the gain of base station in a 4G network by 10 dB based on test results in Japan and more obviously in a 5G environment. This thus requires a unification of the CPRI (common public radio interface) and electrical CPRI.
- The OLT (optical line terminal) have evolved into cloudified mobile engines (MCEs), receiving or transmitting to new 5G core nodes.
- 5G network architecture is trending towards common and dynamic broadband and metro private line networking as well as multi-service transport.
- Connecting devices range from massive capacity premises such as education campus DCs and direct access by academic departments, machine-to-machine communications, fiber [8] to the premises (FTTP), and fiber or optical to the antenna (FTTA/OTTA), as shown in figure 2.8, in which the data communication link is delivering data to the antenna at which the optical signals are received and mixed with a local oscillator laser. The frequency of the LO is tuned to a frequency difference of the mmW center frequency. Thus, there are conversions of signals to the mmW band. The mmW power amplifier delivers to the antenna element, a mm size horn antenna embedded in printed circuit board. This is the optical to MIMO

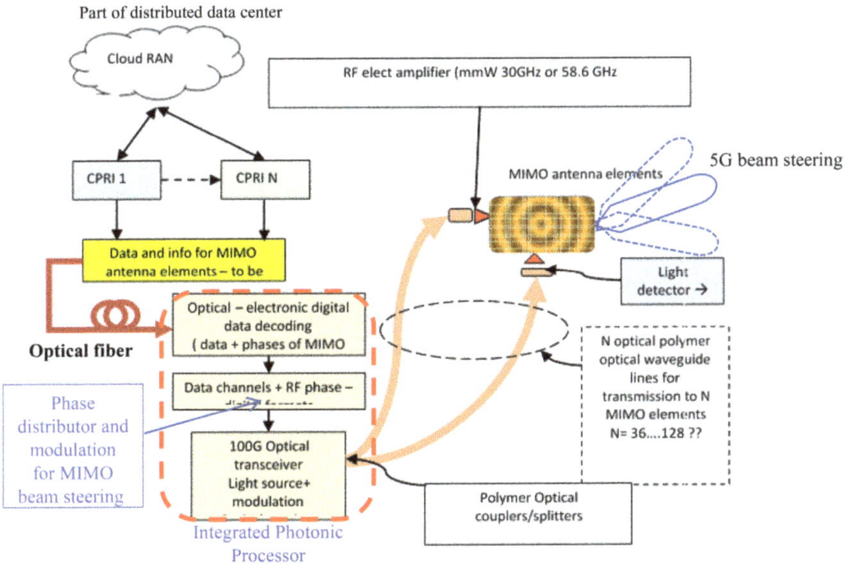

Figure 2.8. Integrated photonic processor in an OTTA optical to the antenna.

(multiple input multiple output) antenna for a mmW band wireless carrier. Furthermore, these OTTA are passive optical networks for distribution and hence sharing of capacity and cost.

The frequency bands for 5G, especially 5G$^+$ besides the 2–6 GHz (FR 1, FR = frequency range 1), are located in the microwave and millimeter wave (mmW) centralized at 18.6 GHz; 28.6 GHz or 56.8 GHz, 90 GHz. The free-license band is of 7.0 GHz of the 58.6 GHz (FR3). Data channels are to be delivered to mmW antennae which is a MIMO integrated horn, via optical guided medium and converted to mmW electronic frequency region and amplified to drive mmW MIMO antenna elements.

Connecting devices range from massive capacity premises such as education campus DCs and direct access by academic departments, machine-to-machine communications, FTTP, OTTA, as shown in figure 2.8, in which the data communication link is delivering data to the mmW antenna elements at which the optical signals are received and mixed with a local oscillator laser in the photo-detector. The frequency of the LO is tuned to a frequency difference with respect to that of the data carrier of mmW center frequency, e.g. 58.6 GHz. Thus, there are opto-electronic conversions of optical signals to the mmW band. The mmW power amplifier delivers to the antenna element, a mm size horn antenna is embedded in the printed circuit board. This is the optical to MIMO antenna for mmW band wireless carrier. Thus, there is a bank of receivers[2] and mixers and mmW RF electronic

[2] See appendix B of the book for details of reception sub-systems, especially on coherent receiver and noise considerations

drivers. A bank of photonic phase shifters is used to generate a pattern of phase delay so that beam steering can be implemented. These phase shifters must be modulators in integrated optic structure so that fast and high-speed beam steering can be done. Furthermore, these OTTA are passive optical networks for distribution and hence sharing of capacity and cost. The integrated photonic processor can be implemented in silicon photonics so that the cost can be sufficiently low. Beam steering phase modulators can be embedded using p–n junction by ion implantation, operating in junction reverse mode whose speed can be in tens of GHz.

With the 7.0 GHz bandwidth of the free-license band and 500 MHz per sub-band for data allocation, high level complex modulation can be used, as in the common RF wireless. Thus more than 10 Gbps can be delivered to MIMO antenna elements. The frequency comb lines and the minimized spectral distribution are shown in figure 2.9. Such a shift left or right of the spectrum can be adjusted and designed based on the modulation of single sideband or double sideband. With an antenna gain of about 22–0 dB for a 56 × 128 elements array of the MIMO antenna we can deliver more than 1.0 Tbps capacity covering an area of radius of 500 m by beam steering, for example, to clients in a crowded stadium.

Thus, the current transport layers have been massively migrating to lower latency and high capacity delivery to users via cloud broadband (Cloud-BB) networks from which direct access to wireless mobile devices can be implemented. Such dense-user environments include shopping centers, sport stadia, social gatherings etc, with direct 3D video transmission/broadcasting for close-up viewing.

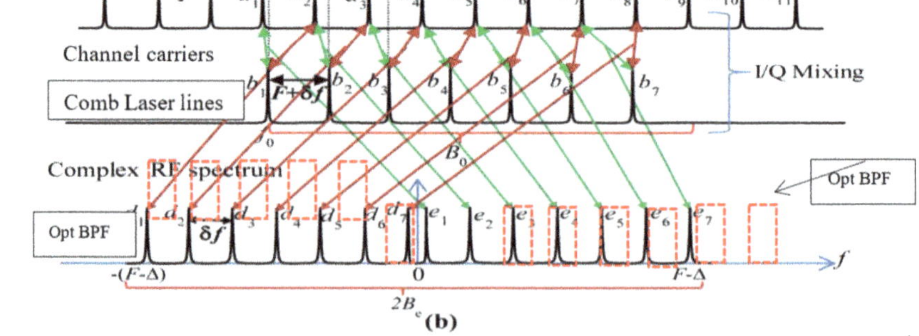

Figure 2.9. Comb sub-carriers and mixing in multi-super-channel transmission with mixing of multi-carrier signal channels and a comb source. Complex RF generated channels under an effective passband of optical filters. (a) Comb laser spectral lines as the source and local oscillator, and (b) beat frequency spectra in the electrical RF domain.

2.2.2 Optical transport technology

Ring and tree networks

The transport of networks for 5G must evolve from the traditional ring networks to tree networks. Infrastructures existing in the current carrier network deployed over the years by the carriers offer significant advantages to the flattening of the traditional telecom carriers. That is, the ring configured structure can be modified to connecting the nodes to nodes to form multi-tree architecture with optimized latency and distributed multi-access edge computing (MEC) pending on capacity demands of 5G mobile users. In such tree optical transport network, the precision of the clock is much less than that accumulating in ring network nodes to nodes at least 4–5 times, hence the stability of the tree networks. The tree network minimizes the latency to least a few times less than that occurring in the ring node by node accumulation. The latency in a tree network can be reduced to less than 10 μs. For distributed RAN the latency can be less than 100 μs for tree networks. This latency is a few times longer for such a 3–4 node ring (figure 2.10).

Much more detail of such evolution of micro-metro transport network can be the subject of another study but is not in the scope of this chapter. We now touch on the modulation techniques and transmission systems required for low cost 5G access networking. The transmission systems for long-haul core and interconnection of DCs and core network nodes can be referred to in several published papers over the last decade, especially on digital signal processing (DSP)-based optical receiving systems either in coherent or direct detection (figure 2.11).

The standardization of 200G/400G[3] enabled PAM4 multi-level format with baud rate of 28G Baud/56 GBaud and multi-lane to form such high bit rate transmission

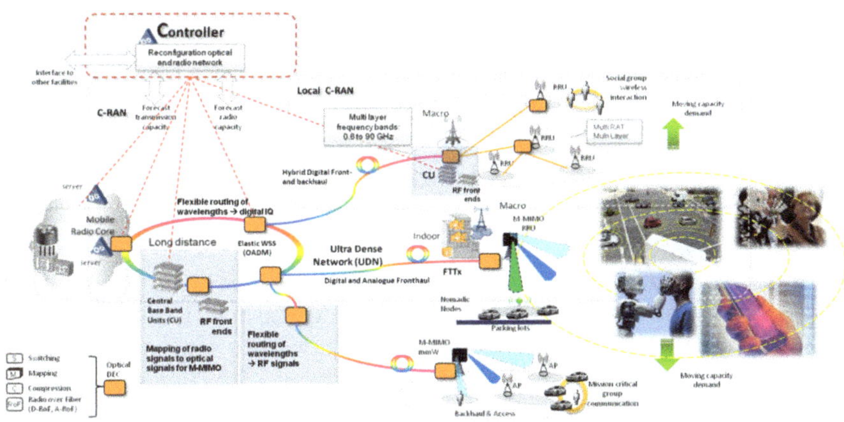

Figure 2.10. Generic view of 5G transport networks in a densely populated city, interconnection long-haul and user-concentrating environments, as well as machine-to-machine IoET (Internet of Every Thing).

[3] LAN/MAN Standards Committee of the IEEE Computer Society 'IEEE Standard for Ethernet: Amendment 10: Media Access Control Parameters, Physical Layers and Management Parameters for 200 Gb/s and 400 Gb/s Operation', 6 December 2017

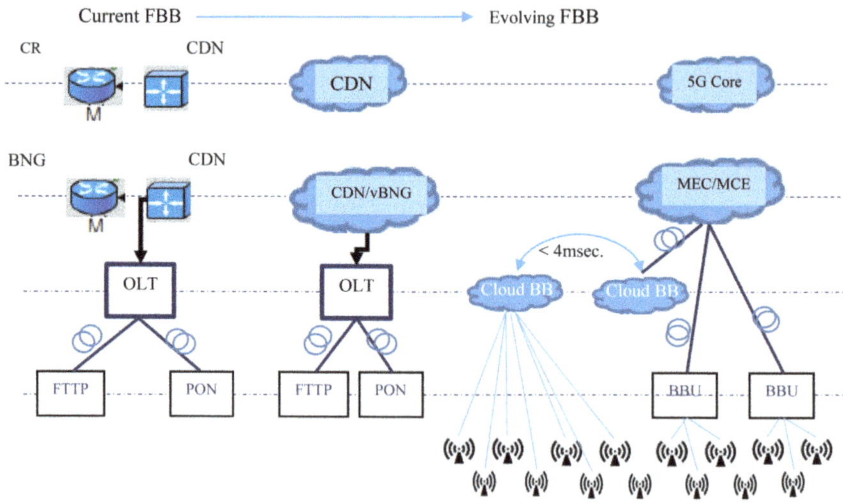

Figure 2.11. Evolution of current transport to cloudified networks for 5G. Labels: CDN = content delivery networks; FBB = fixed broadband; CDN = current delivery network; vBNG = variable business network group; OLT = optical line terminal; MCE = mobile cloud engine; BBU = base-band unit; FTTP = fiber to the premises; PON = passive optical networks.

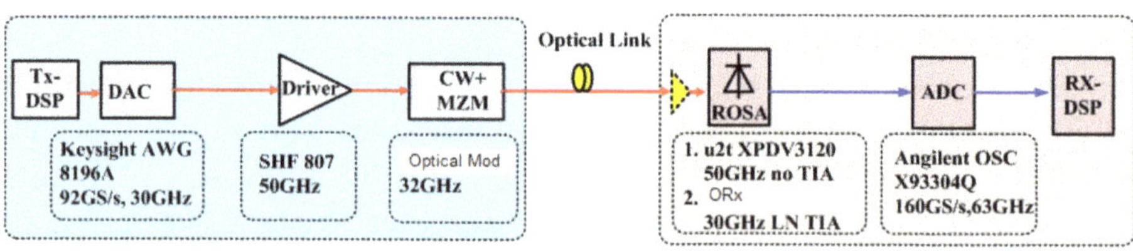

Figure 2.12. Optical transmission link for >200 Gbit s^{-1} per λ (optical carrier) for 5G access capacity delivery.

line. The aggregate bit rate per lamda (wavelength located for each accumulated channel) can be expected to be higher if other modulation formats can be applied.

Data transmission and transmission techniques

The discrete multi-tone (DMT) transmission technique using frequency-domain sub-carriers modulated by higher level modulation can lead to reaching speeds of more than 200 Gbit s^{-1} (Gbps) per λ, thus for four lanes or eight lanes (optical carriers) one can obtain more than 1.0 tera-bit per transmission module over short distance for 5G access networking. Figure 2.12 depicts a transmission set up for more than 200 Gbps per λ in the C-band. Orthogonal multi-sub-carriers are generated from an arbitrary waveform generator of bandwidth higher than 33 GHz and sampling rate of 160 GSamples s^{-1}. The sub-carriers are modulated by PAM4 formats and then fed into a wideband driver to drive the optical modulator, whose bandwidth is 32 GHz, with an extension ratio at biasing point of 7 dB in complete linear region of

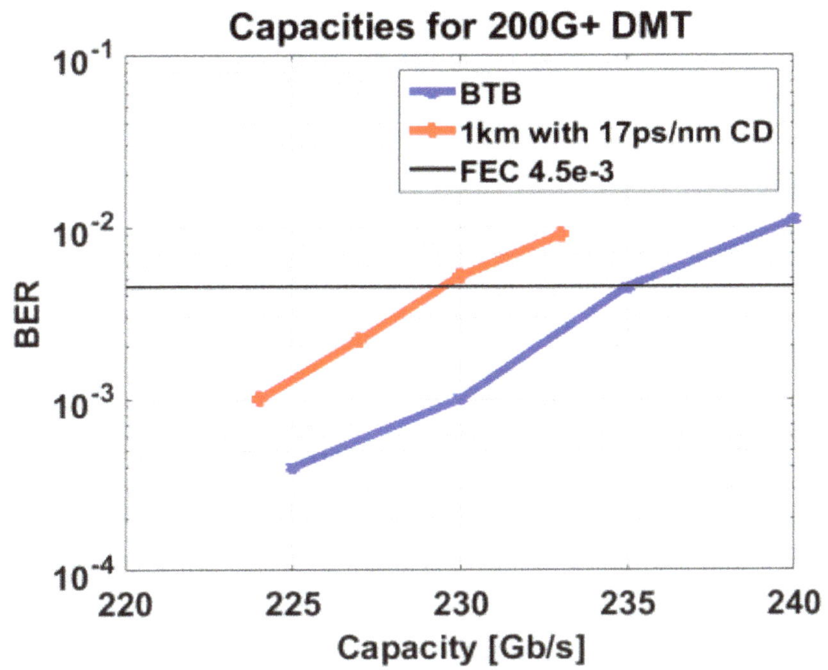

Figure 2.13. Bit error rate versus transmission capacity of a channel (per single wavelength λ) of a Tera-bits s^{-1} transmission link.

the voltage–intensity transfer characteristics of the modulator. Figure 2.13 shows the bit error rate (BER) versus the transmission capacity of an optical carrier channel of an aggregate tera-bits s^{-1} link whose platform is shown in figure 2.12. DSP processing with forward error correction of 7% is able to recover the data channel without error.

Thus, the most common and promising modulation formats are either PAM4 optical domain with direct detection or QAM 16 in coherent DSP-based receiver. For much higher bit rate and direct detection, the discrete multi-tone modulation in frequency domain can offer higher than 200 Gbps with optical and electronic components of limited bandwidth around 34 GHz. Higher capacity can be reached with coherent transmission techniques provided that the costing of optical and electronic components is lower or much more novel coherent receiving and modulation techniques are devices taking advantage of DSP power in processing.

Coherent QAM and DSP-based reception

Frequency-domain orthogonal modulation

Orthogonal frequency division multiplexing (OFDM) is an attractive modulation format that recently received lots of attention in practical and R&D fiber-optic communities [18]. The main advantage of optical OFDM is that it can cope with virtually an unlimited amount of inter-symbol interference (ISI). In high-speed

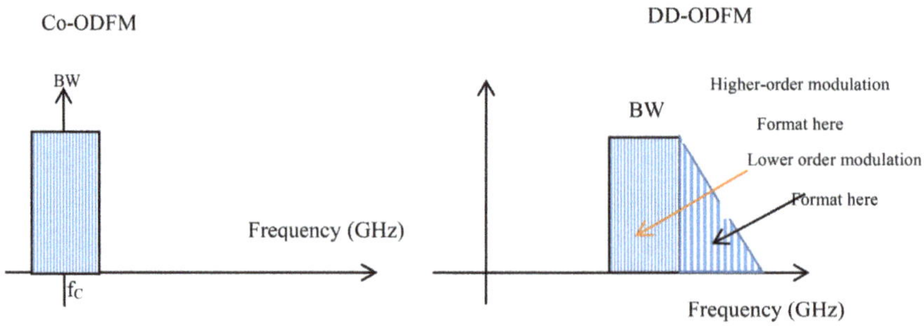

Figure 2.14. Spectra of optical OFDM-Co-OFDM and DD_OFDM. Sharp roll off at spectrum boundary → compact channels multiplexing with lower order of modulation format in the 3 dB to 6 dB band to compensate for lower sensitivity due to band limited components.

optical transmission systems, ISI is caused for instance by chromatic dispersion (CD) and polarization mode dispersion (PMD), which are serious issues in long and short haul systems whose bit rate is higher than 56–100 GBaud, especially at these ultra-high rate and multi-level modulation in optical domain. In this subsection some of the basics of an optical OFDM system are given. Furthermore, high-speed transmission experiments up to 122 Gbps are briefly presented (figure 2.14).

In principle, the basic concept behind OFDM is the division of a high bit rate data stream into several low bit rate orthogonal sub-carrier streams. In general, the subcarriers are generated in the electronic digital domain, therefore these systems typically consist of many sub-carriers (typically more than 64 and should be in a factor of base-2 and exponent N). In these systems, channel estimation is realized by periodically inserting training symbols. In fiber-optic transmission systems, the OFDM systems where the sub-carriers are generated in the optical domain are also created. These systems are sometimes referred to as coherent WDM systems[2]. Coherent WDM systems typically have few sub-carriers and do not use training symbols, but rely on blind channel estimation instead. In optical OFDM systems, the front end of the transmitter consists of an optical modulator, where the OFDM signal is unconverted to the frequency of an optical carrier. The front end at the receiver consists of either a coherent (Co-) or a direct detection (DD-) scheme. DD-OFDM is realized by sending the optical carrier along with the OFDM band so that direct detection with a single photodiode can be used at the receiver. In a Co-OFDM system, the optical carrier is suppressed at the transmitter and a local oscillator (LO) and optical hybrid is required, such as that shown in figure 2.15. The superior performance of Co-OFDM with respect to optical signal-to-noise ratio (OSNR) requirements, PMD (polarization mode dispersion) tolerance and spectral efficiency makes it an excellent candidate for long-haul transmission systems, whereas DD-OFDM, which requires fewer components at the receiver than Co-OFDM, is more suitable for cost-effective short reach applications. The spectra of both these two frequency multiplexing schemes are shown in figure 2.16. The DSP processing of DD-OFDM is much simpler as only the real part of the OFDM signals are required to be generated. The alternative name for DD_OFDM is discrete multi-tone (DMT).

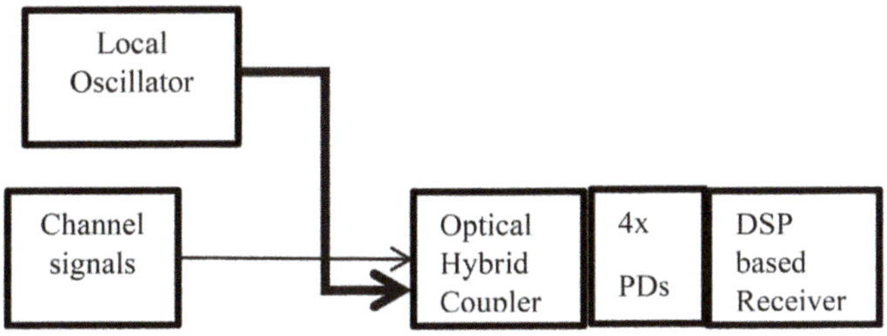

Figure 2.15. Schematic of an optical receiver for Co-OFDM.

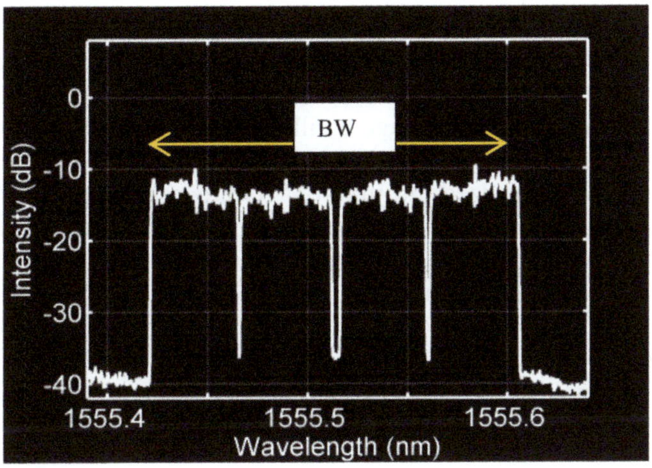

Figure 2.16. Optical spectrum of 122-Gbit s^{-1} OFDM signal, intensity versus wavelength of four channels with a total bandwidth (**BW**) of about 28 GHz or 28*0.8/100 = 0.224 nm.

A recent 100 Gbit s^{-1} transmission experiment has been reported. For both Co-OFDM [19] and DD-OFDM [20, 21] systems, 100 Gbps transmission experiments could be demonstrated. In order to increase the bit rate to 100 Gbps level, a polarization-division-multiplexing (PDM) is indispensable for a reduction of the signal bandwidth which determines the required bandwidth of components in the transmitter and receiver (figure 2.16). An analog-to-digital converter (ADC) and then a DSP can be used in such a PDM-OFDM, and the polarization components can be separated digitally by MIMO processing [22]. In order to apply the MIMO processing in DD-OFDM systems, an optical hybrid at the receiver [18] or a set of two carriers together with specially designed training signal patterns [19] could be employed. This group reported CO-OFDM transmission of 10 × 122 Gbps over 1000 km SSMF without any dispersion compensation[3]. In this experiment, the 122 Gbps OFDM signal consists of two polarization-multiplexed signals with four sub-carrier-multiplexed OFDM bands each. Together with a non-rectangular 8-QAM constellation (figure 2.17(a)) used for symbol mapping, the 122 Gbps

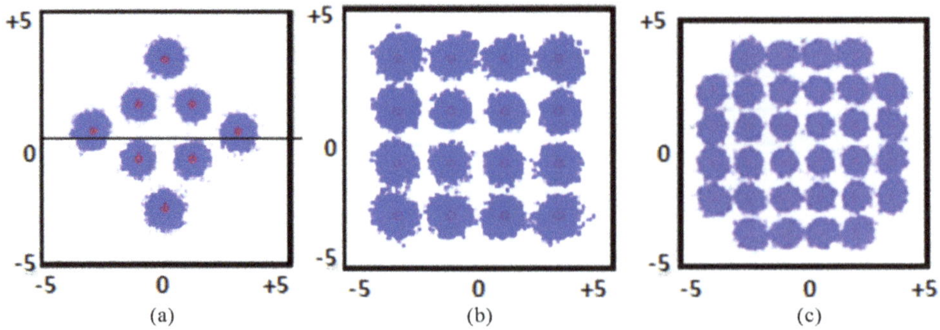

Figure 2.17. Signal constellations (a) 8-QAM; (b) 16-QAM; and (c) 32-QAM.

OFDM signal can be packed in a 23 GHz bandwidth as shown in figure 2.16. At the receiver, the signal is coherently detected with an optical hybrid and an external cavity laser. A real-time digital storage oscilloscope is used to sample the outputs of the optical hybrid and the acquired data is post-processed off-line. Polarization reversed rotating can be realized through MIMO processing and an RF-aided phase noise compensation can also possible [23]. After 1000 km transmission, the obtained BER values for all channels are well below the threshold of a concatenated FEC (forward error coding) code with 7% overhead. Along the whole link, the OFDM signal is continuously detectable demonstrating a dispersion tolerance of more than 18 500 ps nm^{-1}. Such a large dispersion tolerance is attractive for high speed transmission systems because it can eliminate the required inline dispersion compensator.

OTTA

Transport of data channels from Cloud-RAN or D-RAN to massive parallel antenna such as MIMO antenna elements in the mmW regions (58.6 GHz with 7 GHz bandwidth or in the 90 GHz band), shown in figure 2.8, are critical for increasing the wireless delivery capacity, yet to be exploited. One of the techniques we propose is the use of OTTA for beating the optical channels, which is modulated in the base-band with an optical local oscillator whose frequency is different from the optical carrier channel by a millimeter wave frequency, say 58.6 GHz. The 7 GHz free-license band in this 56.8 GHz band is very attractive to deliver multi-Gbps (giga-bits s^{-1}) over the 14 sub-bands each with a bandwidth 500 MHz and multi-level modulation. The number of elements of MIMO antenna can be 16×16 or 8×32. The antenna gain is sufficient to offer several Gbps per user over more than 200-meter wireless transmission.

2.2.3 Basic rates, capacity and server clusters for intra-and inter-DC connections

The transmission rates for interconnecting DCs must be as high as possible and the distance should be at least 40–80 km SSMF. The baud rate can reach 112 GBaud so that for Co-OFDM the aggregate rate for two polarization channels of 16-QAM would lead to 800 Gbps per wavelength with two polarized channels. Thus, for direct detection PAM4 can be employed with shorter transmission distance of 2–10 km for intra-DCN.

Then the frequency domain OFDM or DMT can offer much more compact channels in multiple multiplexed channels with increased number of sub-carriers. The band limited components can be used with their 6 dB band to increase the channel rate to >200 Gbps with only 40 GHz 6 dB bandwidth of optical modulators and receivers.

We note that both the time and frequency domain transmission systems require electronic digital signal processing. At 112 GBaud the sampling rate of these processors is quite high and thus requires significant power consumption. These limitations and difficulties can be avoided by photonic signal processors.

2.3 Exabits s^{-1} integrated photonic interconnection technology for flexible data-centric optical networks

Optical networking is evolving from classical service-provider-based DC-based internetworking environments with massive capacity, hence demanding novel optical switching and interconnecting technologies. We present proposed distributed and concentric DC-based networks and the essential optical interconnection technologies, from the photonic kernels to electronic and opto-electronic server clusters, in both passive and active structures. Optical switching devices and integrated matrices are proposed, composed of tunable (bandwidth and center wavelength) optical filters and switches as well as resonant micro-ring modulators (μRM), i.e. switching and spectral demux/mux, for multi-wavelength flexible-bandwidth optical channels of aggregate capacity reaching Ebps. The design principles and some experimental results are also reported.

2.3.1 Introductory remarks

Recent advances in information technology (IT) and the diffusion of broadband, fixed and radio, connectivity [24, 25], and the significant increase of the transmission bit rates from 100G to Tbps super-channels, are leading to an aggregate DWDM capacity of approaching and possibly surpassing 100 Tbps per fiber, hence dramatic reduction of hardware costs per bit, and the wider availability of open source software, are creating introductions to innovation design and restructuring of the optical and radio access, metropolitan and long haul transmission networks [26–30]. The emerging 5G wireless networks and their integration with cloud-based architectures [31], and demands on data services from smart mobile users put tremendous pressure on the networking of the DC to be structured in both distributed or concentric topologies.

The baud/symbol rate can now reach 56 Gbaud together with multi-level modulation formats and either coherent or direct reception, respectively, enable the aggregate bit rates per single carrier to reach 400 Gbps, hence several Tbps capacity (see figure 2.18) if using multiple wavelength with Nyquist pulse shaping or efficient spectrum packing in both long haul and access networks.

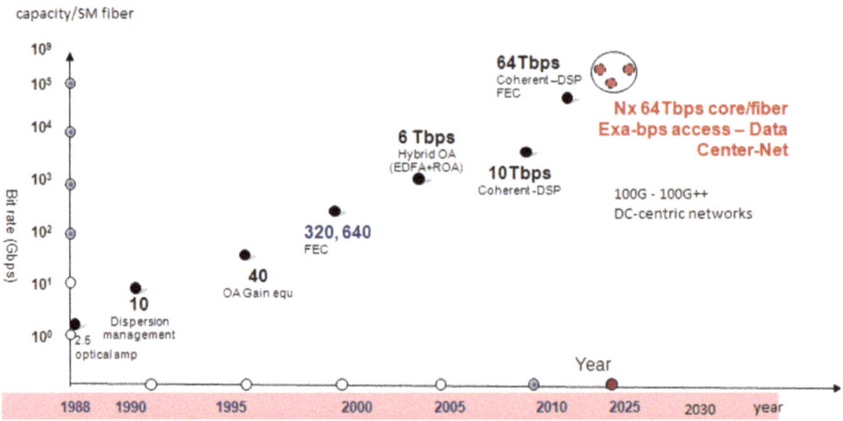

Figure 2.18. Evolution of transmission capacity in the past and predicted for the next 40 years.

Data servicing thus demands optical switching and routing for distributed and concentric topologies so as to overcome the limitations of electronic routers. Thus modern optical switching and routing systems must be able to offer: (i) non-blocking switching; (ii) flexible channel bandwidth filtering at the input and output ports with minimum cross-talk noise; (iii) fast switching speed; and last but not least (iv) the channel counts must be high, reaching a total capacity of about Ebps.

Recent advances in photonic devices permit optical switching system designers to consider this huge possibility in the implementation of compact Ebps optical switching systems, or systems-on-a-chip (SoC). These devices include micro-ring resonators (MRR), X-switches, Mach–Zehnder interferometers (MZI) and cascaded multimicro- rings (MMRs) [32, 33] etc. Micro-ring resonators (μRS) act as wavelength mux/de-multiplexers and switching/modulation simultaneously, especially when the injection and depletion of impurity carriers are employed to manipulate the refractive index and hence phases of the lightwaves guided in Si photonic waveguides. The X-over switches are based on the principle of manipulating the refractive index by carrier injections, hence altering the waveguide structures to diverting the lightwaves in different directions. The MZI is necessary for cascading with ring resonators as these two devices complement each other in terms of offering a zero or depletion of guided lightwaves at a certain phase state different between the parallel paths of the interferometers. On the other hand, the ring resonators produce a maximum intensity when the total accumulated phase of the guided lightwaves reaches a multiple number of 2π, i.e. operating at resonance. This depleted interference and the resonance phenomena are equivalent, respectively, to the zero and pole of an electrical network or in the z-plane of the transfer function (optical transmittance) of a sampled optical network, which is represented in terms of the delay sampled variable z [34].

Optical switching systems have been under extensive research over the years employing guided wave structures such as $LiNbO_3$ [35–38]. Recent advances in Si photonics offering opportunities in the integration of both CMOS electronic analog and digital circuits with photonic devices attract significant interest in the SoC approaches to minimize the cost as well as increase the input and out port counts. This SoC would fulfill the dream of integrated optics that was first proposed by S E Miller in 1969 [39]. The total capacity that a silica SSMF can carry is about 100 Gbps and even higher if a vestigial sideband is used and only in the C-band. So, for Exa-bps we expect 10 000 fiber ports to be employed. It is not hard to estimate this massive capacity requirement with 100 × 100 fiber connections at two main ports of an optical matrix of North–East–West–South ports or an interconnection of smaller matrix size interconnection optical chips. It is this massive interconnection technology that we wish to give an overview on in this section.

In this sub-section we present an overview of the following. (I) Data-centric networks comprising distributed DCs and DC concentric networks and demands on optical networking whose capacity reaches Ebps under flexible multi-wavelength channels of optical transport networks, hence general structures of active optical interconnecting technologies in both spatial and spectral domains with channels of aggregate bit rates ranging from 100G to 400G and to 10 or 32 Tbps. (ii) Integrated photonic technologies including MMR modulators/switches simultaneously performing as multiplexing and de-multiplexing functions, multi X-over switches incorporating MMRs as the basis elements for multi-Tbps and Ebps interconnection matrices. The switched and transmission optical paths can handle the base baud rate variable from 100G to 400G and then Tbps, thus flexible-bandwidth channels or flex optical transport networks. Design of integrated tunable bandwidth optical filters for incorporating with X-switches are presented for forming Exa-bps (Ebps) optical interconnecting matrices. (iii) Finally, the fundamental designs of ultra-compact flexible filters and switching integrated components based on Si photonics for Ebps active interconnection are presented. Experimental results on multi-channels transmissions and performances of optical switching matrices and effects on that of data channels are reported.

2.3.2 Exa-bps optical network topologies

Figure 2.19 shows the evolutionary network structure strategy emerging in the near future. This type of network exhibits a much-simplified control and routing mechanism for servicing of demands to the end users especially in the new broadband Internet. It is expected that all traditional (classical) telecom carrier networks will be flattened to a much-simplified structure in order to meet the challenges of such significant trends in the near future expected in the next coming decades of the 21st century. It is expected that in terms of bandwidth consumption, the average bandwidth of the end user will be up to 1 Gbps (range: 100 Mbps–1 Gbps) with a coverage of 90% of households by high-bandwidth broadband service networking. This will lead to the demand for the transport of information to the multi-Tbps to be delivered to the edges of the networks in the metro core and access for the DCNs and classical flattened carrier networks.

Figure 2.20 shows the topologies of two typical DCs, the distributed and concentric types. The latter topology is very common and expected to offer servicing

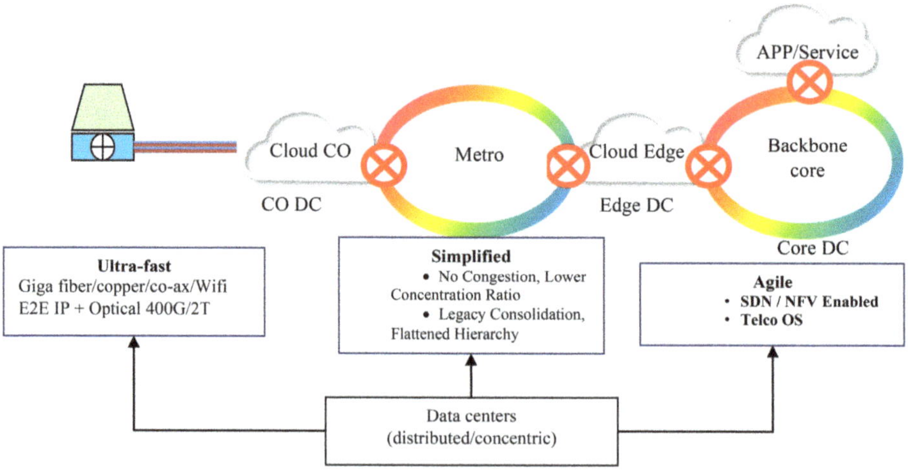

Figure 2.19. Evolutionary network modernization strategy.

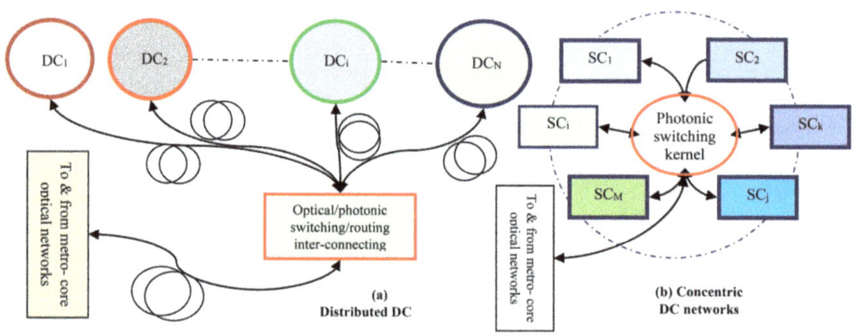

Figure 2.20. Generic structure of (a) distributed DCs interconnected via an optical switching center; (b) concentric DC consisting of server clusters interconnected via a photonic switching kernel.

and cloud operations for storage and servicing. For the former topology, the distributed interconnected DCNs (DIDCNs), there are a number of medium range DCs placed closer to the mobile clouds or cloud radio access networks (CRANs) which are interconnected via an optical switching center. On the other hand, in a concentric DC (CDC) the transport to and from and the capacity of information to be switched or routed is massive, reaching in the order of Ebps, such that the number of server clusters (SCs) to be interconnected can reach to an order of more than 10 000 with about 100 Tbps through each single fiber strand. Similarly, for distributed DCN the capacity can reach this high with the expectation that there are thousands of DCs distributed in a region whose interconnecting distance satisfyies the required dispersion tolerance and OSNR for a certain recoverable BER [40, 41].

The demands on the transmission links are different between the two topological networks. For DIDCN the transmission distance between distributed centre to the

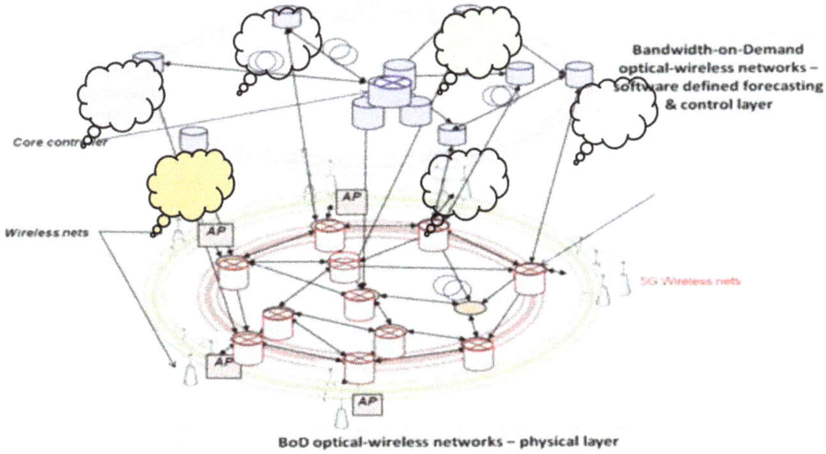

Figure 2.21. Converged flexible-bandwidth optical transport structures and software defined network 'SDN control', a special case with 5G wireless access and optical core networks via several clouds.

optical switching and interconnecting center would be in the range from 2 to 10 km, while this distance can only be no longer than 300 m in a CDC. Furthermore, for both transmission link types, direct modulation and direct detection are used for economic reasons despite coherent detection and recovery of the reception of data channels transmitted from the core fiber networks, as shown in figures 2.20(a) and (b). It is thus expected that the data channels are recovered in the electronic domain and then regenerated and switched and routed in the optical domain, to appropriate server clusters or distributed data centers. This is due to the fact that routing at such ultra-high baud rate (28G or 56G) would be fairly difficult with an unavoidable high probability of error in the electrical domain, due to cable/track frequency dependent losses.

Furthermore, the current evolution of the optical transport is focusing towards bandwidth on demand with flexible bandwidth of the overall super-channel which can employ modulated multiple sub-carriers. The number of sub-carriers can vary and hence its effective bandwidth can be from 75 GHz for 400G to 375 GHz for 1.0 Tbps, as shown in figure 2.21, with the network routing control right down to the optical physical layer, the software define networking (SDN) technique. In SDN, the control layer penetrates down to the physical layer. This control layer must also be matched and adapted to the topology and the capacity of the physical networks. Thus, the photonic switching/routing systems should be simple even at ultra-high capacity operation.

Basic optical switches

Since 1969, integrated optics technology has witnessed intensive research on optical switching based on a number of basic phenomena, including electro-optic (EO) effects via an applied electric field thus the phase changes exerted on the lightwave

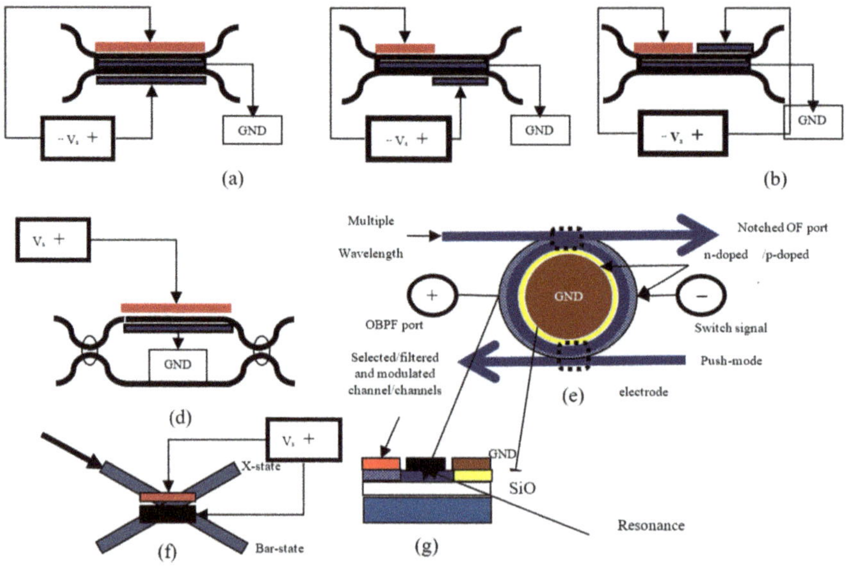

Figure 2.22. Photonic switches employing carrier injection and change of refractive indices of waveguides and interferometers, X-over or resonators: (a–c) directional coupler switches in normal, differential and push–pull modes; (d) Mach–Zehnder interferometric (MZI) (e) RS; (f) X-over switch; (g) cross-sectional view of Si waveguide and n-/p-doped region for carrier injection.

paths, acousto-optic (AO) effects via elastic/perturbed stress/strain on crystal structure, the effective gratings for the Bragg or Raman–Nath diffraction regime for beam deflection/switching, and the magneto-optic (MO) effects to alter the optical phases. These are commonly used for moderate high speed optical switching. Optical switching can be implemented by employing a number of fundamental phenomena in optics, that include interferometers, cross-over waveguides by alteration of guided modes and resonance, and diffraction for deflection/spatial switching. We are interested in the moderately high switching speed so the AO switches are not further presented here, only EO or carrier injection (CJ) types are examined. The structures of these switches are shown in figure 2.22. It is noted that except for the resonant micro-ring of figure 2.22(e), all others are spatial switching with the passband quite wide. Thus they cannot be employed in flexible grid optical switching or routing matrices, but must be integrated with other interferometric filters so that their passband can be tuned to the channel bandwidth as required. The passband of the micro-ring must also be wide enough to accommodate the aggregate bandwidth of the channels or super-channels. On the other hand, the switches shown in figures 2.22(a) and (b) are of directional coupler type whose cross coupling can be implemented by the carrier injection into one of the waveguides to either isolate (by increasing the refractive index) or spread the evanescent field to coupling the TE or TM field into the other waveguide branch. When the total phase reaches quadrature state then a complete switching occurs. For Si doped with electron impurity ΔN_e of $1.0e^{+17}$ cm^{-3}, the change is -8.8×10^{-8} ΔN_e or $1e^{-3}$ [42] at an applied voltage of about 0.9 V and $1e^{-2}$ at 1.2 V, respectively. A similar change is also expected for the doping with holes as the minority carriers. The total change of the refractive index due to both electrons and corresponding holes, is given as

$$\Delta n = -[8.8 \times 10^{-22}\Delta N_e + 8.5 \times 10^{-18}\Delta N_h^{0.8}] \tag{2.1}$$

Hence a total length of about 487 and 244 μm, respectively, for a π/2 phase shift for type (a) and (b), that is $\Delta\beta_{eff}.\ L \cong k_0 \partial n_{eff}.\ L = \frac{\pi}{2}$, with β the propagation constant of the guided wave in the channel/ridge waveguide and L the coupling length. Figure 2.22(f) shows the cross-section of the Si waveguide and the impurity doped regions as well as the electrodes to inject carriers into the guiding volume. For type (c) (figure 2.22(c)) the required phase for cross bar state is π and via the 3 dB coupler at both ends requires the total length of the switch to be twice as long as type (a).

For the μRS the total resonance occurs when the total phase of the lightwaves around the ring equals a multiple of 2π, thus the free spectral range happens by an integer of this change. The ring can be set into resonance or out of this mode by carrier injection into the ring and a switching signal whose voltage level is sufficient to inject the impurities into the active ring waveguide area. Push–pull operation on the μRS can be implemented by differential driving signals, as shown in figure 2.22(e) to reduce the required driving voltage level or minimize the radius of the ring as well as doubling the passband of the ring resonator.

Tunable spectral bandwidth switch structures

Photonic circuits, especially optical filters, can be designed using techniques well known in digital signal processing with discrete variable z representing the delay. In photonics this delay can be represented as $z = e^{-j\beta L} = e^{-j\omega T}$ with β being the effective propagation constant of the guided wave in optical waveguide, L is the propagation length, ω is the angular frequency of the lightwaves in vacuum and T is the delay interval. Thus, the photonic transfer function of an optical circuit can be written as a cascade of networks of transfer function consisting of all poles and all zeros as

$$\hat{H}(z) = \hat{A}\prod_{k=1}^{M}\frac{(z-\hat{z}_k)}{(z-\hat{p}_k)} = \hat{A}\frac{(z-\hat{z}_1)}{(z-\hat{p}_1)}\frac{(z-\hat{z}_2)}{(z-\hat{p}_2)}\ldots\ldots\frac{(z-\hat{z}_M)}{(z-\hat{p}_M)} \tag{2.2}$$

where M is the total number of poles and zeros of the optical networks, which is the same according to network theory. The poles and zeros are given by $\hat{z}_k, \hat{p}_k$. The values of the poles are corresponding to the resonant frequency at which the optical circuit resonates to obtain the greatest energy, such as those in a micro-ring when the total phase of the lightwaves circulating equals a multiple of 2π. Correspondingly, the zeros of an optical circuit occur when the output of the circuit becomes nullified, such as those in optical interferometers when the phases of the optical paths are in phase reversal with respect to one another. Thus we can observe that the filters can be constructed by cascading μRM and Mach–Zehnder delay interferometer (MZDI) incorporating phase shifters as shown in figure 2.23 [34, 43]. Theoretical design using

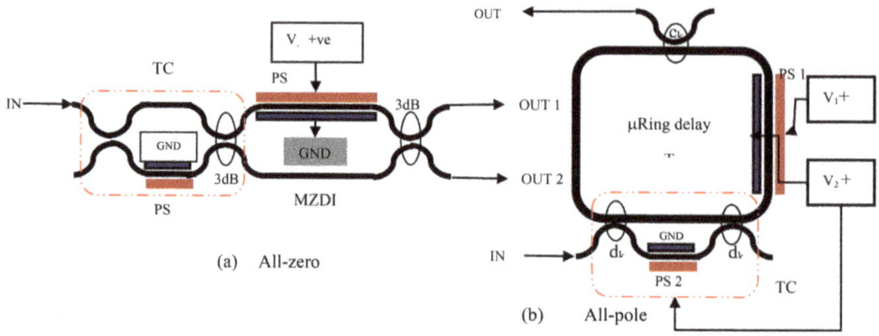

Figure 2.23. Schematic diagram of (a) all zero (interferometric) optical circuit; (b) all-pole (μRS) photonic stage. PS = phase shifter; TC = tunable coupler, MZDI = Mach–Zehnder delay interferometer. Bold lines represent optical waveguides.

the Butterworth filter in discrete domain can be applicable to the photonic domain and the transfer function in parallel connection can be given as

$$H_{\mathrm{BWparallel}}\left(z^{-1}\right) = \frac{1.678\ 719 \times 10^{-11}(1 + z^{-8}) - 6.714\ 876 \times 10^{-11}z^{-2}(1 + z^{-4}) + 1.007\ 231 \times 10^{-10}z^{-4}}{1 + 9453 \times 10^{-5}z^{-1} + 3.989\ 407z^{-2} + 5.968\ 278z^{-4} + 3.968\ 334z^{-6} + 0.989\ 463z^{-8}} \quad (2.3)$$

Similarly as in parallel (interferometric), cascade branches of the interferometeric branches form as

$$H_{\mathrm{BWcascade}}(z^{-1}) = 1.68 \times 10^{-11} + \frac{1 + 0.008z^{-1}}{1 + 0.004z^{-1} + z^{-2}} + \frac{1 - 0.008z^{-1}}{1 - 0.004z^{-1} + z^{-2}} + \frac{1 - 0.005z^{-1}}{1 + .002z^{-1} + z^{-2}} + \frac{1 + 0.005z^{-1}}{1 - 0.002z^{-1} + z^{-2}} \quad (2.4)$$

The magnitude and phase responses as a function of wavelength can be obtained for an eighth-order filter, as shown in figures 2.23(a) and (b) respectively. Thus, there are eight μR and three MZDI for this Butterworth filter. It is noted that a 3 dB passband of about 4 nm is observed for this design. A third-order can similarly be obtained with three μR without any difficulty [44] with a narrower passband. The bandwidth of these filters can be tuned by tuning the phase shifters and the coupling coefficients to accommodate for flexible channels (figure 2.24). Similarly, a multiple passband and sixth-order Chebychev filter can be designed and implemented with the amplitude response shown in figures 2.25(a) and (b). Thus, any photonic filter with specific response can be designed and implemented using cascade MZDI and μR can be obtained with tunable passband to accommodate for flexible channels in Tbps.

The overall occupied area of these cascade optical circuits would be in the range of a few μm^2. Thus, dense optical switching matrix and flexible filters can be implemented in a reasonable area of a Si-based photonic integrated chip.

2.3.3 Photonic switching and routing kernels and systems

The speed of the tunable passband switch can operate at a reasonably high speed, which is contributed to by the rise-time taken for the injection or withdrawal of the impurity carriers from the guiding area. This is an electronic process and the

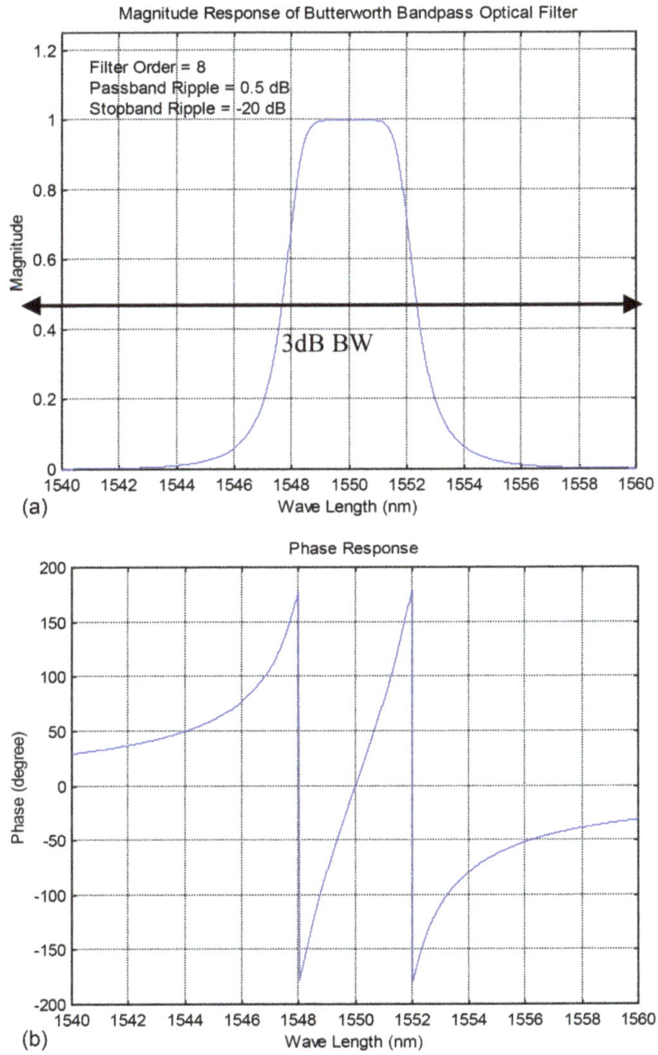

Figure 2.24. (a) Flat-top amplitude response of a linear-phase multi-band half-band photonic filter. Note the linear vertical scale. (b) Phase response of Butterworth bandpass optical filter corresponding to (a).

frequency can reach a few tens of GHz. These filtering switches can be structured into a large matrix form, as shown in figure 2.26(a). The area limitation is due to the size of the pads of the electrode. In this schematic, the central part shows the four-port switches whose bandwidth can be variable, as explained in section 2.3.1.2. The input ports and output ports must be coupled to multiplexers and de-multiplexers, which are most feasible by implementation in silica over Si technology that is also compatible with the CMOS process. Alternatively, the wavelength selective switches can be designed using μR, as shown in figure 2.26(b) in which three μRs are

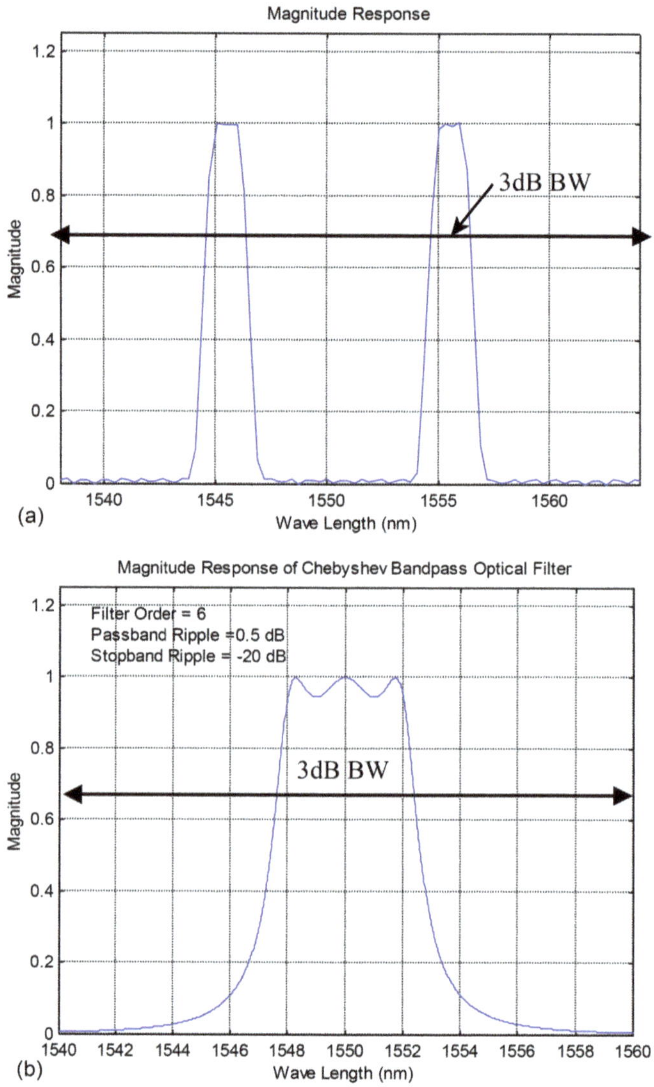

Figure 2.25. (a) Amplitude response of linear-phase multi-band half-band photonic filter. Note the linear vertical scale. (b) Flat-top amplitude response of a linear-phase Chebychev photonic filter with a ripple of 0.5 dB.

employed and each resonator is tuned to a specific wavelength and possibly tunable to a different bandwidth. The operational principles of this switching matrix are based on firstly resonating to circulate the desired channel to a particular ring μR_1/ μR_2/μR_3. The waveguide couplers WC_1 or WC_2 then couple the resonant channel to a specific output port. It is noted that the ring diameters are designed so that the WC can be shared or separated depending on the switching control and designed matrix

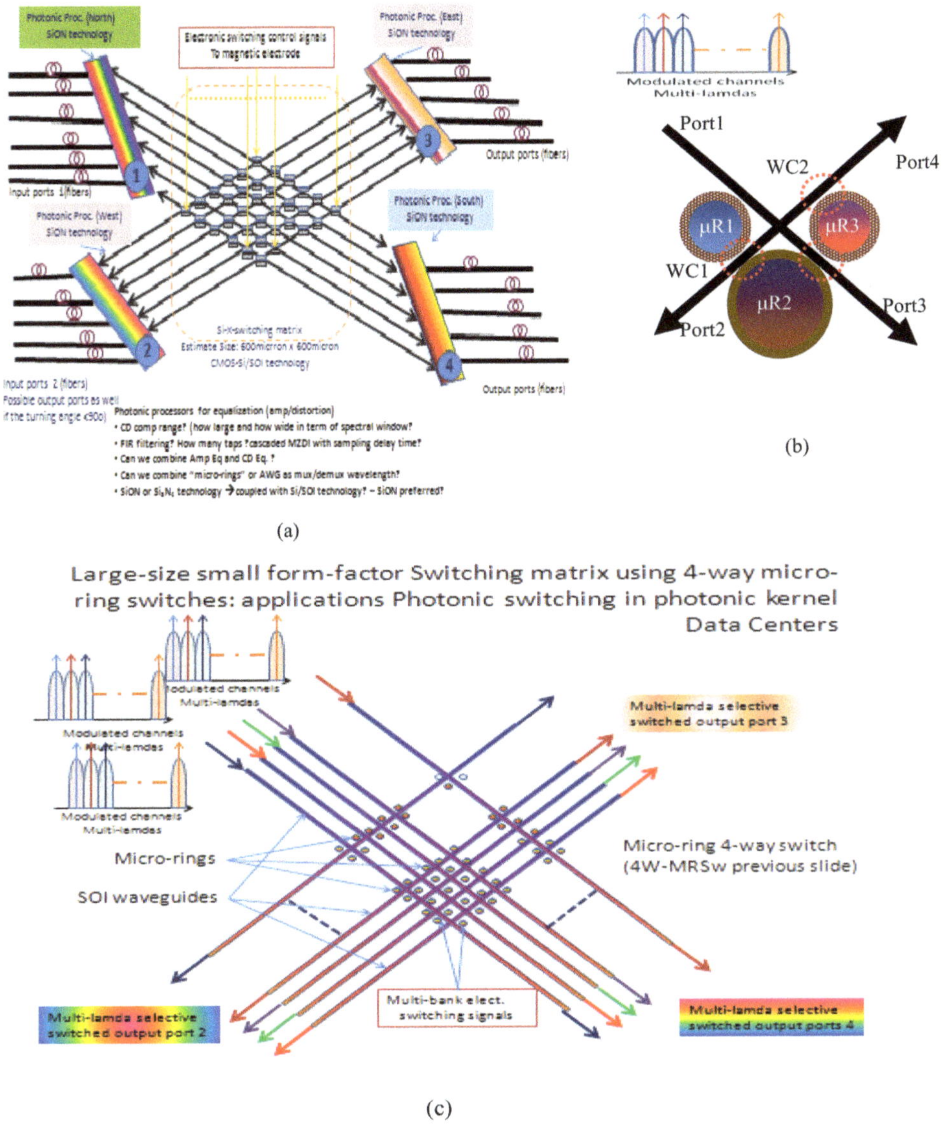

Figure 2.26. (a) Photonic processing-switching matrix using cascaded photonic circuit switches (Si/SOI MZDI and μR under field modulation/switching); (b) four-way resonant selective optical router; and (c) large form factor switching matrix 4-way switching matrix based on MRMs routing. WC = waveguide coupler (c) photonic switching matrix employing μRS.

operations as indicated by the circles of the WC in figure 2.26(b). However, the bandwidth of μR may not be that wide for super-channels and so these switching matrices formed by this type of X-μR of 2 × 2 type can only be used for massive interconnection whose channel bit rate may reach only to 400 Gbps under multi-level modulation.

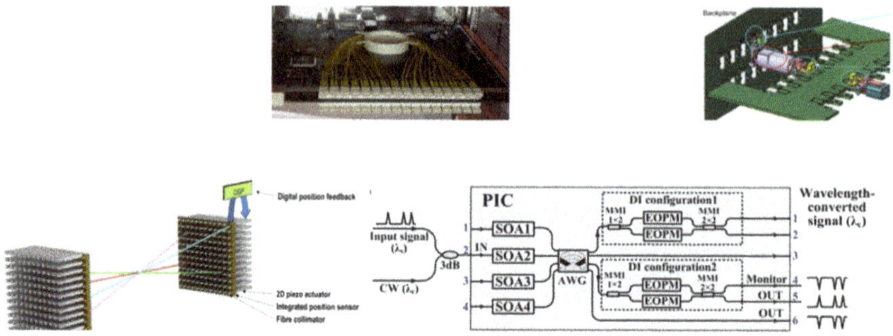

Figure 2.27. Schematic of large-scale optical switching matrix for both space and wavelength channels switching/routing.

2.3.4 Current technologies for optical switching and routing

An all-optical free-space switching and routing matrix is shown in figure 2.27. The matrix size is of dimensions 384 × 384 spatial switching (Huber and Suhner, Switzerland) and 16 × 16 WDM switches in the C- and O-bands (TU Eindhoven) (WDM WSS switch ~6 dB). The spatial optical switching matrices are composed of a base unit with 12 ports in a single row, making an up to 384 × 384 optical switching matrix with only 1 dB loss per port-to-port switching (technical director: Nick Parsons, formerly of Polaris). The physical mechanism is by piezo-electric control of the mirrors deposited at the fiber ends, and rotating and alignment of the fiber to fiber ports. The free space is the medium through which the lightwaves are propagating from one fiber to the other. One of the disadvantages is that one channel fiber (of several colors) can be switched to one other output fiber, hence no multiple distribution. A schematic of an integrated optical switch matrix and routing of multiple wavelengths and spacing is shown in figure 2.28. The integrated switching/routing matrix can be Si photonic type and hence minimum power consumption in contrast to the free-space switching matrix shown in figure 2.27.

2.3.5 Remarks

This section has addressed the challenges and technological developments towards ultra-fast ultra-high capacity optical networks whose capacity may reach of the order of Ebps from the long-distance core to the metro core and access networks. Both core networks in long-distance and metro have employed DSP-based coherent reception and transmitters as well as transponders with bit rates of 100G, 200G, 400G and super-channel Tbps in flexible grid spectra. Metro access networks would require low cost receiving systems, and hence direct detection technologies would be most appropriate. Given that DC centric networks are serving the huge demands of data intense consuming societies, traditional telecoms carriers are flattening their networks so as to simplify and make them agile as well as increasing their content services. The number of wavelength channels can now be planned to about 92 with frequency spacing of 50 GHz due to improvement of optical amplifiers over the entire range of the C-band. Over this band the total capacity transportable can reach

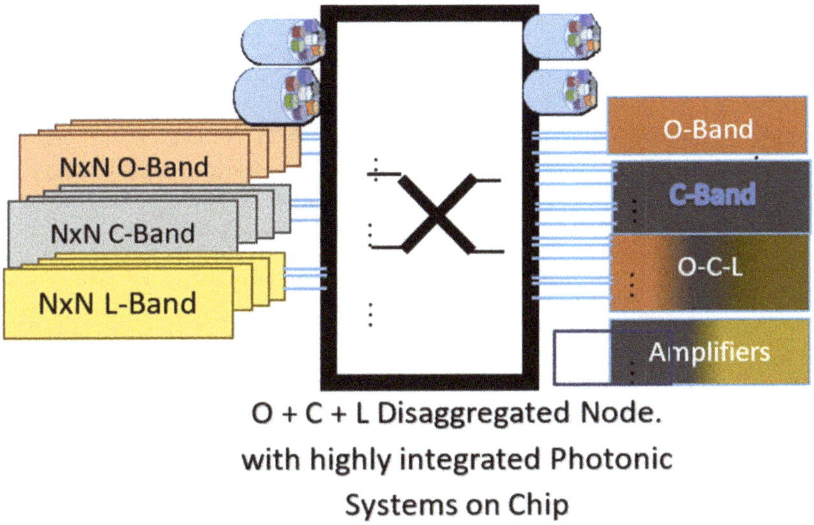

Figure 2.28. Schematic of an integrated switching/routing matrix.

100 Tbps per fiber. Several thousands of fibers are expected to be employed for the Exa-bps capacity and hence the optical switching and interconnection in such networks and intra- and inter-DC connections which are also proposed in section 2.4. These optical routing systems would be deployed throughout the networks acting as flexible ROADMs (reconfigurable optical add/drop multiplexers) and OXCs (optical cross-connect). Optical filters with sharp roll off characteristics given in section 2.3 allow the optical pulse shaping e.g. Nyquist/raise cosine pulse shaping that will allow high density packing of super-channels for such Exa-bps networks.

Integrated Si photonics are expected to offer the integration of optical components, both in the active and passive mode for high speed transmitters and receivers under coherent or non-coherent reception and OXC as well as ROADM. The designs for such OXC/ROADM are given in section 2.3.4. Furthermore, integration of laser sources including comb laser structures on Si substrate will also be very critical to obtain compact and low cost multi-Tbps systems for Exa-bps and even Zeta-bps networks.

2.4 Concluding remarks

This chapter gives a brief description of structures of DCN for both intra- and inter-DCs. The transmission techniques are also given so that the total capacity rate can reach higher than 800 Gbps and even higher than 1.0 Tbps, multi-Tbps capacity. Thus, in chapters 3–5 both DSP-based and PSP-based transmission systems are described in detail. A photonic processor for 5G/6G/7G mmW MIMO antenna beam steering in cloudified networks is given. Chapters 6 and 7 give details of optical switching and cross-connect as well as reconfigurable optical add/drop multiplexers for DCN.

References

[1] Bilal K *et al* 2013 Quantitative comparisons of the state of the art data center architectures *Concurr. Comput.* **25** 1771–83

[2] Noormohammadpour M and Raghavendra C S 2018 Datacenter traffic control: understanding techniques and trade-offs *IEEE Commun. Surv. Tutor.* **20** 1492–525

[3] Al-Fares M, Loukissas A and Vahdat A 2008 A scalable, commodity data center 2 network architecture *ACM SIGCOMM 2008 Conf. on Data 3 Communication (Seattle, WA)* pp 63–74

[4] Guo C, Wu H, Tan K, Shi L, Zhang Y and Lu S 2008 DCell: a scalable and fault tolerant network structure for data centers *ACM SIGCOMM Comput. Commun. Rev.* **38** 75–86

[5] Bilal K, Khan S U and Zomaya A Y 2013 Green data center networks: challenges and opportunities *11th IEEE Int. Conf. on Frontiers of Information Technology (FIT) (Islamabad, Pakistan)* pp 229–34

[6] Cisco 2010 *Cisco Data Center Infrastructure 2.5 Design Guide* (Cisco Press)

[7] Bilal K *et al* 2014 A taxonomy and survey on green data center networks *Future Gener. Comput. Syst.* **36** 189–208

[8] Bilal K, Manzano M, Khan S U, Calle E, Li K and Zomaya A Y 2013 On the characterization of the structural robustness of data center networks *IEEE Trans. Cloud Comput.* **1** 64–77

[9] Guo C *et al* 2009 BCube: a high performance, server-centric network architecture for modular data centers *ACM SIGCOMM Comput. Commun. Rev.* **39** 63–74

[10] Costa P *et al* 2010 CamCube: a key-based data center *Technical Report* MSR TR-2010-74, Microsoft Research

[11] Li D *et al* 2009 FiConn: using backup port for server interconnection in data centers *IEEE INFOCOM 2009*

[12] Singla A *et al* 2012 Jellyfish: networking data centers randomly *9th USENIX Symp. on Networked Systems Design and Implementation (NSDI)*

[13] Gyarmati L and Trinh T A 2010 Scafida: a scale-free network inspired data center architecture *ACM SIGCOMM Comput. Commun. Rev.* **40** 4–12

[14] Manzano M, Bilal K, Calle E and Khan S U 2013 On the connectivity of data center networks *IEEE Commun. Lett.* **17** 2172–5

[15] Bilal K, Khan S U, Madani S A, Hayat K, Khan M I, Min-Allah N, Kolodziej J, Wang L, Zeadally S and Chen D 2013 A survey on green communications using adaptive link rate *Cluster Comput.* **16** 575–89

[16] Heller B *et al* 2010 ElasticTree: saving energy in data center networks *Proc. of the 7th USENIX Symp. on Networked Systems Design and Implementation, NSDI 2010 (April 28–30, 2010, San Jose, CA)*

[17] Matthias S, Gomes N, Juarer A, Lange M, Leiba Y, Mano H, Murata H and Stoehr A 2018 Public field trial in a shopping mall of multi-RAT (60 GHz 5G/LTE/Wifi) mobile networks *IEEE Wirel. Commun. Mag.* Spec. Issue-5G Testing and Field Trials, 2018

[18] Shieh W and Djordjevic I 2009 *OFDM for Optical Communications* 1st edn (New York: Academic)

[19] Jansen S *et al* 2009 121.9-Gb/s PDM-OFDM transmission with 2-b/s/Hz spectral efficiency over 1000 km of SSMF *J. Lightwave Technol.* **27** 177–88

[20] Schmidt B *et al* 2009 100 Gbit/s transmission using singleband direct-detection optical OFDM *Optical Fiber Communication Conf. and National Fiber Optic Engineers Conf., OSA Technical Digest (CD)* paper PDPC3

[21] Qian D *et al* 2010 108 Gb/s OFDMA-PON with polarization multiplexing and direct-detection *Optical Fiber Communication Conf. and National Fiber Optic Engineers Conf., OSA Technical Digest (CD)*

[22] Jansen S *et al* 2008 Long-haul transmission of 16 × 52.5 Gbits/s polarization-division-multiplexed OFDM enabled by MIMO processing *J. Opt. Netw.* **7** 173–82

[23] Jansen S *et al* 2008 Coherent optical 25.8-Gb/s OFDM transmission over 4160-km SSMF *J. Lightwave Technol.* **26** 6–15

[24] Verdone R and Manzaalini A ed. 2014 5G Experimental Facilities in Europe, White Paper, Networld 2020 ETP

[25] Puttmtnam B J *et al* 2015 2.15 Pb/s transmission using 22 core homogeneous single mode multi-core fiber and wideband optical comb, ECOC2015 PDP3.2, ECOC2015, Valencia, Sspain

[26] Binh L N, Mao B N, Xie C S, Stonajovic N and Yang N 2015 Synchronous modulator incorporated re-circulating comb laser sources for superchannel optical transmission *Advanced Solid-State Lasers Congress, M. Ebrahim-Zadeh and I. Sorokina, eds., OSA Technical Digest* paper JTh2A.13

[27] Zhou X *et al* 2010 64 Tb/s (640×107-Gb/s) PDM-36QAM transmission over 320 km using both pre- and post-transmission digital equalization *Proc. OFC 2010 (San Diego)* March 2010 paper PDPB9

[28] Karinou F, Stojanovic N, Xie C S, Ortsiefer M, Daly A, Hohenleitner R, Kögel B and Neumeyr C 2015 28 Gb/s NRZ-OOK using 1530-nm VCSEL, direct detection and MLSE receiver for optical interconnects *Proc OFC 2015 (L.A. USA)* March 2015

[29] Lee J, Kaneda N, Pfau T, Konczykowska A, Jorge F, Dupuy J-Y and Chen Y-K 2015 Serial 103.125-gb/s transmission over 1 km SSMF for lowcost, short-reach optical interconnects *Proc. OFC 2015 (Los Angeles, CA)* March 2015

[30] Rahman T, Rafique D, Spinnler B, Pincemin E, Bouëtté C L, Jauffrit J and Calabro S 2015 Record field demonstration of C-band multi-terabit 16QAM, 32QAM and 64QAM over 762 km of SSMF *Opto-Electronics and Communications Conf. (OECC) (Shanghai, June 28–July 2)*

[31] Liu J, Zhao T, Zhou S, Cheng Y and Niu Z 2014 Concert: a cloud-based architecture for next generation cellular systems *IEEE Wirel. Commun.* **21** 14–22

[32] Ngo N Q and Binh L N 2006 Synthesis of tunable optical waveguides filters using digital signal processing techniques *IEEE J. Lightwave Technol.* **24** 3520–30

[33] Binh L N Tunable photonic filters: a digital signal processing design approach *Appl. Opt.* **48** 2799–810

[34] Binh L N 2007 *Photonic Signal Processing* (Boca Raton, FL: CRC Press)

[35] Deng K-L, Runser R J, Toliver P, Glesk I and Prucnal P R 2000 A highly scalable, rapidly-reconfigurable, multicasting-capable, 100-Gbit/s photonic switched interconnect based upon OTDM technology *IEEE J. Lightwave Technol.* **18** 1892

[36] Prucnal P R, Baby V, Rand D R, Wang B C, Xu L and Glesk I 2002 All optical processing in switching networks *LEOS* http://photonicssociety.org/newsletters/oct02/prucnal.html

[37] Prucnal P R, Blumenthal D J and Perrier P R 2006 *Self-Optical Switching with Optical Processing* (Berlin: Springer)

[38] Prucnal P R 1987 Integrated optics and optical switching *Proc. ECOC/LAN 87*

[39] Miller S E 1969 Integrated optics: An introduction *Bell Syst. Tech. J.* **48** 2059–69

[40] Karinou F, Prodaniuc C, Stojanovic N, Ortsiefer M, Daly A, Hohenleitner R, Kögel B and Neumeyr C 2015 Experimental performance evaluation of equalization techniques for 56 Gb/s PAM-4 VCSEL-based optical interconnects *Proc. ECOC 2015, Paper (Valencia)*

[41] Karinou F, Prodaniuc C, Stojanovic N, Ortsiefer M, Daly A, Hohenleitner R, Kögel B and Neumeyr C 2015 Directly PAM-4 modulated 1530-nm VCSEL enabling 56 Gb/s/λ data-center interconnects *IEEE Photon. Technol. Lett.*

[42] Soref R A and Bennett B R 1987 Electro-optical effects in silicon *IEEE J. Quant. Electron.* **23** 123–9

[43] Ngo N Q and Binh L N 2006 Synthesis of tunable optical waveguides filters using digital signal processing technique *IEEE J. Lightwave Technol.* **24** 3520–31

[44] Poon A P, Luo X, Xu F and Chen H 2009 Cascaded microresonator-based matrix for silicon on-chip optical interconnection *Proc. IEEE* **97** 1216–37

IOP Publishing

Transmission, Processing, and All-Optical Routing for Ultra-High Capacity Data Center Networking (Second Edition)

Le Nguyen Binh

Chapter 3

Access and DC networking transmission technologies

3.1 Introductory remarks

Communications and switching between servers and racks of servers in data center networking (DCN) are critical in the operation or servicing and storage of massive data. Over the last decade, the transmission rates have been increased gradually and exponentially from 10 Gbps to 1.6 Tera-bps per wavelength carrier over the short distance of a few hundred maters to a few kilometers of single-mode fibers, even up to 180 Gbps over multi-mode fibers and much higher over multi-core fibers. It is important for DC engineers to gain a basic understanding of transmission technology. The switching of massive data streams is also treated in chapter 6 which covers the add and drop of wavelength channels as well switching, so that dynamic and agile DCN can be achieved.

This chapter focuses on the description of the functional principles and operations of the modulation formats and receiving techniques of incoherent and coherent reception so as to deliver the highest speed with the limit transmission distance (figure 3.1). A schematic of the optical pre-processing of the coherent reception of polarization multiplexed optical channels is shown in figure 3.2, all of which can be implemented in integrated photonic structure, especially Si-photonic technology [1, 2].

Under coherent reception the aggregate bit rate can reach 800 Gbps per optical carrier with both polarizations of guided waves in association with the digital signal processing in the electronic digital domain, or effectively the baud rate is at 112 GBd. Coherent orthogonal frequency division multiplexing (OFDM) can also be employed to reach this high rate and beyond. Such 5G optical networking in distributed topology is shown in figure 3.1.

Then, for low cost and a high rate under the limited bandwidth of optical, optoelectronic and electrical components, the use of discrete multi-tone (DMT) or

doi:10.1088/978-0-7503-5849-1ch3

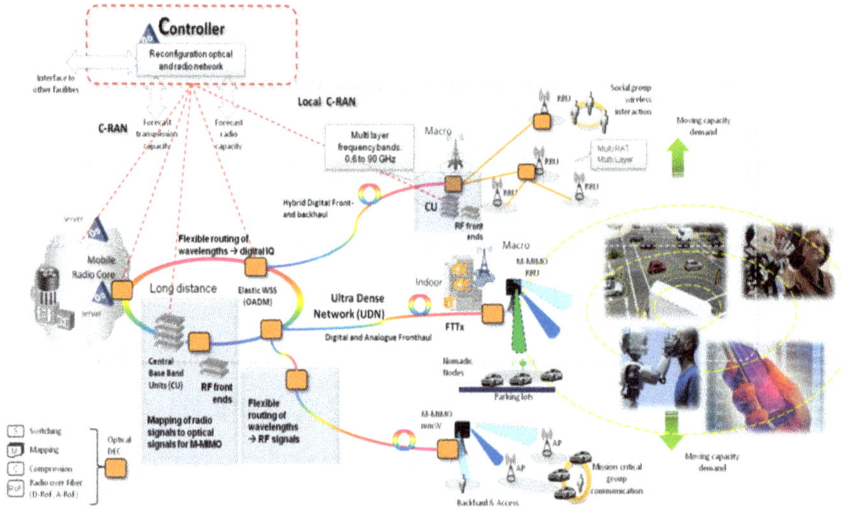

Figure 3.1. DCN in the evolution of optical networks on core and metropolitan to 5G distributed networks.

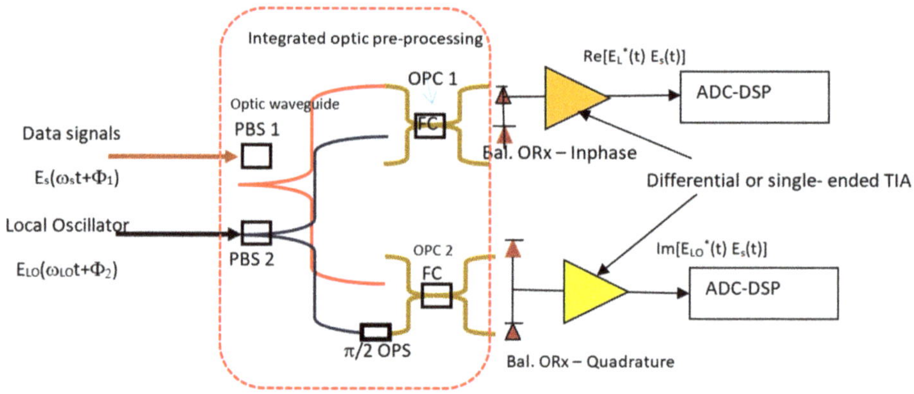

Figure 3.2. Schematic of a coherent receiver using balanced detector and an integrated optic pre-processor for I–Q components phase and amplitude recovery, incorporating DSP and ADC. Labels: Im = imaginary part or quadrature phase component; Re = real part or in-phase components; PBS = integrated optic polarization beam splitter; ADC = analog-to- digital converter DSP = digital signal processor; TIA = trans-impedance amplifier. ω_s, ω_{LO} = signal carrier and local oscillator frequencies; OPS = optic phase shifter; OPC = optical polarization combiner (same pol); TE = transverse electric (horizontal), TM transverse magnetic (vertical).

direct detection (DD-)OFDM, with a sufficient number of sub-carriers and the extended bandwidth of 0–6 dB and lower order modulation formats, can extend the bit rate per optical carrier to higher than 200 Gbps with additional capacity employed in the 3–6 dB bandwidth region and a lower order of modulation. Furthermore, we also describe the PAM4 technique in such transmission to 112 Gbps or 56 GBd per optical carrier with use of a minimum DSP sampling speed.

The aggregation of four to eight optical carriers leads to a total bit rate in Tbps that allows the ultra-high capacity link in the data center (DC) networking and access layer of the evolved 5G optical Internet. The typical distance is 2–10 km and the spectral window can be in the O-band (1310 nm window) or 1550 nm spectral window, or even the 850 nm window by a vertical-cavity surface-emitting laser (VCSEL). A typical commercial SFP module for 1.25 Gbps is shown in figure 3.3 at a cost per bit of about 10$/1.25 Gbps or ~10 $/billion bits. This cost is expected to rise to 10 times but for 100 Gbps in the near future. It is noted that the working spectral region is in the 1310 nm window using VCSEL, at which the dispersion effect is almost negligible. Table 3.1 shows some current typical interfaces for

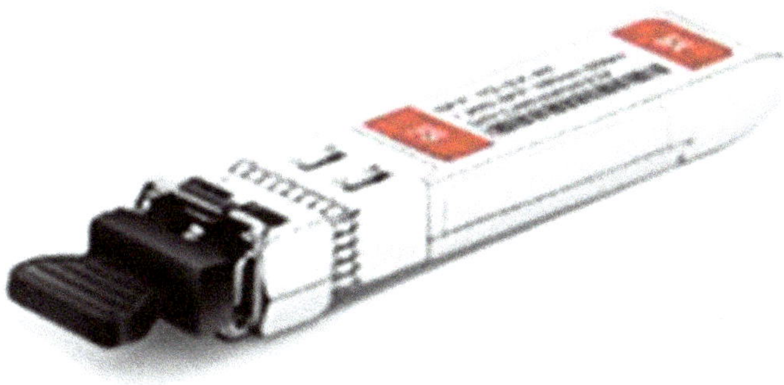

Figure 3.3. A commercial Gbps SFP-size compatible device operating at 1310 nm with a 2 km transceiver module. Its operating data rate is up to 1.25 Gbps. Technology: 850 nm VCSEL laser transmitter; fiber: 550 m with 50/125 μm MMF 300 m on 62.5/125 μm MMF; operating conditions: Single 3.3 V power supply and LVTTL logic interface; pluggable SFP Footprint Duplex LC connector interface; Class 1 FDA; commercial temperature range: 0 to +70 °C; compliant with SFP MSA specification; built-in digital diagnostic functions, including optical power monitoring (2019 price ~$10). See also table 3.1 or more types and standards.

Table 3.1. Typical table of compact transmission with high capacity up to 5 Gbps. 100G modules are expected to be the same size with integrated optic devices.

Type	Wavelength	Transmission distance	Fiber waveguide link	IEEE-802.3-Standard, remarks
1000BASE-SX	850 nm	500 m	50 μm/125 μm Fiber pair—MMF	Clause 38
		220 m	62.5 μm/125 μm Fiber pair—MMF	
1000BASE-LX	1310 nm	5 km 10 km	9 μm/125 μm Fiber pair SSMF	Clause 38 Clause 59,

(Continued)

Table 3.1. (*Continued*)

Type	Wavelength	Transmission distance	Fiber waveguide link	IEEE-802.3-Standard, remarks
1000BASE-LX101000BASE-LH				'1000BASE-LX'
1000BASE-EX	1310 nm	40 km	9 μm/125 μm SSMF Fiber pair	no IEEE-802.3-Standard, negligible dispersion
1000BASE-ZX	1550 nm	80 km	9 μm/125 μm SMF Fiber pair	no IEEE-802.3-Standard, negligible dispersion
1000BASE-EZX	1550 nm	120 km	9 μm/125 μm SMF Fiber pair	no IEEE-802.3-Standard
1000BASE-BX101000BASE-BX	1490 nm (-D) 1310 nm (-U)	10 km	9 μm/125 μm SMF Fiber pair	Clause 59, CWDM with 1490 nm 'downstream' and 1310 nm 'upstream'
1000BASE-T	electrical	100 m	Twisted Pair Cat-5e	Clause 40, SFP module

non-coherent transmission links for DCN and short haul low cost access networking. Note that any DSP would be in a separate electronic module in association with these modules. The principles of coherent transmission systems can be found in [3] and are given in appendix A. Appendix B illustrates the operational principles of balanced reception.

3.2 DSP-based coherent optical transmission systems

A flow chart of the digital signal processing in a coherent DSP-based transmission system is shown in figure 3.4. This diagram is now introduced here to include an optical processing section and the sampled signals are then digitally processed in a DSP, including the clock/timing recovered signals fed back into the sampling unit of the analog-to-digital converter (ADC) so as to obtain the best correct timing for sampling the incoming data sequence for processing in the DSP. Any errors made at this stage of timing lead to high deviation of the bit error rate (BER) in the symbol decoder, shown in figure 3.4. A dual polarized channel is propagating in the same optical guided beam and arriving at the coherent receiver consisting of an optical pre-processor in which a local oscillator (LO) laser is mixing with the signals in a photodetector (PD), a square-law detector. The beating signals resulting from the square of the summation of the LO and the small signal level of the arrived channels, comprises three parts: (i) a DC term due to the intensity of the LO as the dominated

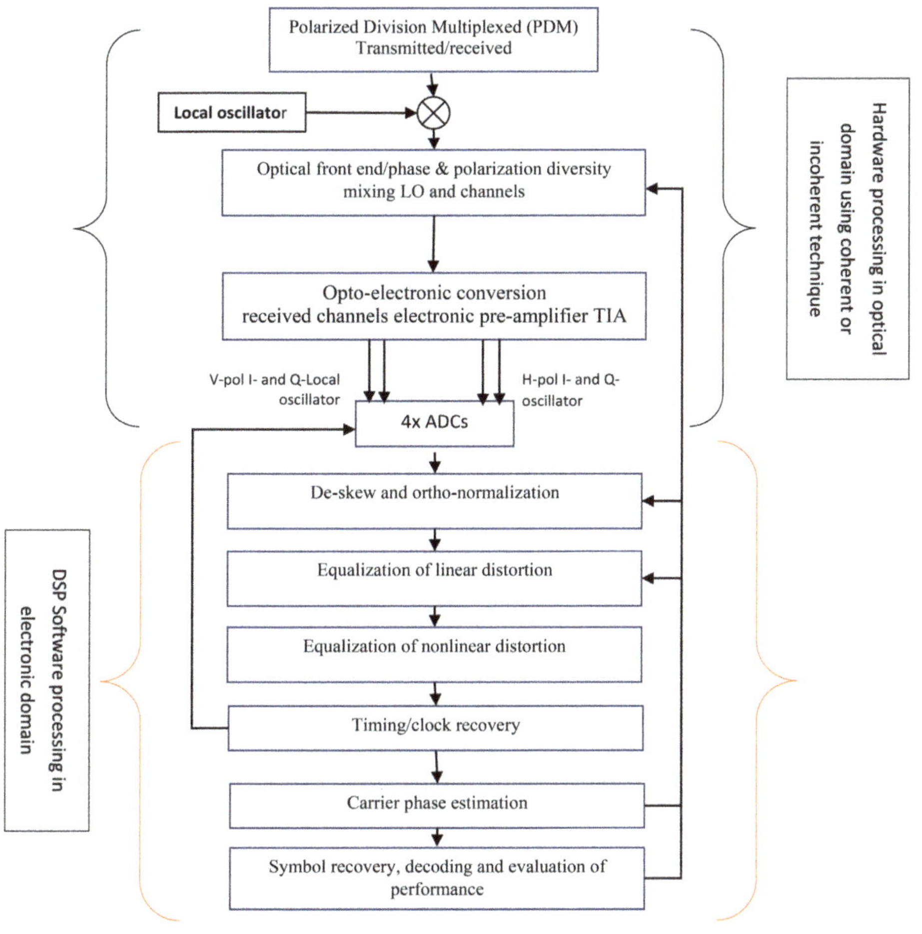

Figure 3.4. Flow of functionalities of DSP processing in a coherent optical receiver of coherent transmission system with feedback path diverted to ADC.

term; (ii) an RF term as a multiplication of the amplitude of LO and the signal with the frequency term as the difference between the carried signals and the LO frequency; and (iii) the frequency term as the amplitude of the summation of the frequencies of the LO and the signal carrier. The frequency band of the third term is too far away from the sensitivity of the PD, and is thus eliminated. The DC term is also filtered by the low frequency roll-off of the optoelectronic (O/E) components, hence is negligible.

It is also noted that the vertically polarized channel (V-pol) and horizontally polarized (H-pol) channels are detected in the integrated optic pre-processing (IOPP) section, a hybrid coupler (HC) and a balanced PD pair (PDP). Their in-phase (I-) and quadrature (Q-) components are produced in the electrical domain with signal voltage conditioned for the conversion to digital domain by the ADC.

When several channels are transmitting in association with multi-carrier optical sources and compact bandwidth shaping, they are called superchannels. This technique is also important for carrying massive capacity in DC networking. The processing of a sampled sequence from the received optical data and photo-detected electronic signals passing through the ADC relies on the timing recovery from the sampled events of the sequence. The flowing stages of the blocks given in figure 3.4 may be changed or altered accordingly, depending on the modulation formats and pulse shaping, for example the Nyquist pulse shaping in Nyquist superchannel transmission systems.

This section describes the performance and impact of processing algorithms in optical transmission systems, short distance or long distance, employing coherent reception techniques over highly dispersive optical transmission lines, in particular the multi-span or single-span optically amplified non-DCM long distance. First, the QPSK homodyne scheme is examined, and then the 16-QAM incorporating both polarized division multiplexing (PDM) in the optical domain, hence the terms PDMQPSK or PDM-16-QAM. We then expand the study to superchannel transmission systems in which several sub-channels are closely spaced in the spectral region so as to increase the spectral efficiency so that the total effective bit rate must reach at least 1.0 Tbps. Due to the overlapping of adjacent channels, there is the possibility that modifications of the processing algorithms are to be made. Further, the nonlinearity impairments on transmitted sub-channels would degrade the system performance and the application of DSP-based processing algorithms, such as back propagation algorithms, must be combined with linear and nonlinear equalization schemes so as to effectively combat the degrading.

A schematic of a balanced optical receiver is shown in appendix B.

3.3 Quadrature amplitude modulation (QAM)

Homodyne coherent reception requires a perfect match of the frequency of the signal carrier and the LO. The phase difference contributes significantly to the degradation of QAM systems as the phase orthogonality is the main feature of these modulation schemes. Any frequency difference will lead to phase noise of the detected signals. This is the largest hurdle for the first optical coherent system initiated in the mid-1980s. In DSP-based coherent reception systems the recovery of the carrier phases and hence the frequency is critical to achieve the most sensitive reception with maximum performance in the BER, or evaluation of the probability of error. This section illustrates the recovery of the carrier phase for QPSK and 16-QAM optical transmission systems. The constellations of these quadrature amplitude modulation schemes can be represented by a set of geometrical circles with different amplitudes. How would the DSP algorithms perform under the physical impairment effects on the recovery of the phase of the carrier?

Currently, several equipment manufacturers are striving to provide commercial advanced optical transmission systems at 100 Gbps employing coherent detection techniques for long distance backbone networks and metro networks as well. Since the 1980s, it has been well known that single-mode optical fibers can support such

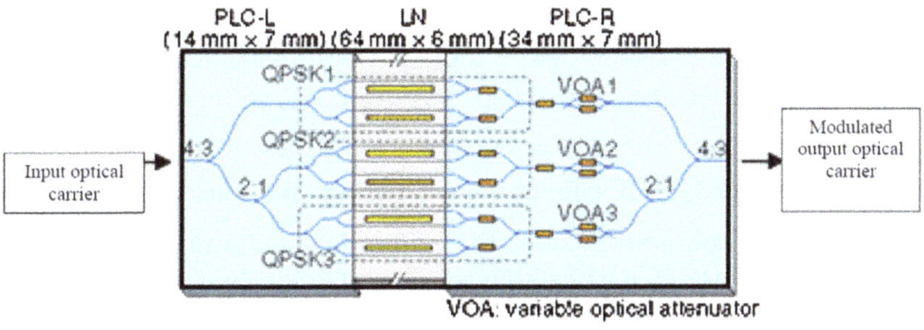

Figure 3.5. Structure of a complex signal generator using dual polarized modes and in-phase and quadrature components via two parallel MZIMs per component and $\pi/2$ phase shifters. Variable optical attenuator (VOAs) are used to adjust the quadrature amplitude of the optical fields. PLC = planar lightwave circuit, L = left; R = right; LN = lithium niobate ($LiNbO_3$). See also schematics given in figures 3.7 and 3.8. VOA = variable optical attenuator: QPSK(n = 1,2..4) optical modulator employing QPSK format, number 1 to 4.

transmission due to the preservation of the guided modes and their polarized modes of the weakly guiding linearly polarized (LP) electromagnetic waves [4]. Naturally both transmitters and receivers must satisfy the coherency conditions of narrow linewidth sources and coherent mixing with a local oscillator, an external cavity laser, to recover both the phase and amplitude of the detected lightwaves. Both polarized modes of the LP modes can be stable over long distances so far, in order to provide the polarization division multiplexed (PDM) channels, even with polarization mode dispersion (PMD) effects. All linear distortion due to PMD and CD can be equalized in the DSP domain, employing algorithms in either offline processing or real-time.

It is thus very important to ensure that these sub-systems are performing the coherent detection and transmitting functions. This section thus presents a summary of the tests conducted with a back-to-back and transmission of QPSK PDM channels. The symbol rate of the transmission system is 28 GBd under the modulation format PDM-CSRZ-QPSK. We note that the differential QPSK (DQPSK) encoder and the bit pattern generator are provided in the optical transmitter. A QAM modulator using $LiNbO_3$ integrated optics and two planar lightwave circuits (e.g. silicon nitrate) is shown in figure 3.5. The electrical broadband signals are applied to the electrodes of the Mach–Zehnder interferometric modulators (MZIMs). Two polarized optical channels are guided through sections of MZIM so that the in-phase and quadrature components are generated and then summed up at the output. Two PLC sections are connected at interfaces of the LiNbO3 so that the phase shifts via thermal effects and variation of the amplitude attenuation. PLC Si3N4 of low propagation loss is used to minimize the total insertion loss. Lightwaves modulated by this QAM modulator reserve their coherence and are commonly employed in high speed broadband coherent and intensity-modulated transmission systems. The analog signals are fed into an RF amplifier driver and then fed into the electrodes of the MZIMs. Digital waveforms are generated in a DSP system of either frequency domain, e.g. OFDM, or time domain multi-level modulation format, e.g. PAM-n, complex M-ary QAM.

Pre-distortion or pre-emphasis of the signals can be implemented in this DSP-based transmitter so as to compensate for the transmission fiber lines or any band-limited sub-systems of the transmission systems.

3.3.1 112G—800 Gbps QPSK coherent transmission systems

It is expected that the coherent polarization multiplexed channel QAM formats with a baud rate of 100 GBaud can offer high sensitivity 800 GBps per carrier. 25 GBaud × 4 lanes for 100 Gbps modules are quite popular now, but still not high enough for DCN in the near future and DC access networks. This section gives a brief outline of the conceptual design of the optical transmitter and receiver for 400 Gb s^{-1} bit rate whose spectrum would look like 28 Gb s^{-1} symbol rate.

The spectral distribution of the channel consists of two sub-channels separated by 50 GHz symmetrically allocated from the laser central frequency. The modulation is a multi-level QPSK based on the existing QPSK 100G structures. The multi-level 16-QAM is used in order to increase the capacity by a factor of 2 and thus with the frequency division mux contained within the allowable spectral spacing would allow us to generate and transmit a 400G capacity optical channel. The generic optical Tx and Rx of these advanced sub-systems are briefly described.

The setting and testing of the currently researched 100G transmission systems are outlined, in particular the noise contribution under the superposition of the differentially amplified and coherently detected noise on the phase modulated and balanced receiving structure. The detection of the combined constellation would suffer some penalty pending on the Euclidean distance between the constellation points. This is still to be explored. The golden ratio of the 16-QAM may offer some advantages. However, we expect at least a 3 dB reduction in the sensitivity of the receiver and thus it may require higher effective number of bits (ENOB) of the DSP sub-system.

The roles of digital signal processing in the recovery of received signals is also remarked. This proposed system is certainly is discussed for reference.

3.3.1.1 Modulation format 16 QAM

QPSK is considered as a basis/kernel for the multi-level advanced modulation format. Other options are to be considered depending on the noise levels contributed by optical amplifiers and quantum shot noise of the optoelectronic front end. In addition, two frequency sub-channels are used whose carriers are generated using an external modulator with suppression of the lightwave carrier.

The formation of 100 Gbps can be calculated as 100 Gbps = [2 bits/symbol (DQPSK)]*2 [polarizations (Pol-Mux)[1]]*28 GBaud (symbol rate). Thus, for 400G = 4 × 100G—the factor of 4 is obtained by frequency division multiplexing (FDM) and one-order higher level modulation (M-ary-16-QAM at 50 GSymbols s^{-1}) plus two polarizations. The frequency multiplexing with the channel frequency carriers is generated by an external optical modulator which is driven by a single microwave

[1] Pol-Mux = polarization multiplexing

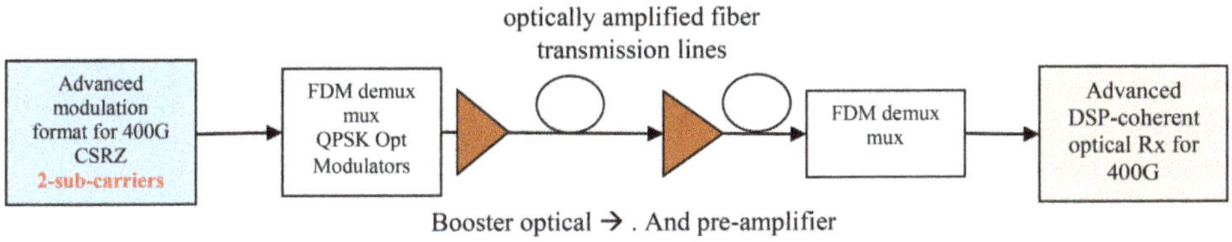

Figure 3.6. 400G–800G optical transmission systems.

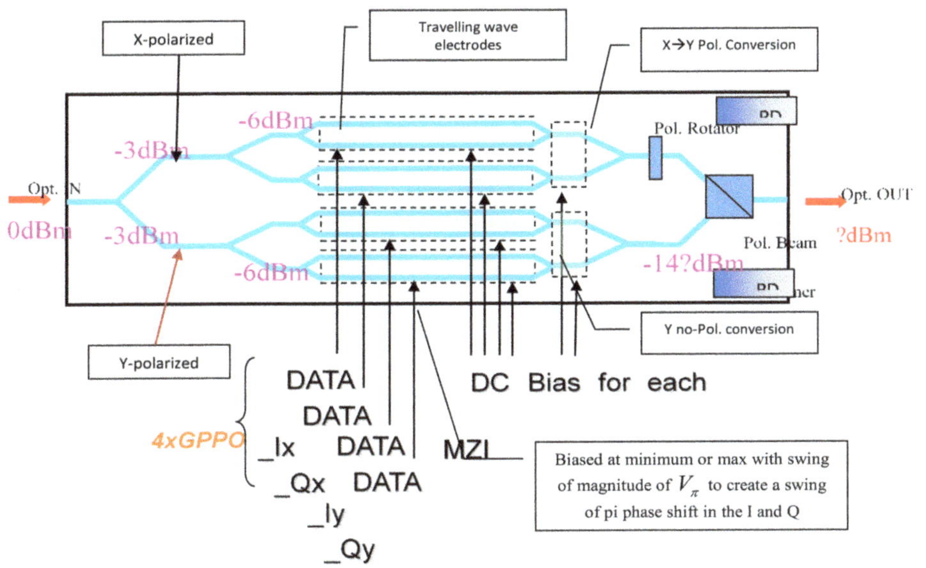

Figure 3.7. Schematic of a Fujitsu I–Q modulator with integrated polarization mux employed in a 100G Tx coherent transmission system (data spec supplied by Fujitsu). Data sequence applied to I and Q electrical ports are 28 GS s^{-1}. Note: there is a possibility of biasing point drift due to electro-optic effects generated by static charges accumulated on the surface of the travelling wave electrodes.

sinusoidal wave to create the double sideband carriers shifted by 25 GHz from both sides of the lightwave carrier. The modulator is based symmetrically on the electrical–optical transfer characteristics of the MZIM at $V_\pi/2$ and the amplitude swing is equal to $V_\pi/2$ so that the carrier can be suppressed. Thus, the optical transmission is arranged as shown in figure 3.6. Naturally the receiver for 400 Gbps at 50 GSymbols s^{-1} requires wider bandwidth, almost doubling that for 25 GSymbols s^{-1} for 100 Gbps PDM-QPSK.

3.3.1.2 Optical modulators

Figure 3.7 shows the existing I–Q modulator employed in the 100G Pol-Mux optical transmitter for generating 28 GS s^{-1} with two orthogonal polarized guided lightwave channels. We note the following: (i) the total insertion loss per polarized channel is at least 14 dB; (ii) the modulation is by single-electrode driving signals,

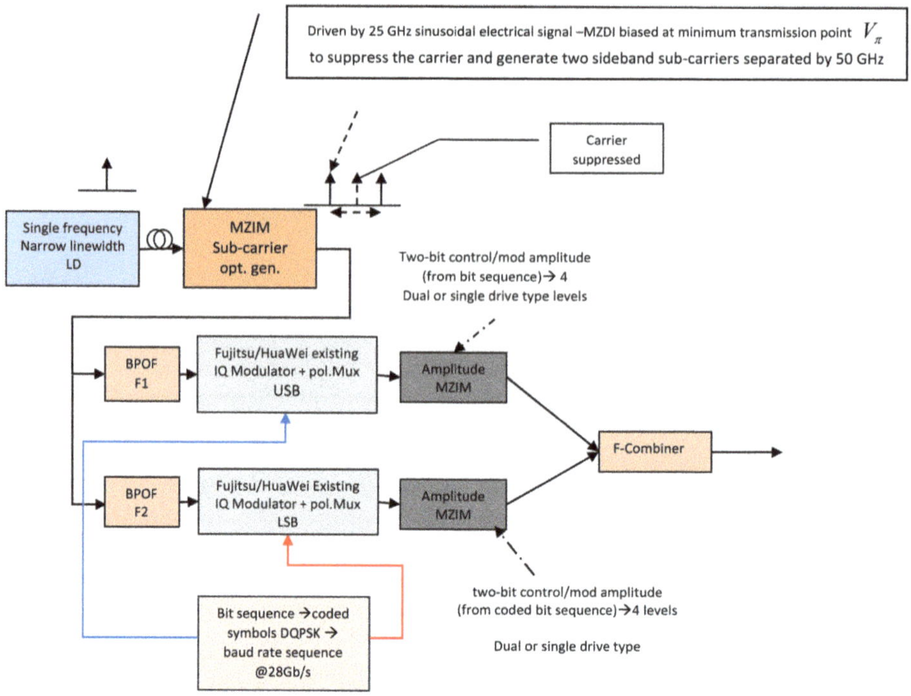

Figure 3.8. Generic structure of 400 Gb s^{-1} with 28 GS s^{-1}-based Pol-Mux FDM 16-QAM optical transmitter. Notes: de-skewing (phase shifters) components must be used to match the electrical paths wherever necessary.

meaning that the amplitude of the RF electrical signal is quite high. However, this is compensated for by the simplicity of the driving signal requirements and RF connecting; (iii) the biases are implemented by integrated T-bias built in the optical IC. Note that there would be drift of the DC biasing point.

Two carriers can be generated using the CSRZ modulators and then optical filtering, as shown in figure 3.8. However, it is much simpler if two MZIM modulators are used with dual-drive electrodes and the sinusoidal RF waves are $\pi/2$ phase shifted with each other to generate upper side band single band and lower sideband instead of using optical filters and one CSRZ optical modulator. This structure is shown in figure 3.9. It is quite straightforward that when the two RF sinusoidal waves are shifted by $\pi/2$ then the lower sideband are cancelling each other, resulting one sideband in the optical domain (figure 3.10), and similarly, for the lower sideband generation. The frequency spectra at different position of the optical FDM modulators are shown in figure 3.11.

It is noted that if this FDM and dual constellation multiplexing are used for a 100G transmitter then the symbol (baud) rate can be halved, thus the dispersion tolerance and hence transmission distance can be quadrupled. It is assumed that the digital optical coherent receiver is the same and that the detection performance offers the same for both QPSK and DQPSK. At worst the penalty for 8PSK

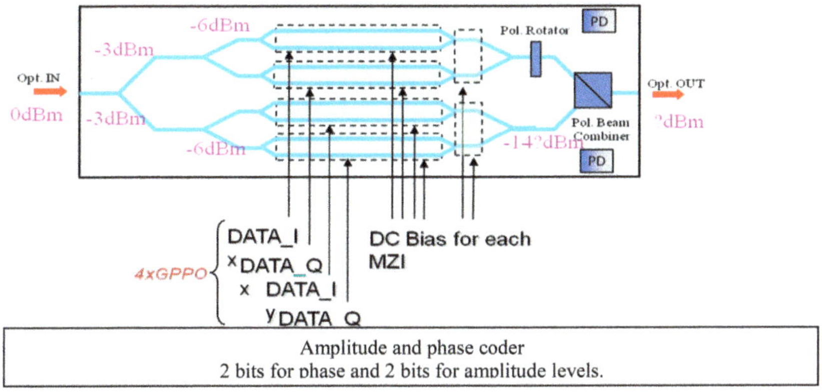

Figure 3.9. Structure of optical transmitter using four-level APSK with amplitude modulation and Fujitsu I–Q PDM modulator shifted.

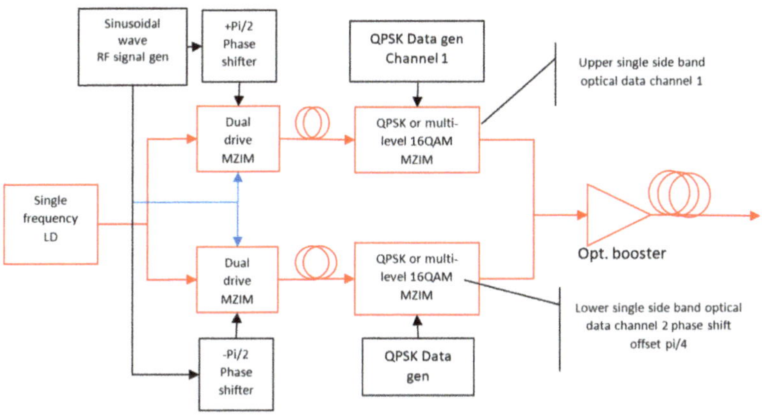

Figure 3.10. Schematic diagram of the optical transmitter using 2 × FDM (upper and lower single sideband) and $\pi/4$ phase shift between QPSK constellation.

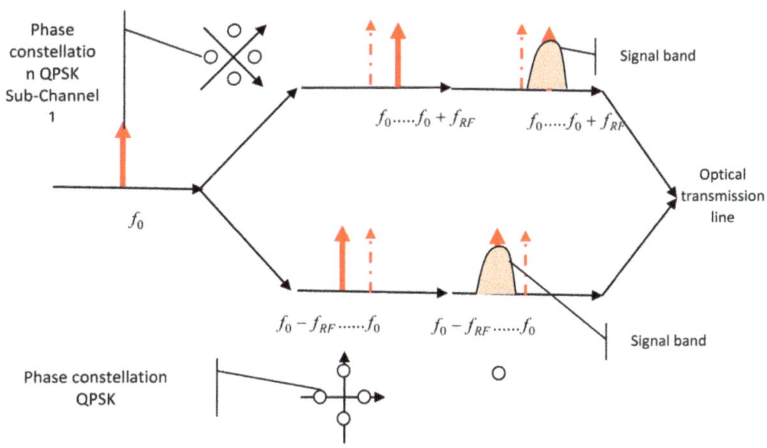

Figure 3.11. Optical frequency spectrum and constellation at different positions of the transmitter described in figure 3.10.

(multiplexed constellation) could be about 3 dB for the 8PSK as compared with the QPSK case.

Using the multiplexed constellation optical coherent receiver, we can simplify the receiver configuration by using only one receiver rather than two digital optical receivers for each constellation.

3.3.1.3 Optical receivers

The receiver consists of: (i) an FDM demultiplexer for each 400G channel, array waveguide grating (AWG) if DWDM operation is employed; (ii) a polarization demux to separate two polarized lightwave channels. Then a demodulator for multi-level detection including an intensity detection and DQPSK demodulators at 100G (Hua Wei 100G receiver); (iii) a digital processing unit for decision making of the phase states of the received I and Q channels.

3.3.1.4 Operating principles

Optical transmitter

The structure of the optical transmitter is shown in figure 3.8 except that the optical band pass filters are not used and only one channel is employed. Figure 3.8 shows a 400 Gb s^{-1} optical transmitter. The modulation scheme is DQPSK with polarization multiplexing and de-multiplexing with forward error coding (FEC). Thus for 100 Gb s^{-1}, this modulation scheme would result in 28 GS s^{-1} (baud/symbol rate).

The CW light wave is generated by a narrow linewidth laser and fed into a Mach–Zehnder amplitude modulator biased at the minimum transmission point of the electrical–optical transfer curve so that a sinusoidal electrical signal at a frequency equal to half of the bit rate and a swing voltage of $2V\pi$ would give an optical clock sequence at the output. This is the pulse carver. This optical clock pulse sequence would have the original optical carrier suppressed.

A Fujitsu optical modulator shown in figure 3.7 is used to modulate and generate two polarized lightwave channels. Each polarized optical path is fed into an I–Q optical modulation section in which two parallel MZIM sub-sections of $\pi/2$ optically delayed with each other, are employed and accepting the electrical RF data sequence inputs to modulate the phase of the optical carriers.

Following the modulation, the polarized channels are then combined and launched into the transmission fiber line.

Note that if 400 Gbps is to be designed then another factor of 4 can be achieved by 16-ADQPSK (amplitude differential quadrature phase shift keying) and frequency division multiplexing. A proposed structure of this transmitter is shown later in this document. The multi-level modulation format of order N would give another factor to reduce the bit rate to the base baud rate of 28 GS s^{-1}. In this case, $N = 16$ gives another factor of 2 higher than that of the DQOPSK scheme. Thus with the FDM and the 4 bits/symbol of the high level modulation scheme, 400G would give a base baud rate of 28 GS s^{-1} transmission sequence.

Optical receiver
At the receiver the optical differential signals are fed into a balanced receiver which consists of two PDs connected back to back (BTB) as differential inputs into an electronic differential amplifier with the differential optical-voltage transfer gain of 2800 V W^{-1}. This differential amplifier then amplifies the differential current and converts to electrical voltage output, the RF output and its complementary. The total equivalent noise density is 80 pA/$\sqrt{\text{Hz}}$ and a 3 dB bandwidth typically specified at 30 GHz. This gives a total equivalent noise current $i_{\text{Neq.}}$ of

$$i_{\text{Neq.}} = \sqrt{S_{\text{Neq.}}B} = \sqrt{80^2.10^{-24} \times 30.10^9} = 8.\sqrt{3} \times 10^{-6}\text{A} = 8.\sqrt{3}\ \mu\text{A} \tag{3.1}$$

This is the total current represented at the input of the differential amplifier. If this total equivalent noise is brought forward to the input of the amplifier then this amplifier is now considered to be noiseless.

Superimposing on top of this total equivalent electronic noise current, there are a number of noise sources presented at the input of the electronic differential amplifier, as follows.

Quantum shot noises is generated by the large power local oscillator mixing with the signal optical power. Assuming that this LO is about 10 dB above the signal power level then this quantum shot noise dominates over that of the signal.

The signal dependent shot noise is thus considered to be much smaller than that of the LO quantum shot noise. Thus, the total noise superimposed on top of the signal level is the total current formed by both the quantum shot noise of the LO and the total equivalent electronic noise of the differential amplifier.

The signal current and hence the output differential voltage level can be estimated by considering the differential power of the state |0| and state |1| or 0 and π radians of the phase states of the modulated symbols. For example, if an average power of −10 dBm signal power is launched into the U2T receiver then the differential amplitude level is −0.1 mW and +0.1 mW or a total differential power is 0.2 mW. This would give a differential output voltage level of 2800 V W^{-1} × 0.2 mW = 4.8 mV[2].

Thus, if the power of the LO laser is 13 dBm fed into the 90° hybrid coupler whose total insertion loss is 13 dB then the LO power is 0 dBm per polarized channel, which generates a quantum shot noise of

$$i_{\text{NLO}} = \sqrt{2qP_{\text{LO}}\Re B} = (2 \times 1.6 \times 10^{-19} \times 10^{-3} \times 1.30 \times 10^9)^{1/2} \approx 1 \times 10^{-6} = 1\mu\text{A} \tag{3.2}$$

This current is about nine times less than that of the total equivalent electronic noises as seen at the input port of the differential amplifier. For a differential optical power of 0.2 mW and a transfer coefficient of the differential amplifier of 2800 V W^{-1} we would expect an output voltage level of 5.6 mV. Alternatively, the equivalent signal current is 0.75 A W^{-1} gives a total signal differential current of 0.3 mA. Thus, a signal-to-noise ratio (SNR) can be estimated to be

[2] Technical specification of U2T optical receiver 43 Gb/s DPSK balanced optical receiver, Product Code: BPRV2125(A), www.U2T.com

$$\text{SNR} = \frac{\langle i_s \rangle^2}{i_{\text{NLO}}^2 + i_{\text{Neq.}}^2} = \frac{300}{1 + 8 \times 1.732} = 20 \text{ or SNR} = 13 \text{ dB}. \tag{3.3}$$

This is sufficient for optical receivers to operate successfully. Note that the receiver sensitivity is specified at −10 dBm. Thus, the LO level boosts the signal power to the appropriate level for the receiver to generate optical pulse sequence for digital sampling and processing.

3.3.2 ENOB and clipping effects of real ADC and automatic gain controller on transmission performance of a coherent QAM system

The roles of ADC ENOB and automatic gain controller (AGC) clipping effect are investigated in a coherent QAM optical transmission system by experiments and simulations, under different transmission effects. Potential implementation risks and optimization for AGC and ADC are presented in this section.

3.3.2.1 ENOB and optical systems

Since the development of 100 Gbps optical transmission systems, coherent detection and digital signal processing (DSP) technologies seem to be indispensable as well as for systems beyond 100G [5–9]. In a coherent system, an ADC plays special role, which converts data from the analog into the digital domain. There are two critical ADC issues, the effective number of bits (ENOB) and the clipping effect. The requirement for ADC bit resolution in coherent systems was discussed in [6]. A more realistic description of ADC noise assumes the Gaussian distribution of electrical noise rather than the uniform distribution of noise generated by the quantization process [10]. An AGC implemented in front of the ADC should be considered and optimized together with the ADC. The AGC output signal strength is monitored and compared to the reference voltage (RV) that may be fixed or derived from some optimization algorithms. Besides clipping, a nonlinear AGC gain transfer function introduces additional system penalties. It becomes more critical for higher modulation formats, e.g. 16-QAM systems, since it requires higher ADC resolution.

In transmission systems with different channel impairments the AGC optimization and required ENOB should be analyzed in detail [11]. They strongly depend on the amount of channel distortion and considered transmission scenario. Assuming realistic ADC and AGC behavior, we studied the ENOB and the clipping effect in the coherent 112G PDM-DQPSK and 224G PDM-16-QAM optical transmission systems by experiments and simulations. The required optical signal-to-noise ratio (OSNR) at a BER of 10^{-3} was estimated over ENOB and RV ranges for real transmission systems and simulated systems including linear effects (a) first-order polarization mode dispersion (PMD), which is characterized by state of polarization angle (SOPA) and differential group delay (DGD), and (b) chromatic dispersion (CD).

3.3.2.2 Experimental and simulation configuration

The 112 Gb s^{-1} PDM-DQPSK was experimentally investigated in the system with the channel spacing of 50 GHz and a span loss of 22 dB. In the DWDM transmission, eight channels of 112 G PDM-DQPSK signals were transmitted over a 1500 km link consisting of 20 spans. Each span consists of a 75 km standard single-mode fiber (SSMF) and a two-stage erbium-doped fiber amplifier (EDFA), in which a section of dispersion compensation fiber (DCF) can be inserted. In the hybrid transmission of 112 G and 40 G, one 112 G channel was set in the middle of eight 40 G QPSK channels. In this case, the CD was fully compensated using DCFs. In the hybrid transmission of 112 G and 10 G, the scenario was the same except that the link length was 900 km. In all cases, the 112 G signal was detected by a coherent receiver that includes a conventional polarization diversity $\pi/2$ optical hybrid coupler together with the local laser and balanced photo diodes outputting four electrical signals. A $\pi/2$ hybrid coupler for PDM-QAM signals consists of a polarization splitter in integrated optic structure, this also accepts the input LO oscillator. Both the polarized LO and channel signals are then combined in the balanced PD. Prior to this combination, each polarized optical channel *X*- and *Y*- are shifted by $\pi/2$ to distinguish the in-phase and quadrature components so that they can then form the phase states of the modulated and detected constellation state.

These electrical signals were sampled by either a real-time oscilloscope or a sampling oscilloscope with 50 GSamples/symbol (GSamples s^{-1}) and 20 GHz bandwidth. The offline data were then processed by the DSP program based on a Matlab© platform, including resampling, CD compensation, clock recovery, PMD compensation, carrier frequency compensation, carrier phase recovery, decision, and BER estimation (figure 3.12).

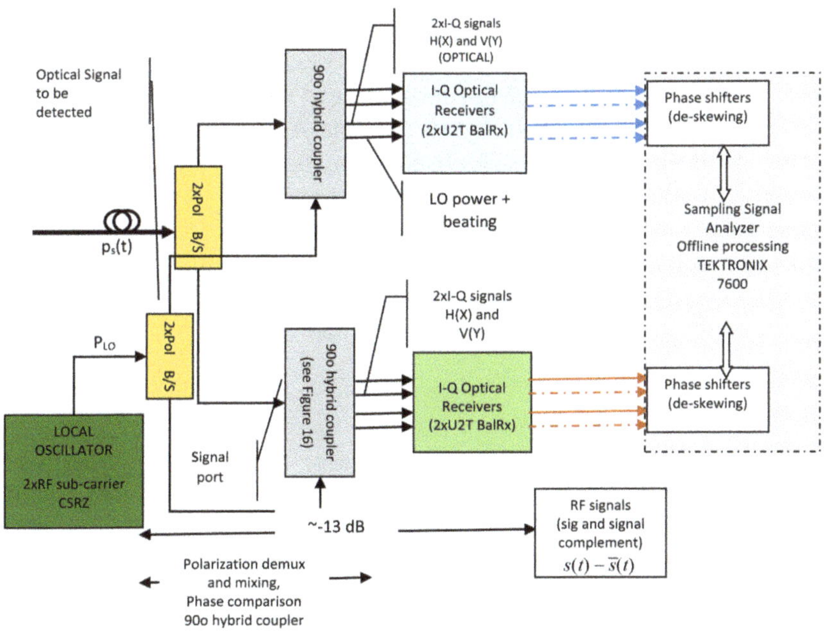

Figure 3.12. Coherent optical receivers including polarization de-multiplexing.

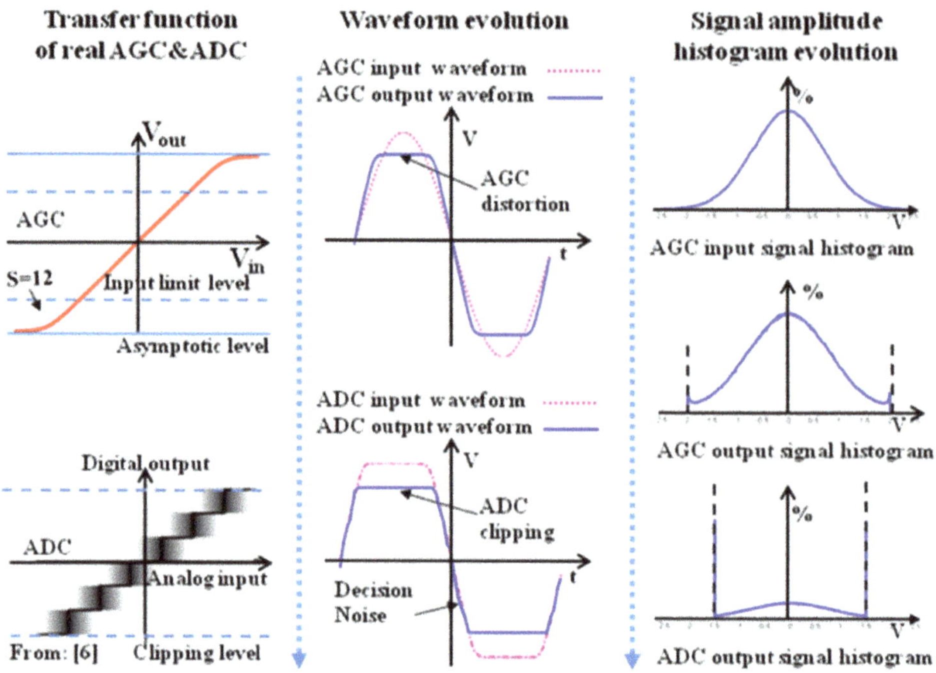

Figure 3.13. Demonstration of the signal waveform and amplitude histogram evolution with the transfer functions of real AGC and ADC.

The AGC and ADC models were implemented before data resampling. A realistic ADC with an AGC can be modeled by two ideal ADCs, additive noise source and an AGC amplifier [12]. Assuming Gaussian distribution of electrical noise presented at the ADC decision levels, the ADC was modeled with variable ENOB. Regarding the AGC model, the absolute AGC input limit level was set to 1.5 (clipping starts at this point), the absolute asymptotic level was set to 2 (clipped signal within ±2), and the knee sharpness parameter was set to 12 [13]. The 8-bit ADC input limit was set to 1.5, which corresponds to the AGC linear region. Figure 3.13 shows the transfer functions of realistic AGC and ADC along with the time domain waveforms and amplitude histograms of signals before and after these circuits. Large clipping can be observed in the outer ADC bins when AGC gain is set to high value.

3.3.2.3 Simulation and experimental results

Simulation results for DQPSK modulation format are presented in figures 3.14(a)–(c). As expected, channels with CD and PMD are more sensitive to ENOB value. Even with the appropriate selection of RV value the ADC with ENOB = 3 bits results in very bad performance. With ENOB of 3 and under the back-to-back (B2B) case, DQPSK may achieve acceptable performance with optimum performance AGC. However, under channel distortions, the performance is decreased by almost 1.0 dB. At higher ENOB values the system performance is less sensitive to the RV parameter. Low AGC gain can be compensated by finer ADC resolution. However, when AGC gain is too high the ADC input signal is clipped, and it cannot be compensated by finer ADC resolution. These penalties are rapidly growing with RV value. In contrast to distorted channels, the B2B can accept larger RV range. The ENOB value of 5 provides good DQPSK performance although carefully optimized AGC enables acceptable performance (0.2 dB loss) with ENOB = 4. The most critical case for ADC resolution is the channel with high CD. The optimum RV value also depends on the transmission impairments. In the B2B case, this sensitivity is not emphasized. However, when distortions are presented, the optimum value depends on ENOB. With the current ADC technology, the ENOB is close to 5 bits. Looking at simulation results one can observe that the optimum RV value varies between 0.5 and 1.

Similar effects are observed with 16-QAM systems such as presented in figures 3.14(d)–(f), with the exception that ENOB of 3 generates

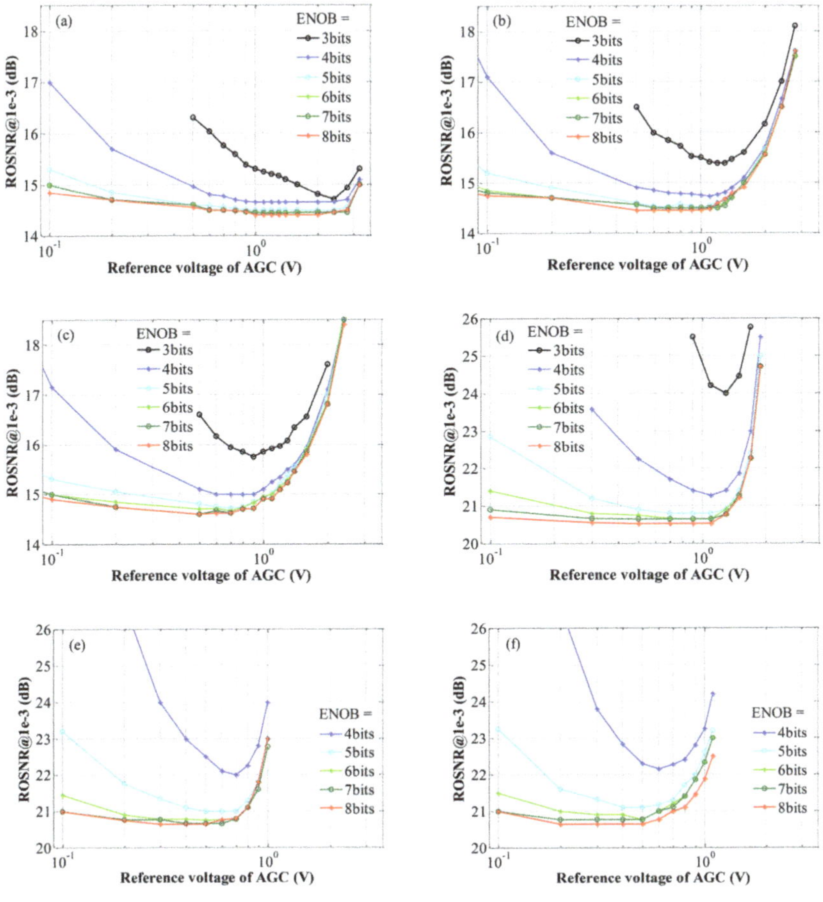

Figure 3.14. Simulation results of the 112 G PDM-DQPSK system (a) B2B; (b) DGD = 30 ps; SOPA = 45°; (c) CD = 10 000 ps nm^{-1}; DGD = 30 ps; SOPA = 45°; and of the 224 G PDM-16-QAM system (d) B2B; (e) DGD = 30 ps; SOPA = 45°; (f) CD = 10 000 ps nm^{-1}; DGD = 30 ps; SOPA = 45°.

unacceptable penalties. Compared to DQPK, one bit more is required to approach optimum working regime. The AGC clipping degrades performance for high values of RV parameter. The AGC dynamic range is more limited by low ENOB than in the DQPSK case. Also, the optimal RV value is much smaller in 16-QAM systems. Additionally, the optimum AGC gain more strongly depends on transmission impairments. It can vary for more than 50% with and without dispersion. That indicates the need for accurate AGC control based on the monitoring of channel performance and degradation (OSNR, CD, PMD, etc). According to the rapid penalty increase due to the clipping effect at high RV, the possibility of RV drifting to higher voltage levels should be considered and controlled properly.

The results shown in figure 3.14 are confirmed by the experimental results presented in figure 3.15. Since the measured ENOB was close to 5, we were only able to decrease ENOB in simulations from 5 to 4 and 3. More degradation comes

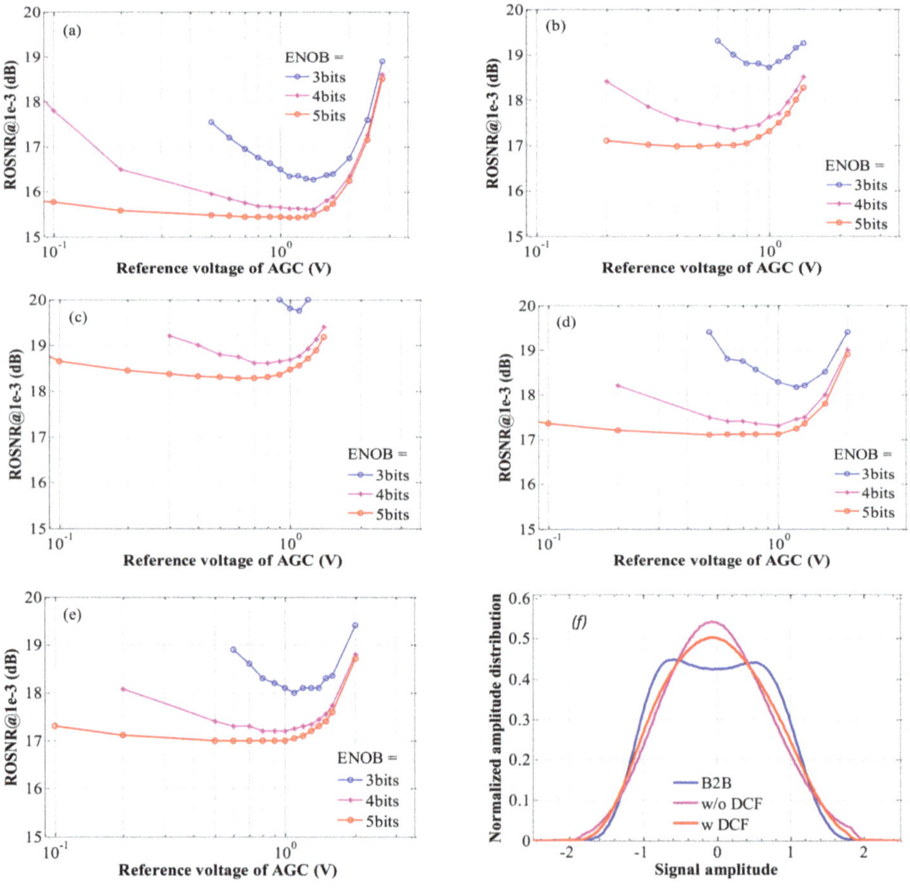

Figure 3.15. Experimental results of 112 G PDM-DQPSK (a) B2B; (b) DWDM transmission without DCF; (c) DWDM transmission with DCF; (d) hybrid transmission with 40 G; (e) hybrid transmission with 10 G; and (f) signal amplitude histogram before AGC.

from nonlinear effects. Also, in the B2B case, a larger difference between 3 and 4 ENOB likely comes from component imperfections and electrical bandwidth limitations. Anyhow, for ENOB of 5, the RV equal to 0.7 provides a good choice for uncontrolled AGC reference value.

In figure 3.15(f) amplitude histograms of some channels are presented for the same RV value. A large difference between B2B and long transmission is observed. In the B2B case, two levels pop up while only noise and electrical bandwidth influence the signal. With distortion the histograms become more Gaussian while the signal is corrupted by nonlinear effects, CD and PMD. This is a reason why lower AGC RV is preferable when distortion is presented.

In future optical transmission systems utilizing higher level modulation formats and digital-to-analog converters (DAC) at the transmitter side (OFDM, nQAM, nPSK with n as the order integer), this problem seems to become more critical. Besides higher sensitivity to ENOB, the AGC nonlinear amplitude function and precise AGC control will be key factors for maximizing the transmission performance.

3.3.2.4 Remarks

The limited ENOB and clipping effects of real AGC and ADC cause penalties that can be partly compensated using sophisticated AGC design and transfer functions correction. Furthermore, the penalties are enhanced by transmission impairments like CD, PMD, nonlinear effects etc. Therefore, the AGC and ADC dynamics must be optimized with appropriate parameters. The problem becomes more serious in future optical networks utilizing tight channel packaging.

3.3.3 Quantization errors and noises

ENOB is a very important parameter for the evaluation of a digital processing system, especially at the ADC and vice versa. Effectively, it is the measure of the corruption of the noise levels on the resolution of the useful signal level in terms of the natural log scale. This parameter ENOB should be considered and evaluated for digital optical receivers which normally consist of an optical receiver (coherent or non-coherent) followed by ADC and DSP sub-systems. Thus, the common definition of the ENOB should now be extended to include the optical receivers at the front end. This optical receiver may include a PD or pair of detectors for balanced receiving techniques and then followed by a wideband electronic amplifier.

The noise of such a digital optical receiver would consist of the following: (i) electronic noises of the ADC and DSP; (ii) electronic noise or equivalent noise spectral density of the electronic wideband amplifier following the photodetectors; (iii) optical noise including accumulated optical amplification noise, signal dependent noise (quantum shot noise), quantum shot noise due to high power of the local oscillator if coherent detection is used (see figures 3.16 and 3.17); (iv) nonlinear inference or distortion of the data signals, e.g. SPM, XPM and FWM impairments; (v) linear distortion effects, e.g. chromatic dispersion, PMD etc.

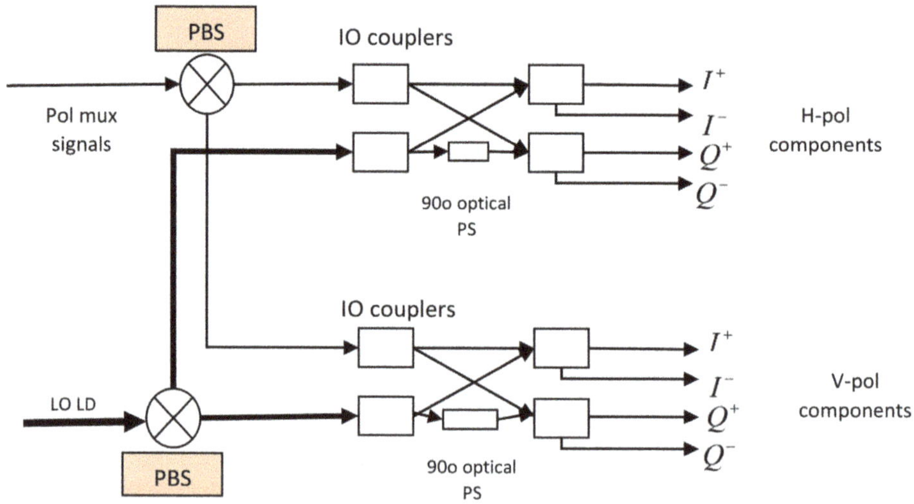

Figure 3.16. 90° hybrid optical coupler with RF output ports of differential difference in optical phase of the in-phase and quad-phase optical waves. Refer to the differential optical gain conversion factor of the receiver (extracted from NTT hybrid coupler). Legends: PBS = polarization beam splitter, IO = integrated optic, I = in-phase, Q = quadrature phase, PS = phase shifter, LOLD = local oscillator laser diode. H—horizontal, V—vertical.

This noise must be evaluated in terms of the number of least significant bits (LSB), which would make it ineffective, thus the ENOB of the digital signal processing sub-system. If this is determined to be important then we should conduct a detailed study of this parameter and its impact on receiver performance (figure 3.18).

Furthermore, if the two 16APSK constellations are shifted by $\pi/4$ with respect to each other, the homodyne detected constellation would look like those given in figure 3.16. The penalty of OSNR would not be much more than that of a single 16APSK due to the phase resolution of the $\pi/2$ separation of the 16APSK (figure 3.19).

3.4 Optical pre-processing reception and transmitter

The transmission system is arranged as shown in figure 3.20 with the integrated optic pre-processor given in figure 3.2 in association with the coherent reception sub-system. The CSRZ-QPSK (carrier-suppressed return to zero-QPSK) transmitter consists of a CSRZ optical modulator which is a single-drive or dual-drive MZIM. For the single-drive case, the MZIM is biased at the minimum transmission point of its electrical-power transfer characteristics or depletion of the carrier and driven by a sinusoidal signal whose frequency is half of the symbol rate, e.g. 14 GHz (one side band) for 28 GBaud. A WDM Mux (multiplexer) is employed to multiplex other wavelength channels located within the C-band (1530–1565 nm). An optical amplifier (EDFA type) is employed at the front end of the receiver so that noises

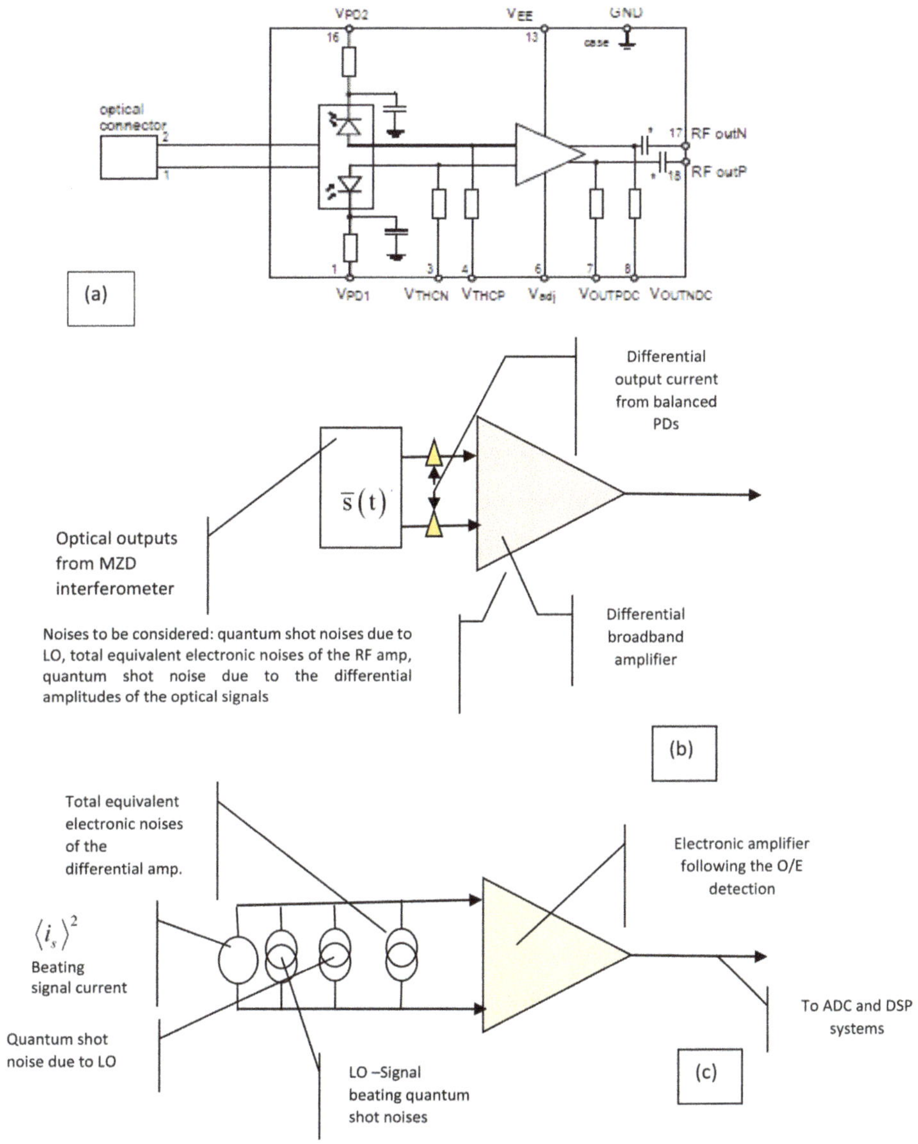

Figure 3.17. Detailed structure of the optical receiver by U2T; (a) general structure and (b) noise representation.

can be superimposed on the optical signals to obtain the OSNR. The DSP processed signals in the digital domain are carried out offline and the BER is obtained.

The transmitter lightwave source can be a narrow linewidth external cavity laser (ECL), an Encore type, a polarization splitter coupled with a 45° aligned ECL beam, two separate CSRZ external $LiNbO_3$ modulators and then two I–Q optical modulators. The linewidth of the ECL is specified at about 100 kHz and with external modulators we can see that the spectrum of the output modulated

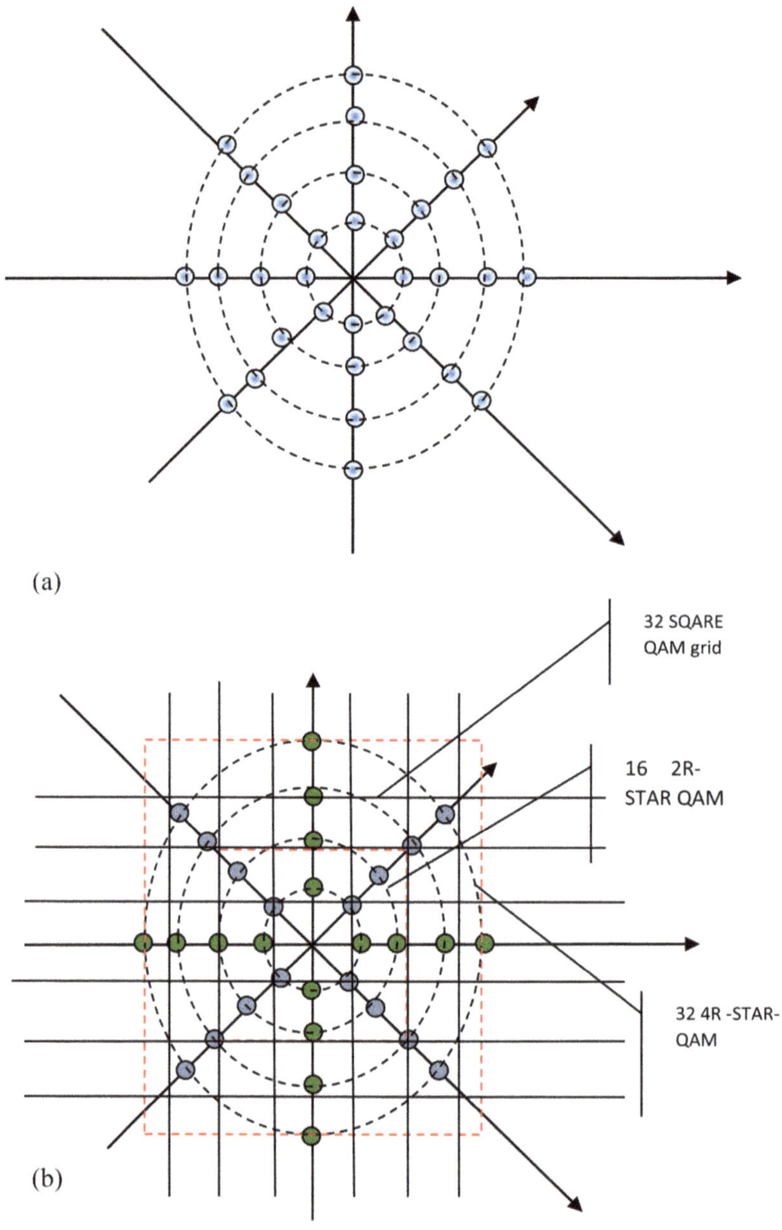

Figure 3.18. Combined 2 × 16APSK to form 4R-32-APSK by homodyne detection using two sc 16APSK (a) without and (b) with grid of square QAM.

lightwaves is dominated by the spectrum of the baseband modulated signals. However, we observed that the laser frequency is oscillating about 300 MHz due to the integration of a vibrating grating reflector so as to achieve stability of the optical frequency.

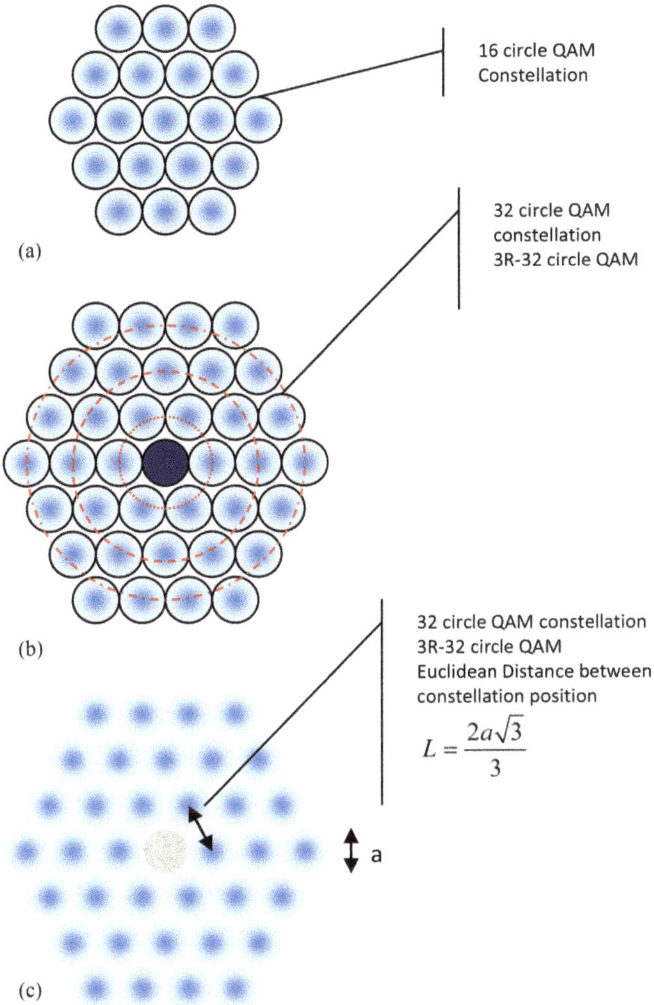

Figure 3.19. Signal constellation of circle QAM: (a) 3R circle QAM; (b) 16 circle QQAM with distance $2a\sqrt{3}/3$; and (c) 3R-32 circle QAM.

The receivers employed in this system are two types. One is a commercialized type, Agilent/Keysight N4391 in association with an Agilent/Keysight external LO, the second one consists of a PD balanced pair connected B2B in a push–pull manner and then followed by a broadband electronic trans-impedance amplifier (TIA). Further, prior to the electronic reception part, $\pi/2$ hybrid optical coupler is employed as the optical mixing sub-system at the front end of the receiver. An optical hybrid coupler can be constructed by a polarization splitter, $\pi/2$ phase shift and polarization combiner that mixed the signal polarized beams and those of the LO (an ECL type identical to the one used in the transmitter). The LO mixed polarized beams (I–Q signals in the optical domain) are then detected by balanced receivers. I–Q signals in

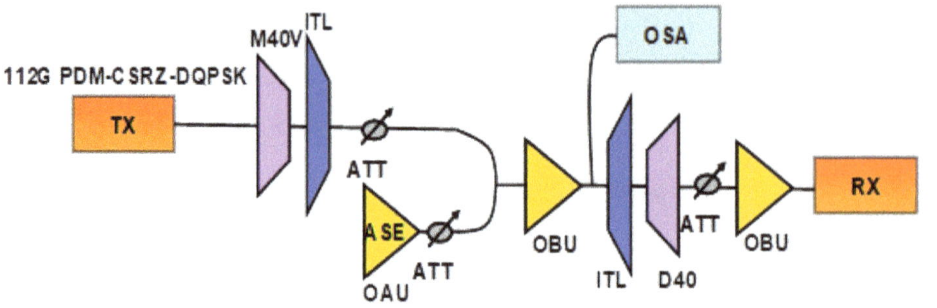

Figure 3.20. Set-up of the PDM QPSK optical transmission system. Tx= transmitter, PDM = polarization division multiplexing, ITL interleaved transmission line, return to zero, OSA=optical spectrum analyzer, Rx receiver, OBU optical boosting unit, D40 = optical demux type 40, Att= attenuator, OAU = optical amplifier unit, ASE = amplification stimulated emission (noise).

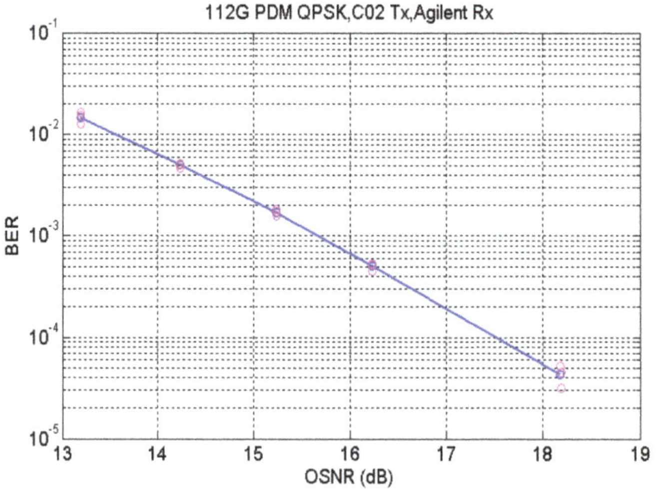

Figure 3.21. B2B OSNR versus BER performance with Keysight Rx in 112 Gbps PDM-CSRZ-DQPSK system.

the electrical domain are then sampled and stored in the real-time oscilloscope (Keysight Digital Processing Osciloscope). The sampled I–Q signals are then processed offline using the algorithms provided in the scope or own developed algorithms such as the evaluation of EVM described above for *Q*-factor and thence the BER.

Both the transmitters and receivers are functioning with the required OSNR for the B2B of about 15 dB at a BER of 2×10^{-3}. It is noted that the estimation of the amplitude and phase of the received constellation is quite close to the received signal power and the noise contributed by the balanced receiver with a small difference, due to the contribution of the quantum shot noise contributed by the power of the LO. Figure 3.21 shows the BER versus OSNR for B2B QPSK PDM channels. As

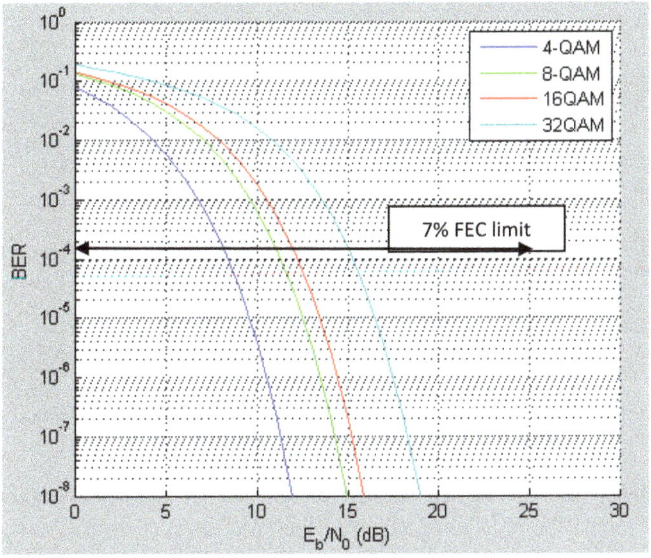

Figure 3.22. Theoretical BER versus SNR for different level QAM schemes obtained using 'bertool.m' of MATLAB.

shown in figure 3.22 the variation of BER as a function of the signal energy over noise for AWGN noise, for 4-QAM or QPSK coherent then the SNR is expected at about 8 dB for BER of 1×10^{-3}. Experimental processing of such a scheme in B2B configuration shows an OSNR of about 15.6 dB. This is due to 3 dB split by the polarized channels and then additional noises contributed by the receiver, hence about 15.6 dB OSNR required. The forward error coding (FEC) is set at 1×10^{-3}. The receiver is an integrated digital coherent optic type (DCO). These DCO receivers are, presently known as CFP2-DCO or DP-DCO-QSFP formats, which will be described in chapter 6. These compact formats are essential for saving spaces in DC networking.

A brief analysis of the noises at the receiver can be as follows. The noise is dominated by the quantum shot noise generated by the power of the local oscillator which is about at least 10 times greater than that of the signals. Thus, the quantum shot noises generated at the output of the PD is

$$i_{\text{N-LO}}^2 = 2qI_{\text{LO}}B$$
$$I_{\text{LO}} = 1.8 \text{ mA} \ @ \ 0.9 - \text{PD\_quantum\_efficiency}$$
$$B = 31 \text{ GHz} - \text{BW\_U2t\_Agilennt\_Rx}$$
$$i_{\text{N-LO}}^2 = 2 \times 10^{-19} \times 1.8 \times 10^{-3} \times 30 \times 10^9 = 10.8 \times 10^{-12}\text{A}$$
$$\longrightarrow i_{\text{N-LO}} = 3.286\,\mu\text{A} \longrightarrow \text{shot-noise-current}$$

The bandwidth of the electronic pre-amplifier TIA (transfer impedance amplifier) of 31 GHz is considered. This shot noise current due to the LO imposed on the PD pair, is about compatible with that of the electronic noise of the electronic receiver

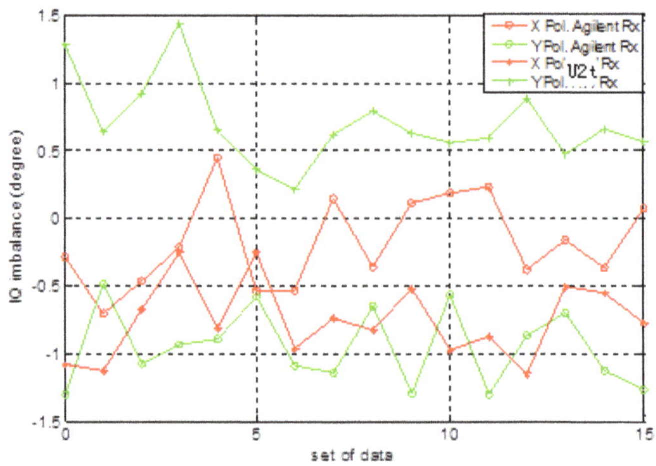

Figure 3.23. I–Q imbalance estimation results for both Rx. Note the maximum imbalance phase of ±1.5°.

given that the noise spectral density equivalent at the input of the electronic amplifier of the U^2t balanced receiver is specified at 80 pA/$\sqrt{\text{Hz}}$, i.e.

$$(80 \times 10^{-12})^2 \times 30.10^9 = 19.2 \times 10^{-12}\ \text{A}^2 \underset{\text{noise\_current}}{\longrightarrow} i_{\text{Neq.}} = 4.38\ \mu\text{A}$$

Thus, any variation in the LO would affect this shot noise in the receiver. It is thus noted that with the trans-impedance (Z_T) of the electronic pre-amplifier estimated at 150 Ω, a 10 dBm difference in the LO would contribute to a change of the voltage noise level of about 0.9 mV in the signal constellation obtained at the output of the ADC. Of further note is that the noise contributed by the electronic front end of the ADC has not been considered. We note that differential TIA offers at least 10 times higher transfer impedance of around 4500 Ω over 30 GHz mid band. These TIAs offer much higher sensitivities as compared to single input TIA type[3] [14].

3.4.1 I–Q imbalance estimation

There is an imbalance due to the propagation of the polarized channels and the I- and Q-components. They must be compensated in order to minimize the error. The I–Q imbalance of the balanced receiver (Rx) is less than 2 degree angle of arc, which can be negligible for the system, as shown in figure 3.23. This imbalance must be compensated for in the DSP domain.

3.4.2 Skew estimation

In addition to the imbalance of the I- and Q- due to optical coupling and electronic propagation in high frequency cables, there are propagation delay time differences

[3] Linear Circuits Inc. From single ended to different input trans-impedance amplifier http://circuits.linear.com/267

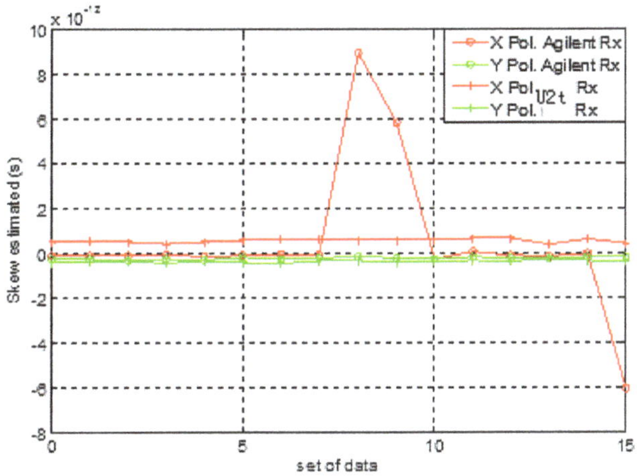

Figure 3.24. Skew estimation results for both types of Rx.

between these components that must be compensated. The skew estimation is shown in figure 3.24 obtained over a number of data sets.

Abnormal skew variation from time to time was also observed, which should not happen if there is no modification of the hardware. Considering the skew variation happened with the Agilent Receiver, which only has very short RF cable and tight connection, there is a high probability that the skew happened inside the Tektronix oscilloscope more than at the optical or electrical connection outside.

Figure 3.25 shows the BER versus the OSNR when the skew and the imbalance between I- and Q-components of the QPSK transmitter are compensated. The OSNR of DQPSK is improved by about 0.3–0.4 dB at 1e−3BER compared with the result without I–Q imbalance and skew compensation as shown in the constellation obtained in figure 3.26. In the time domain, the required OSNR of DQPSK at a BER of 1e−3 is about 14.7 dB, which is 0.1 dB improved. The required OSNR at 1e −3BER of QPSK is about 14.7 dB, which is about the common performance of the state of the art for QPSK. For a BER = 1e−3 an imbalanced CMRR = −10 dB would create a penalty of 0.2 dB in the OSNR for the Agilent receiver and an improvement of 0.7 dB for a commercial balanced receiver employed in the Rx sub-systems.

Figure 3.27 shows the structure of the external cavity laser incorporating a reflection mirror, which is vibrating at a slow frequency of around 300 MHz. A control circuit would be included to indicate the electronic control of the vibration and cooling of the laser so as to achieve stability and elimination of Brillouin scattering effects.

3.4.3 Fractionally spaced equalization of CD and PMD

Ip and Kahn [18] employed the fractional spaced equalization scheme with MSE (mean square error) to evaluate its effectiveness of PDM ASK (amplitude shift keying) with NRZ (non-return to zero) or RZ (return to zero) pulse-shaping

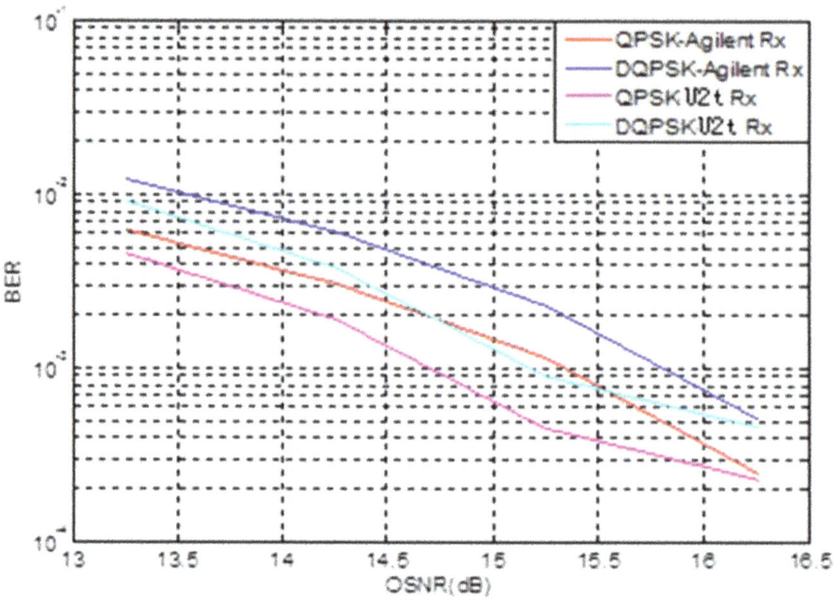

Figure 3.25. OSNR versus BER for two types of integrated coherent receiver after compensating I–Q imbalance and skew.

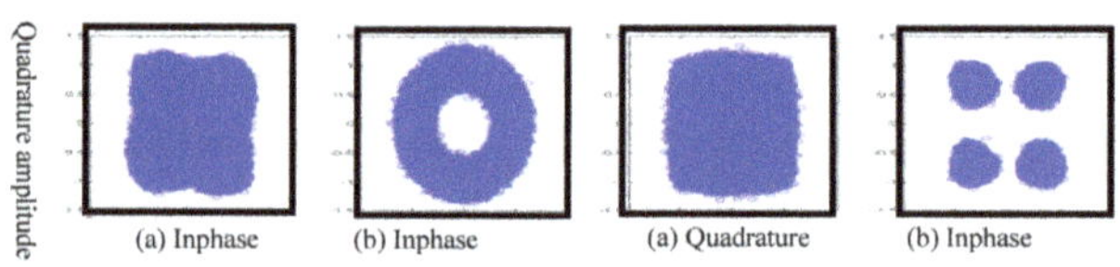

Figure 3.26. Constellation after PMD module (a) and after CPE algorithm module (b).

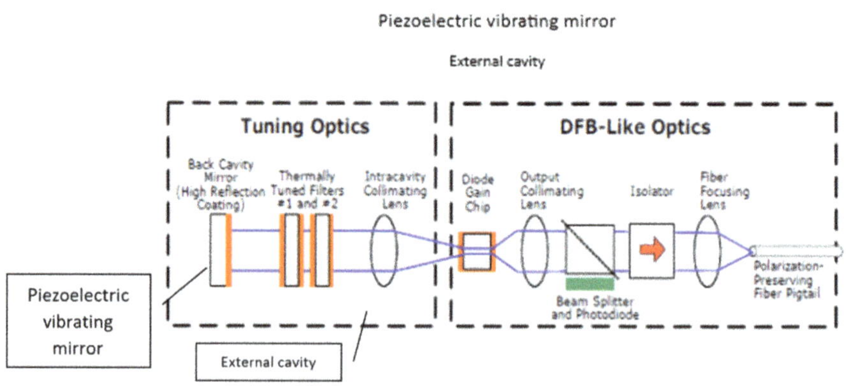

Figure 3.27. (a) Typical noisy QPSK constellation of PDM channels, (a) V-pol (b) H-pol. (c) External optical cavity structure of DFB laser with the back-mirror vibration used as narrow linewidth transmitting source. DFB = distributed feedback.

transmission system. Their simulation results were depicted for a maximum allowable CD (normalized in ratio with respect to the dispersion parameter of the well-known SSMF) versus the number of equalizer taps N for a 2 dB power penalty at a launched OSNR of 20 dB per symbol for ASK RZ and NRZ pulse shapes, using a Bessel anti-aliasing filter with sampling rate $1/T = M/KT_s$; T_s = symbolperiod, with M/K as the fractional ratio.

3.4.4 Electronic digital equalization

Ip and Kahn [15] first developed and applied the back propagation to equalize the distortion due to nonlinear impairment of optical channels transmission through single-mode optical fibers. The back-propagation algorithm is simply a reverse phase rotation at the end of each span of the multi-span link. The rotating phase is equivalent to the phase exerted on the signals in frequency domain with a square of the frequency dependence. Thus, this back propagation is efficient in the aspect that the whole span can be compensated so as to minimize the numerical processes, hence less processing time and central processing unit time of the digital signal processor is required.

Figure 3.28 shows the equalized constellations of a 21.4 GSy s^{-1} QPSK modulation scheme system after transmission through 25 × 80 km non-DCF spans under the equalization using: (a) linear compensation only, (b) nonlinear equalization, and (c) combined back propagation and linear equalization. Obviously, the back propagation contributes to the improvement of the performance of the system.

Figure 3.29 shows the phase errors of the constellation states at the receiver versus launched power of 25 × 80 km multi-span QPSK 21.4 GSy s^{-1} transmission system. The results extracted from[1] shows the performance of back propagation phase rotation per span for 21.4 Gb/s 50 % RZ-QPSK transmitted over 25 ×80 km spans of SSMF with 5 ROADMs (Reconfigurable Optical Add-Drop Module), with 10% CD under-compensation. The algorithm is processed offline of received sampled data after 25×80 km SSMF propagation via the use of the nonlinear Schrodinger equation (NLSE) and coherent reception technique. It is desired that the higher the launched

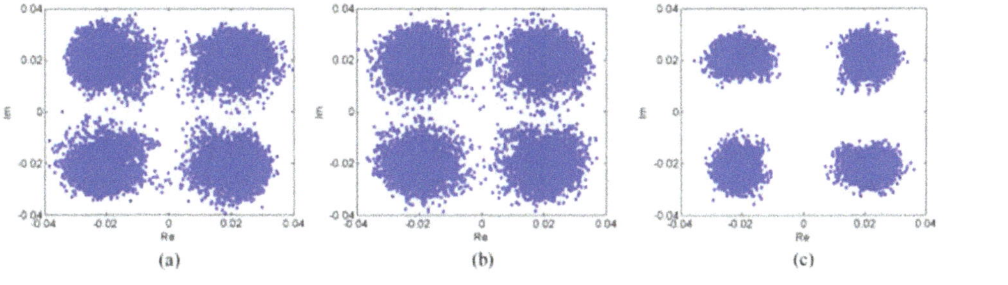

Figure 3.28. Simulated constellation of QPSK scheme as monitored: (a) with linear CD equalization only; (b) with nonlinear phase noise compensation, and (c) after back propagation processing combining with DSP-based linear equalization. Copyright 2008 IEEE. Reprinted, with permission, from [18].

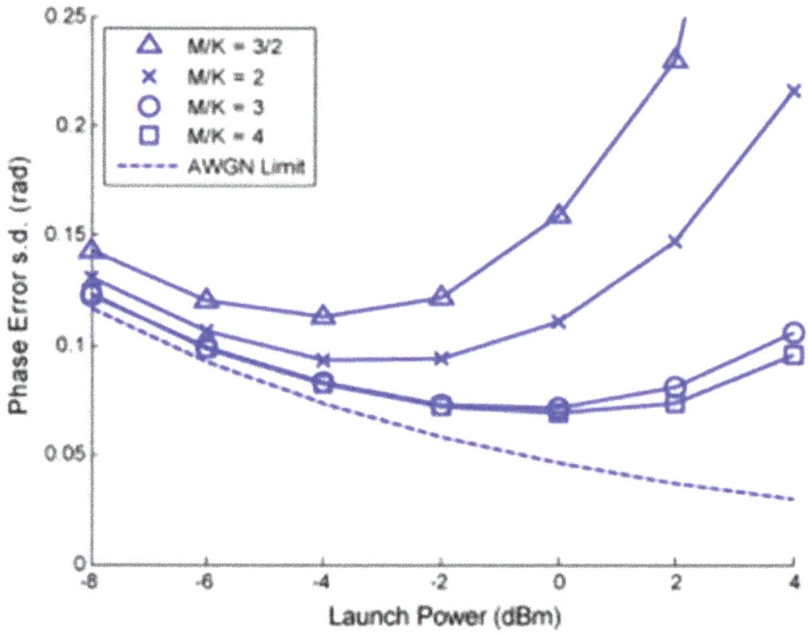

Figure 3.29. Phase errors at the receiver versus launched power or 25 × 80 km multi-span QPSK at 21.4 GSy s^{-1}. Performance of back propagation phase rotation per span for 21.4 Gb s^{-1} 50% RZ-QPSK transmitted over 25 × 80 km spans of SMF with 5 ROADMs, with 10% CD under-compensation using fractionally spaced equalizer with *M*/*K* factor of 1.5–4. (Simulation results extracted from[1] copyright 2008 IEEE. Reprinted, with permission, from [18].)

power the better the OSNR that can be employed for longer distance transmission. So, for fractional space ratio of 3 and 4 offer higher launched power and thus be the preferred equalization scheme as compared with equal space or sampling rate equals to that of the symbol rate. The ROADM is used to equalize the power of the channel under consideration as compared with other DWDM channels.

3.5 16-QAM systems

Consider the 16-QAM received symbol signal whose phase Φ denotes the phase offset. The symbol d_k denotes the magnitude of the QAM symbols and n_k is the noise superimposed on the symbol at the sampled instant. The received symbols can be written as

$$r_k = d_k e^{j\phi} + n_k; \; k = 1, 2, \ldots. L \tag{3.4}$$

Using the maximum likelihood sequence estimator (MLSE), the phase of the symbol can be estimated as

$$l(\phi) = \sum_{k=1}^{L} \ln\left\{ \sum_{d} e^{-\frac{1}{2\sigma^2}|r_k - de^{j\phi}|^2} \right\} \tag{3.5}$$

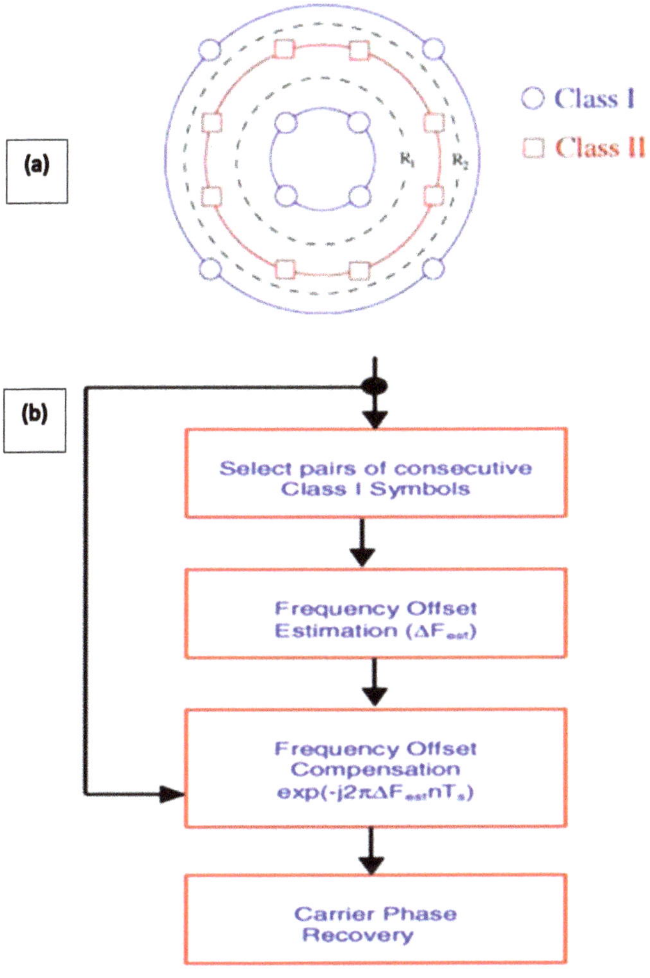

Figure 3.30. Processing of 16-QAM for carrier phase estimation (a) constellation of 16-QAM. (b) Processing for carrier phase recovery with classes I and II of circulator sub-constellation.

Effectively one would take the summation of the contribution of all states of the 16-QAM on the considered symbol measured as the geometrical distance in natural logarithmic scale with the noise contribution of a standard deviation σ.

The frequency offset estimation for 16-QAM can be conducted by partitioning the 16-QAM constellation into a number of basic QPSK constellations, as shown in figure 3.30. There two QPSK constellations in the 16-QAM whose symbols can be extracted from the received sampled data set. They are then employed to estimate the phase of the carrier as described in the previous section on carrier phase estimation for QPSK modulated transmission systems. First, a selection is made of the inner most QPSK constellation, classified as Class I symbols, then an estimation of the frequency offset of the 16-QAM transmitted symbols. Then an FO

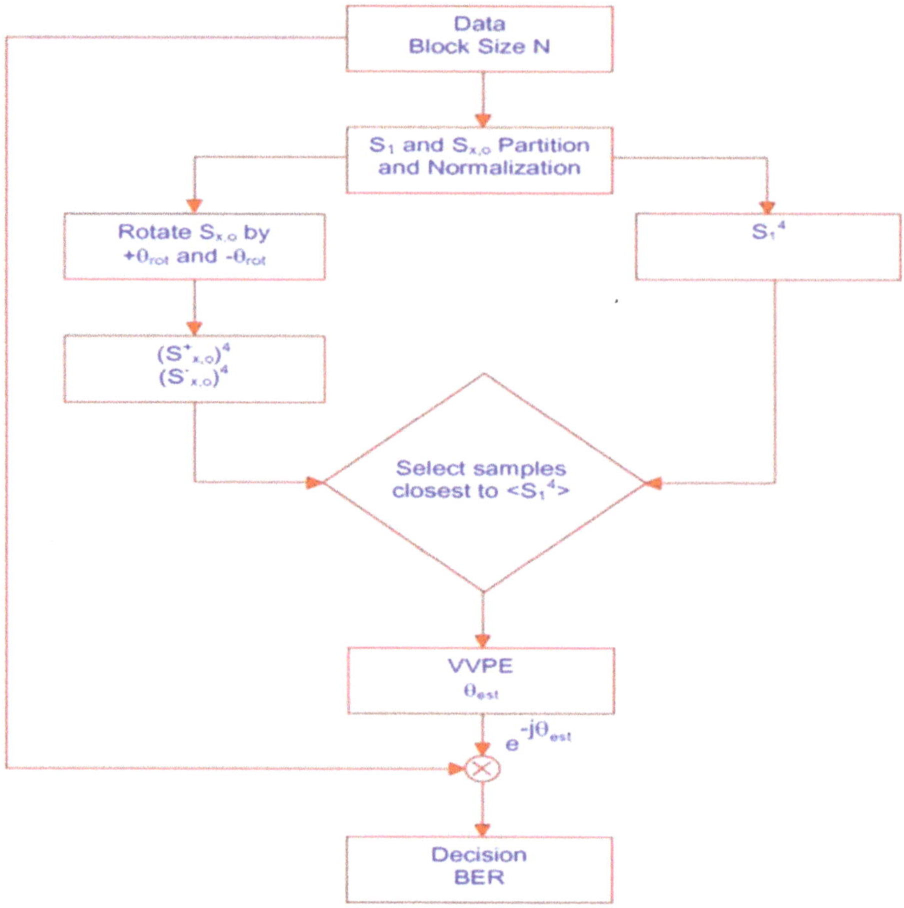

Figure 3.31. Refine carrier phase recovery of 16-QAM by rotation of Class I and Class II sub-constellations.

compensation algorithm is conducted. Finally, the phase recovery of all 16-QAM symbols can be derived. Further confirmation of the difference of carrier phase recovery or estimation can be conducted with the constellation of Class I, as indicated in figure 3.30(a).

Carrier phase recovery based on the Viterbi–Viterbi algorithm on the Class I QPSK sub-constellation of the 16-QAM may not be sufficient and so a modified scheme has been reported by Fatadin *et al* in [16]. Further refining of this estimation of the carrier phase for 16-QAM by partition and rotating is made so as to match certain symbol points to those of Class I constellation of the 16-QAM. The procedures are as shown in the flow diagram of figure 3.31.

Conduct the partition into different classes of constellation and then rotate Class 2 symbols with an angle in either clockwise or anticlockwise of $\pm\theta_{\text{rot}} = \pi/4 - \tan^{-1}(1/3)$. In order to avoid opposite rotation with respect to the real direction, the estimation of the error in the rate of changes of the phase variation or frequency estimation can be found by using the fourth power of the argument of the angles of two consecutive symbols, given by

$$\Delta F_{\text{est.}} = \frac{1}{8\pi T_s}\arg\left(\sum_{k=0}^{N} S_{k+1}S_k\right)^4 \tag{3.6}$$

to check their quadratic mean, then selecting the closer symbol, then apply the standard Viterbi–Viterbi procedure. Louchet *et al* [17] also employed a similar method and confirmed the effectiveness of such a scheme. The effects on the constellation of the 16-QAM due to different physical phenomena are shown in figure 3.32. Clearly the FO would generate the phase noises in (a) and influencing both the I- and Q-components by the CD of a small amount (so as to see the constellation noise) and the delay of the polarized components on the I- and Q- components. These distortions of the constellation allow practical engineers to assess the validity of algorithms which are normally separate and independent and implemented in serial mode. This is in contrast to the constellations illustrated for QPSK as shown in figure 3.26.

Figure 3.33 also shows the real-time signals which result from the beating of the two sinusoidal waves of FO beating in a real-time oscilloscope.

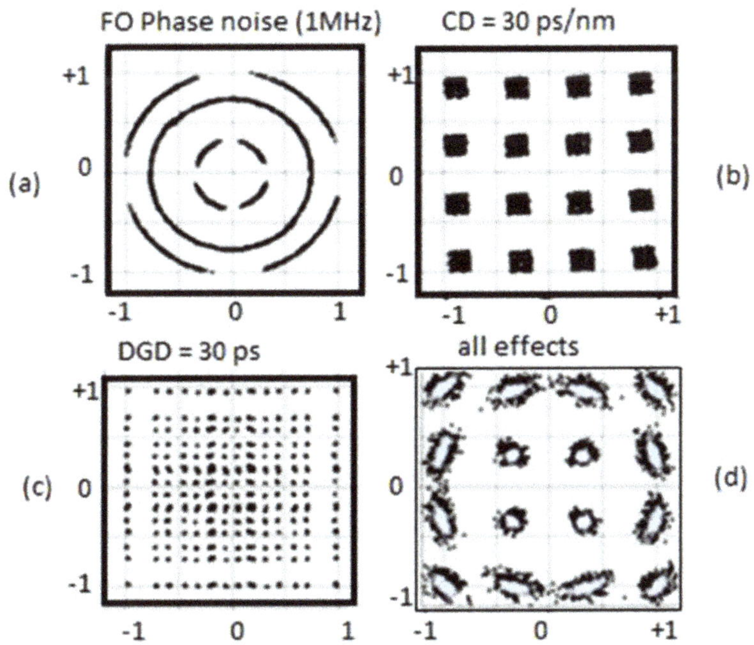

Figure 3.32. 16-QAM constellation under influence of (a) phase rotation due to FO of 1 MHz and no amplitude distortion; (b) residual CD impairment; (c) DGD of PMD effect; and (d) total phase noise effect. Copyright IEEE. Reprinted with permission from [7].

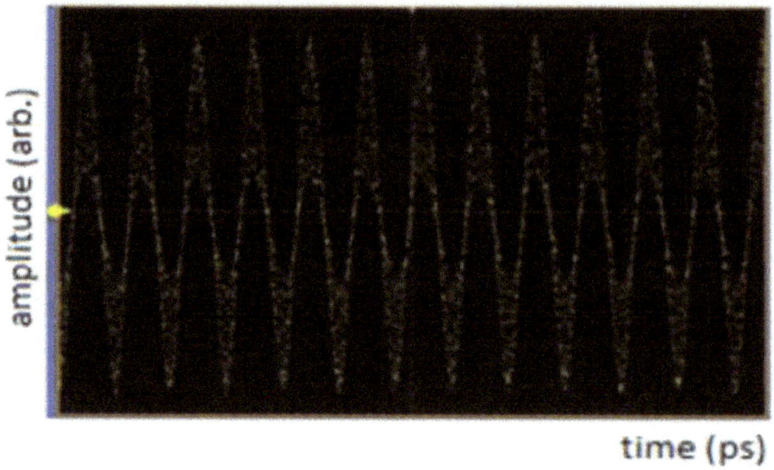

Figure 3.33. Beating signals of the two mixed lasers as observed by a real-time sampling oscilloscope.

Noe *et al* [18] simulated the carrier phase recovery for QPSK with polarization multiplexed (PDM) channels under constant modulus algorithm (CMA) and decision-directed (DD) without and with modification in which the error detected in each stage would be updated. The transmission system under consideration is B2B with white Gaussian noise and phase noise superimposition on the signals. For an OSNR of 11 dB the CMA with modification offers a BER of 1×10^{-3} while it is 4×10^{-2} for CMA without modification. This indicates that the updating of the matrix coefficients is critical to recovering the original data sequence. The modified CMA was also recognized to be valid for 16-QAM.

3.6 Tera-bits/s superchannel transmission systems

3.6.1 Overview

PDM-QPSK has been exploited in 100 Gb s^{-1} long haul transmission commercial systems, and the optimum technologies for 400GE/1TE transmission for next-generation optical networking have now attracted significant interest for deploying ultra-high capacity information over the global internet backbone networks. Further intense research on Tbps transmission systems have also attracted several research groups as the logical rate to increase from 100 Gb s^{-1}. The development of hardware platforms for 1 to *N* Tbps is critical for proving the design concept. The Tbps superchannels can be defined as optical channels comprising a number of sub-rate sub-channels whose spectra would be the narrowest allowable. Thus, in order to achieve efficient spectral efficiency, phase shaping is required and one of the most efficient techniques is Nyquist pulse shaping. Thus, Nyquist QPSK can be considered the most effective format for delivery of high spectral efficiency and effective in coherent transmission and reception as well as equalization at both transmitting and reception ends.

Thus, in this section we describe detailed design and experimental platform for delivery of Tbps using Nyquist-QPSK at a symbol rate of 28 to 32 GSa s^{-1} and 10 sub-carriers. The generation of sub-carriers has been demonstrated using either recirculating frequency shifting (RFS) or nonlinear driving of an I–Q modulator to create five sub-carriers per main carrier, thus two main carriers are required. Nyquist pulse shaping is used to effectively pack multiplexed channels whose carriers are generated by comb generation technique. A DAC with sampling rate varied from 56G to 64 GSa s^{-1} is used for generating Nyquist-shaped pulses including the equalization of the transfer functions of the DAC and optical modulators.

3.6.2 Nyquist pulse and spectra

The raised-cosine filter is an implementation of a low-pass Nyquist filter, i.e. one that has the property of vestigial symmetry. This means that its spectrum exhibits odd symmetry about $1/2T_s$, where T_s is the symbol period. Its frequency domain representation is a 'brick-wall-like' function, given by:

$$H(f) = \begin{cases} T_s & |f| \leqslant \frac{1-\beta}{2T_s} \\ \frac{T_s}{2}\left[1 + \cos\left(\frac{\pi T_s}{\beta}\left\{|f| - \frac{1-\beta}{2T_s}\right\}\right)\right] & \frac{1-\beta}{2T_s} < |f| \leqslant \frac{1+\beta}{2T_s} \\ 0 & \text{otherwise} \end{cases} \quad \text{with} \quad 0 \leqslant \beta \leqslant 1 \tag{3.7}$$

This frequency response is characterized by two values: β, the roll-off factor, and T_s, the reciprocal of the symbol rate in Sym s^{-1}, that is, $1/2T_s$ is the half bandwidth of the filter. The impulse response of such a filter can be obtained by analytically taking the inverse Fourier transformation of (equation 3.7), in terms of the normalized sinc function, as

$$h(t) = \operatorname{sinc}\left(\frac{t}{T_s}\right)\frac{\cos\left(\frac{\pi\beta t}{T_S}\right)}{1 - \left(2\frac{\pi\beta t}{T_S}\right)^2} \tag{3.8}$$

where the roll-off factor, β, is a measure of the excess bandwidth of the filter, i.e. the bandwidth occupied beyond the Nyquist bandwidth as from the amplitude at $1/2T$. Figure 3.34 depicts the frequency spectra of the raised-cosine pulse with various roll-off factors. Their corresponding time domain pulse shapes are given in figure 3.35.

When used to filter a symbol stream, a Nyquist filter has the property of eliminating ISI, as its impulse response is zero at all nT (where n is an integer), except when $n = 0$. Therefore, if the transmitted waveform is correctly sampled at the receiver, the original symbol values can be recovered completely. However, in many practical communications systems, a matched filter is used at the receiver, so

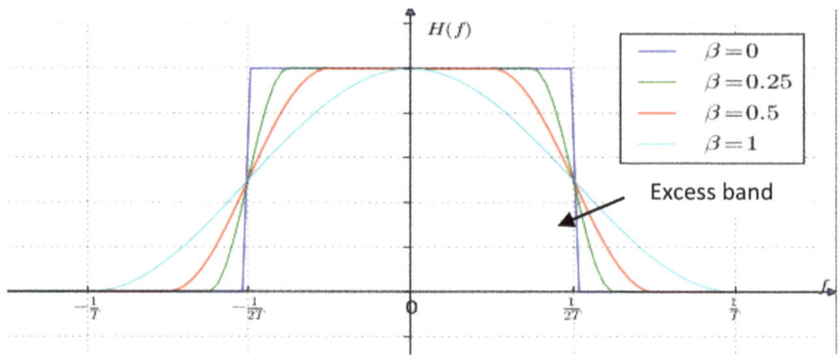

Figure 3.34. Frequency response of raised-cosine filter with various values of the roll-off factor β.

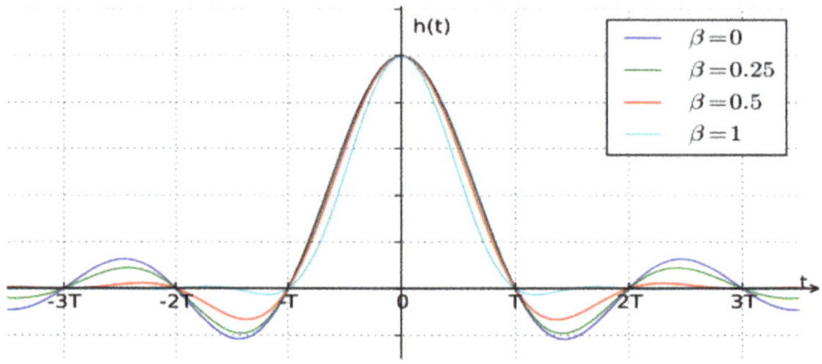

Figure 3.35. Impulse response of raised-cosine filter with the roll-off factor β as a parameter.

as to minimize the effects of noises. For zero ISI, the net response of the product of the transmitting and receiving filters must equate to $H(f)$, thus we can write:

$$H_R(f)H_T(f) = H(f) \tag{3.9}$$

Or alternatively we can rewrite that

$$|H_R(f)| = |H_T(f)| = \sqrt{|H(f)|} \tag{3.10}$$

The filters which can satisfy the conditions of (equation 3.10) are the root-raised-cosine (RRC) filters. The main problem with RRC filters is that they occupy a larger frequency band than that of the Nyquist sinc pulse sequence. Thus for the transmission system we can split the overall raised-cosine filter with RRC filter at both the transmitting and receiving ends, provided the system is linear. This linearity is to be specified accordingly. An optical fiber transmission system can be considered to be linear if the total power of all channels is under the nonlinear SPM threshold limit. When it is over this threshold a weakly linear approximation can be used (figure 3.36).

The design of a Nyquist filter influences the performance of the overall transmission system. Oversampling factor, selection of roll-off factor for different

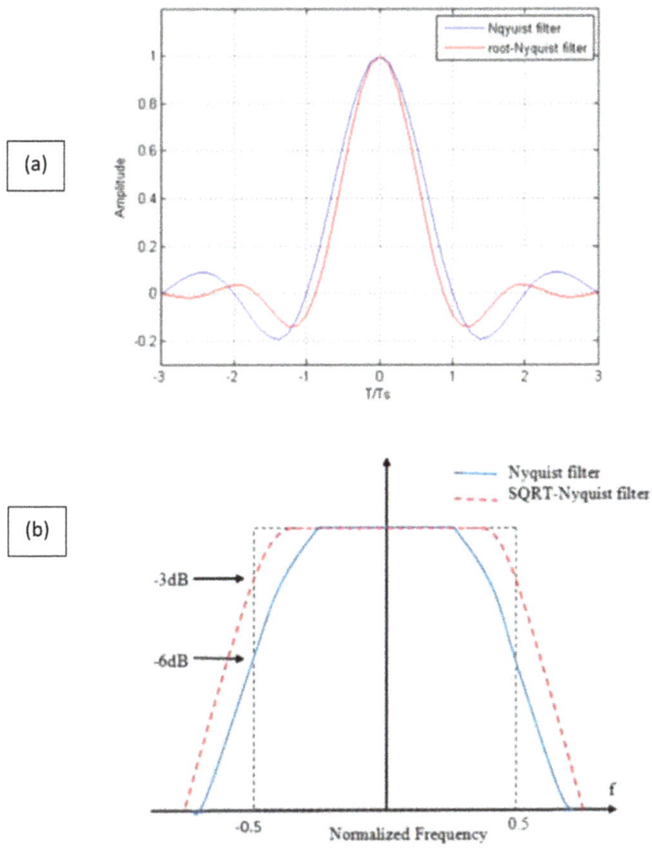

Figure 3.36. (a) Impulse and (b) corresponding frequency response of sinc Nyquist pulse shape or root-raised-cosine (RRC) Nyquist filters.

modulation formats, and FIR Nyquist filter design are key parameters to be determined. If considering the transfer functions of the overall transmission channel including fiber, WSS and the cascade of the transfer functions of all O/E components, the total channel transfer function is more Gaussian-like. To compensate this effect in the Tx-DSP, one would thus need a special Nyquist filter to achieve the overall frequency response equivalent to that of the rectangular or raised-cosine with roll-off factor shown in figure 3.37.

3.6.3 Superchannel system requirements

Transmission distance: as the next generation of backbone transport, the transmission distance should be comparable to the previous generation, namely 100 Gbps transmission system. As the most important requirement, we require the 1 Tbps transmission for long haul to be 1500–2000 km, for metro application ~300 km.

CD tolerance: as SSMF fiber CD factor/coefficient 16.8 ps nm^{-1} is the largest among the current deployed fibers, CD tolerance should be up to 30 000 ps nm^{-1} at

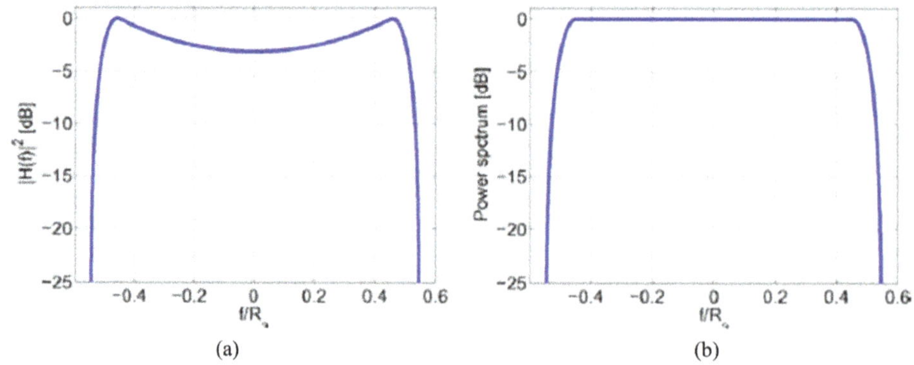

Figure 3.37. (a) Desired Nyquist filter for spectral equalization; (b) output spectrum of the Nyquist filtered QPSK signal.

the central channel whose wavelength is approximated at 1550 nm. At the edge of C-band this factor is expected to increase by about 0.092 ps nm^{-2} km^{-1}) or about 32 760 ps nm^{-1} at 1560 nm and 26 400 ps nm^{-1} at 1530 nm [19].

PMD tolerance: the worst case of deployed fiber with 2000 km would have a DGD of 75ps or about 3 symbol period for 25 GSy s^{-1} per sub-channel. So, the PMD (mean all-order DGD) tolerance is 25ps.

SOP rotation speed: according to the 100 Gbps experiments, SOP rotation can be up to 10 KHz, so we take the same specification as 100G system (figure 3.37).

Modulation format: PDM-QPSK for long haul transmission; PDM-16-QAM for metro application.

Spectral efficiency: compared to a 100G system but increased by a factor of 2. Both Nyquist WDM and Co-OFDM can fulfill this. However, it depends on technological and economical requirements that would determine the suitability of the technology for optical network deployment. Table 3.2 tabulates the system specifications of various transmission structures with parameters of sub-systems, especially when the comb generators are employed using either recirculating or nonlinear generation techniques. The DSP-reception and offline digital signal processing is integrated in these systems.

3.6.4 System structure

3.6.4.1 DSP-based coherent receiver

A possible structure of a superchannel transmission system is depicted in figure 3.38. At the transmitter the data inputs can be inserted into the pulse shaping and individual data streams can be formed. A DAC can be used to shape the pulse with Nyquist equivalent shape, with the raised-cosine forms whose spectra also follow a raised cosine with the roll-off factor β varying from 0.1 to 0.5. If this off factor takes the value of 0.1 the spectra would follow an approximate shape of a rectangular. A comb generator can be used to generate equally spaced sub-carriers for the super-channel from a single-carrier laser source, commonly an external cavity laser of very

Table 3.2. 1 Tbps offline system specifications.

Parameter	Superchannel	Superchannel	Some Specs	Remarks
Technique	**Rcfs Comb Gen.**	**Nonlinear Comb Gen**		
Bit rate	1,2..N Tbps (whole C-band)	1 Tbps, 2,…*N* Tbps	~1.28 Tbps @28–32 GB	20% OH for OTN, FEC
Number of ECLs	1	*N* x 2		
Nyquist roll-off	0.1 or less	0.1 or less		DAC pre-equalization required
Baud rate (GBauds)	28–32	28–32	28, 30 or 31.5 GBaud	Pending on FEC coding allowance
Transmission distance	2500	2500	1200 (16 span~2000 km) (25 spans)	20% FEC req. for Long haul application
			2500 km (30 spans)	
			500 km	Metro application
Modulation format	QPSK/16-QAM	QPSK/16-QAM	Multi-carrier Nyquist WDM PDM-DQPSK/ QAM	For long haul
				For long haul
			Multi-carrier Nyquist WDM PDM-16-QAM	For metro
Channel spacing			4×50 GHz	For long haul
			2 × 50 GHz	For long haul
Launch power	≪0 dBm if 20 Tbps is used		~ −3 to 1d Bm lower if *N*>2	Depending on QPSK/16-QAM and Long haul/metro can be different
B2B ROSNR @2e−2 (BOL) (dB)	14.5	14.5	15 dB for DQPSK	1 dB hardware penalty
			22 dB for 16-QAM	1 dB narrow filtering penalty

(*Continued*)

Table 3.2. (*Continued*)

Parameter	Superchannel	Superchannel	Some Specs	Remarks
Fiber type	SSMF G.652 Or G.655	SSMF G.652 (or 655)	G.652 SSMF	
Span loss	22	22	22 dB (80 km)	
Amplifier	EDFA (G > 22 dB); NF < 5 dB		EDFA (OAU or OBU)	
BER	2e−3	2e−3	Pre-FEC 2e−2 (20%) or 1e−3 classic FEC (7%)	
CD penalty (dB)			0 dB @ +/−3000 ps nm^{-1} <0.3 dB @ +/−30 000 ps nm^{-1};	16.8 p s nm^{-1} km^{-1} and 0.092 ps nm^{-2} km^{-1})
PMD penalty (DGD)			0.5 dB @ 75ps, 2.5 symbol periods	
SOP rotation speed	10 kHz	10 kHz	10 kHz	OPLL may require due to oscillation of the LO carrier
Filters cascaded penalty			<1 dB @ 12pcs WSS	
Driver linearity	required	required	THD <3%	16-QAM even more strict

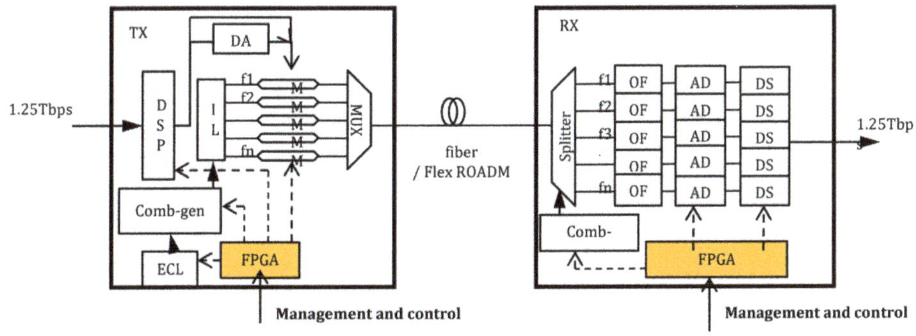

Figure 3.38. The structure of a superchannel Tbps transmission system.

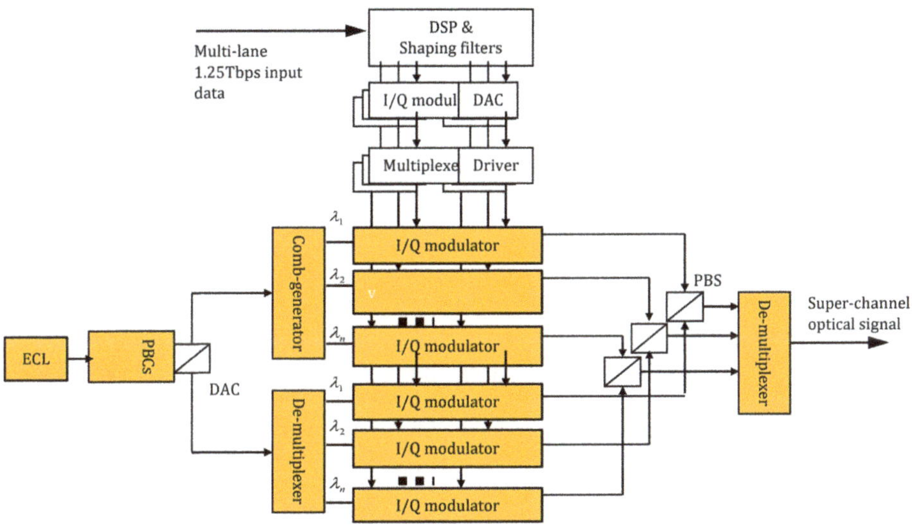

Figure 3.39. Generic detailed architecture of a superchannel transmitter. PBC = polarization beam combiner; PBS = polarization beam splitter; I/Q = in-phase/quadrature phase; DAC = digital-to-analog convertor; ECL = external cavity laser.

narrow band of linewidth of about 100 KHz. These comb-generated sub-carriers (see figure 3.40) are then demultiplexed into sub-carriers and fed into a bank of I–Q optical modulators and Nyquist pulse shape pulse sequence as output of the DAC are then employed to modulate these sub-carriers to form the superchannels at the output of an optical multiplexer, shown in the block at the left side of figure 3.38. More details of the transmitter for superchannels are shown in figure 3.39. It is noted that the generation of a comb source can be recirculating of the shifting of the original carrier around a closed loop. The frequency shift is the spacing frequency between the sub-carriers. So, the *N*th sub-carrier would be the *N*th time circulation of the original carrier. There would be superimposing of noises due to the ASE incorporated in the

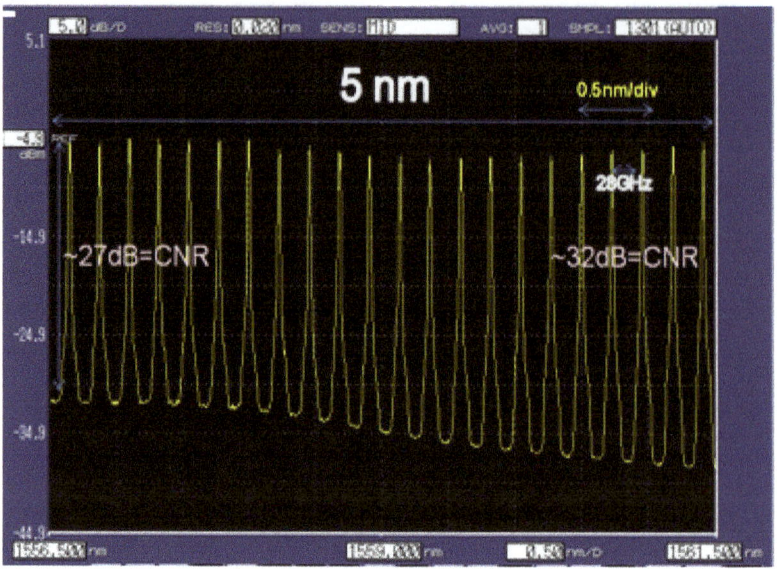

(a) Spectrum of superchannel carrier lines

Figure 3.40. Block diagram of a recirculating frequency shifting comb generator, (a) and a typical generated spectrum of the conb generator, (b) with 28 GHz spacing between channels over more than 5 nm in spectral region and about 30 dB carrier-to-noise ratio (CNR). VOA = variable optical attenuator.

loop, this would be minimized by inserting into the loop an optical filter whose band width would be the same as or wider than that of the superchannel.

The fiber transmission line is an optically amplified optical fiber multi-span without incorporating any dispersion compensating fibers. Thus, the transmission is highly dispersive. The broadening of a 40 ps width pulse would spread across at least an 80–100 symbol period after propagating over 3000 km of SSMF. Thus, one can assume that the pulse launched into the fiber of the first span would be considered to be an impulse as compared to that after 3000 km SSMF propagation.

After the propagation over the multi-span non-DCF line, the transmitted sub-channels are demuxed via a wavelength splitter into individual sub-channels, with minimum crosstalk. Each sub-channel is then coherently mixed with an LO which is generated from another comb source incorporating an OPLL to lock the comb into that of the sub-carriers of the superchannel. Thus, a comb generator is indicated in the right side, the reception system of figure 3.38. The coherently mixed sub-channels are then detected by a balanced receiver, then electronically amplified and fed into the sampler and ADC. The digital signals are then processed in the DSP of each sub-channel system or parallel and interconnected DSP system. In these DSPs the sequence of processing algorithms is employed to recover the carrier phase and hence the clock recovery, compensating for the linear and nonlinear dispersion, and the evaluation of the BER versus different parameters such as OSNR etc. Figure 3.41 shows the modulated spectra of five channels whose sub-carriers are

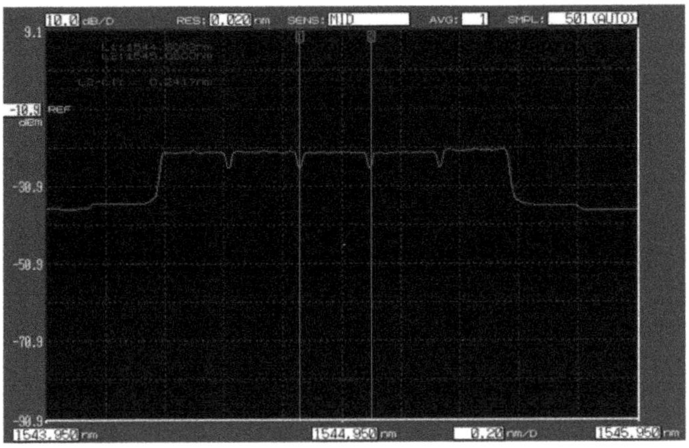

Figure 3.41. Selected five sub-carriers with modulation.

selected from the multiple sub-carrier source of figure 3.40. The modulation is QPSK with Nyquist pulse shaping.

3.7 Timing recovery in the Nyquist QAM channel

Nyquist pulse shaping is one of the efficient methods to pack adjacent sub-channels into a superchannel. The timing recovery of such a Nyquist sub-channel is critical for sampling the data received and improving the transmission performance. Timing recovery can be done either before or after the PMD compensator. The phase detector scheme is shown in figure 3.42, a Godard type [20] which is a first-order linear scheme. After CD compensation (CD^{-1} blocks), the signal is sent to a state-of-polarization (SOP) modifier to improve the clock extraction.. The azimuth of 45° and DGD of a half symbol/unit interval (UI) completely destroys the clock tone. Therefore, the SOP modifier is required for enabling the clock extraction. In practical systems, a raised-cosine filter is used to generate Nyquist pulses. A filter pulse response is defined by two parameters, the roll-off factor (ROF) β and the symbol period T_s and described by taking the inverse Fourier transform. The Godard phase detector cannot recover the carrier phase even with small β, thus the channel spectra is close to rectangular. A higher order phase detector must be used to effectively recover the timing clock period, as shown in [21]. A fourth power law PD (4PPD) with pre-filtering, presented in [22] and shown in figure 3.42(b), can deal with small β values.

The 4PPD operates by first splitting and forming the combination of these components X- and Y-polarized channels, then conducting the frequency domain detection and regenerating, through the VCO, the frequency shift required for the ADC to ensure the sampling timing is correct with the received sampled for processing.

The use of such a phase detector in coherent optical receivers requires a large hardware contribution. Due to PMD effects the direct implementation of this method before PMD compensation is almost impossible. Therefore, 4PPD

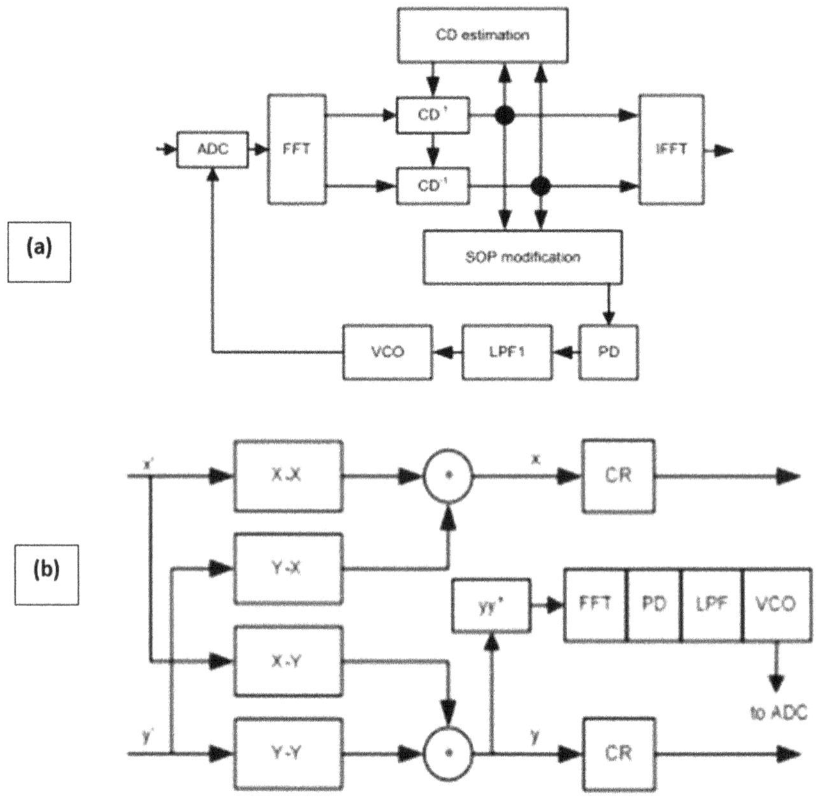

Figure 3.42. (a) Godard phase detector algorithm. (b) Fourth order 4PPD. VCO = voltage control oscillator; FFT = fast Fourier transform; IFFT = inverse FFT; CD = chromatic dispersion; SOP = state of polarization.

implementation in the frequency domain after the PMD compensation is proven to be the most effective and performs well even in the most extreme cases with ROF equal to zero.

3.8 128 Gb s^{-1} 16-QAM superchannel transmission

An experimental set-up by Dong *et al* [23] for the generation and transmission of a six-channels carrying 128-Gbit s^{-1} under modulation format and polarization multiplexing PDM-16-QAM signal. The two 16-Gbaud electrical 16-QAM signals are generated from the two arbitrary waveform generators. Laser sources at 1550.10 nm and the second source with a frequency spacing of 0.384 nm (48 GHz) are generated from two external cavity lasers, each with a linewidth less than 100 kHz and output power of 14.5 dBm, respectively. Two I/Q MODs are used to modulate the two optical carriers with I and Q components of the 64-Gb s^{-1} (16-Gbaud) electrical 16-QAM signals after the power amplification using four broadband electrical amplifiers/drivers, respectively. Two phase shifters with the bandwidth of

5 k–22.5 GHz provide a two-symbol extra delay to de-correlate the identical patterns. For the operation to generate 16-QAM, the two parallel Mach–Zehnder modulators (MZMs) for I/Q modulation are both biased at the null point and driven at full swing to achieve zero-chirp and phase modulation. The phase difference between the upper and the lower branch of I/Q MZIM is also controlled at the null point. The data input is shaped so that about 0.99 roll-off factor of raised-cosine pulse shape could be generated.

The power of the signal is boosted using polarization-maintaining EDFAs. The transmitted optical channels are mixed with an LO and polarization demultiplexed via a $\pi/2$ hybrid coupler, then I and Q components are detected by four pairs of balanced photodetectors (PDPs). They are then trans-impedance amplified and resampled.

The sampled data are then processed in the following sequence, CD compensation, clock recovery, then resampling and going through classical CMA, a three-stage CMA, frequency offset compensation, a feed-forward phase equalization, LMS equalizer (LMSE) and then differentially detected to avoid cycle slip effects.

Furthermore, the detailed processing [24] for the electrical polarization recovery is achieved using a three-stage blind equalization scheme: (i) Firstly, the clock is extracted using the 'square and filter' method, and then the digital signal is resampled at twice the baud rate based on the recovery clock; (ii) secondly, a T/2-spaced time-domain finite impulse response (FIR) filter is used for the compensation of CD, where the filter coefficients are calculated from the known fiber CD transfer function using the frequency domain truncation method; (iii) thirdly, adaptive filters employing two complex-valued, 13-tap coefficients and partial T/2-space are employed to retrieve the modulus of the 16-QAM signal.

The two adaptive FIR filters are based on the classic CMA and followed by a three-stage CMA, to realize multi-modulus recovery and polarization de-multiplexing. The carrier recovery is performed in the subsequent step, where the fourth power is used to estimate the frequency offset between the LO and the received optical signal. The phase recovery is obtained by feed-forward and the LMS (least-mean-square) algorithms for LO frequency offset (LOFO) compensation. Finally, differential decoding is used for BER calculating after the decision.

The spectra of the six Nyquist channels under BTB and 1200 km SSMF transmission are shown in figure 3.43. It is noted that even with 0.01 roll-off of Nyquist channels, we do not observe the flatness of individual channels in the diagram. The constellation as expected would require a significant amount of equalization and processing.

On average, the achieved BER of 2e−3 with a launched power of −1 dBm over 1200 km SSMF non-dispersion compensation link was demonstrated by the authors of [15]. The link is determined by a recirculating loop consisting of four spans with each span length of 80 km SSMF and an in-line EDFA. A WSS (wavelength selective switch) is employed wherever necessary to equalize the average power of the sub-channels. The optimum launched power is about −1 dBm for the six-sub-channel superchannel transmission.

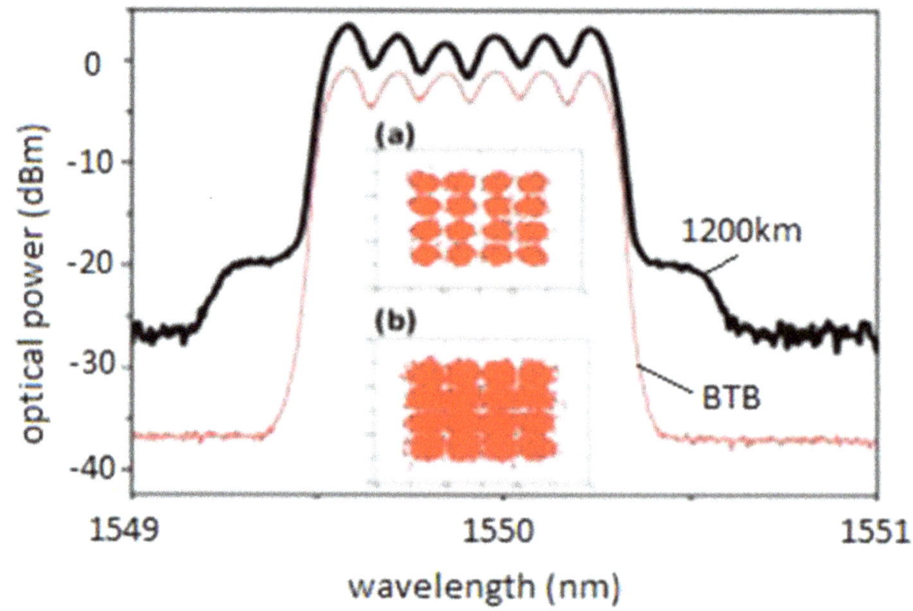

Figure 3.43. Spectra of Nyquist channels, 6 × 128G PDM 16-QAM B2B and 1200 km transmission.

3.9 450 Gb s^{-1} 32-QAM Nyquist transmission systems

Further spectral packing of sub-channels in a superchannel can be done with Nyquist pulse shaping and pre-distortion or pre-equalization at the transmitting side. Zhou *et al* [25] demonstrated the generation and transmission of 450 Gb s^{-1} wavelength-division multiplexed (WDM) channels over the standard 50 GHz ITU-T grid optical network at a net spectral efficiency of 8.4 b s^{-1} Hz^{-1}. This result is accomplished by the use of Nyquist-shaped, PDM 32-QAM, or 5 bits/symbol × 2 (polarized modes) × 45G Sy s^{-1} to give 450 Gb s^{-1}. Both pre- and post-transmission digital equalization techniques are employed to overcome the limitation of the DAC bandwidth. Nearly ideal Nyquist pulse shaping with roll-off factor of 0.01 allows guard bands of only 200 MHz between sub-carriers. To mitigate the narrow optical filtering effects from the 50 GHz grid reconfigurable optical add-drop multiplexer (ROADM), a broad-band optical pulse-shaping method is employed. By combining electrical and optical shaping techniques, the transmission of 5 × 450 Gb s^{-1} PDM-Nyquist 32-QAM on the 50 GHz grid over 800 km and one 50 GHz grid ROADM was proven with soft DSP equalization and processing. The symbol rate is set at 28GSy s^{-1}.

It is noted that the transmission SSMF length is limited to 800 km due to the reduced Euclidean geometrical distance between constellation points of 32-QAM and by avoiding the accumulated ASE noises contributed by EDFA in each span, Raman optical amplifiers with distributed gain are used in a recirculating loop of 100 km ultra-large area fibers. The BER performance for all five sub-channels is shown in figure 3.44(a) with insert of the spectra of all sub-channels. Note the

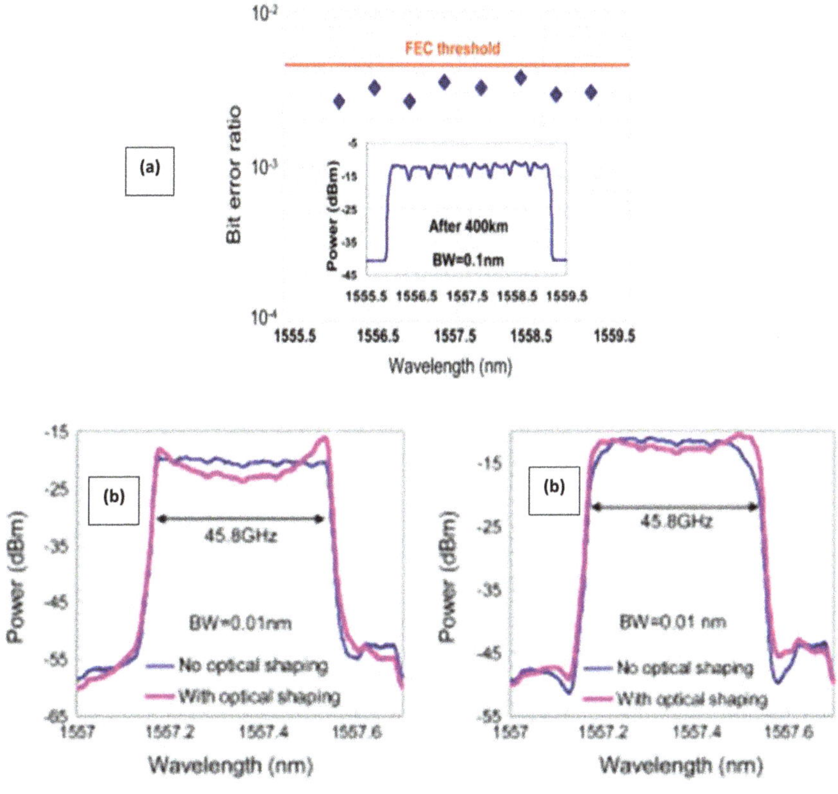

Figure 3.44. (a) Spectra of five sub-channels of 450 Gb s^{-1} channel, that is 5 × 450 Gb s^{-1} superchannels after 400 km (one loop circulating) of 5 × 80 km plus Raman amplification. BER versus wavelength and FEC threshold at 2.3e−3; (b) spectrum of a single sub-channel before and (c) after WSS with and without optical shaping by optical filters (reproduced from [16] with permission). Note: dark line for original spectra, light line for spectra after spectral shaping.

near flat spectrum of each sub-channel that indicates the near Nyquist pulse shaping. Figures 3.44(b) and (c) show the spectra of a single sub-channel before and after WSS with and without optical filtering that performs as spectra shaping. The original pulse shape can be compared with that displayed in figure 3.15. Further to this published work, Zhou *et al* have also time multiplexed a 64-QAM and transmitted over 1200 km [26], that is three circulating around the ring of 400 km of 100 km span incorporating Raman pumped amplification. About 5 dB penalty between 32-QAM and 64-QAM in receiver sensitivity at the same FEC BER of 6e−3 is obtained. Note that electrical time division multi-uplexing of digital sequences from the arbitrary waveform generators was implemented by interleaving so that higher symbol rate can be achieved. Furthermore, 20% soft decision FEC using quasi-cyclic LDPC code is employed to achieve BER threshold 2.4e−2, thus the 20% extra-overhead is thus required on the symbol rate.

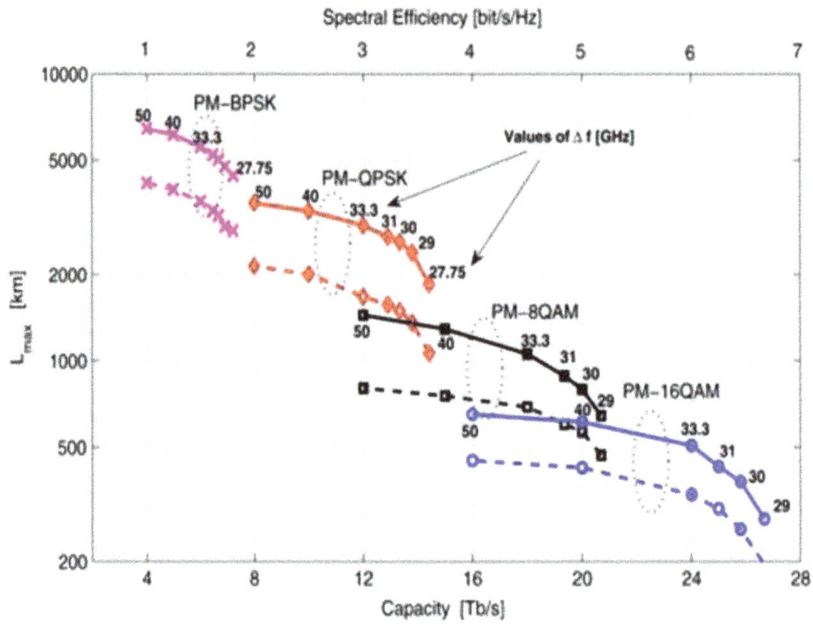

Figure 3.45. Transmission reach distance variation with respect to spectral efficiency and total capacity of supechannels. Different QAM schemes are indicated PM = polarization multiplexing. The numbers are for bandwidth of the sub-channel (reproduced from [21] with permission).

Simulated results by Bosco *et al* [27] without using soft FEC also show that the variation of the maximum reach distance with a BER of 2e−3 for capacity in the C-band (bottom axis) and spectral efficiency for PM-BPSK, PM-QPSK, PM-8QAM and PM-16-QAM, as shown in figure 3.45, for SSMF non-DCF optical transmission lines as well as non-zero dispersion shifted fibers (NZ-DSFs) indicated by dashed lines.

3.10 DSP-based heterodyne coherent reception systems

We have so far described optical transmission systems under homodyne coherent reception, that is, when the frequencies of the lightwave carriers and that of the LO are equal. The original motivation of using homodyne detection is to eliminate the 3 dB degradation as compared with heterodyne technique, and this is possible under DSP-based reception algorithm to avoid the difficulties of locking of the LO and the channel carrier. Under the classical heterodyne reception, the 'at least' 3 dB loss comes from the splitting of the received signals in electrical domain and then multiplied by a sinusoidal cosine and sine RF oscillator to extract the in-phase and quadrature (I–Q or I/Q) components.

So far, we have discussed homodyne reception DSP-based optical transmission systems, which are considered for extensive deployment in commercial coherent communication systems for 100G, 400G, 1T or beyond. However, with the development of large-bandwidth and high-speed electronic ADCs and PDs, once again, coherent detection with DSP has allowed the mitigation of impairments in optical transmission to be compensated by equalization in the electrical domain. For homodyne detection in PDM systems, the I/Q components of each polarization state should be separated in the optical domain with full information. Thus, four balanced PD pairs incorporated with a photonic dual-hybrid structure and four-channel time-delay synchronized ADCs are required.

By up-converting I and Q components to the intermediate frequency (IF) at the same time, not only can heterodyne coherent detection halve the number of the balanced PDs and ADCs of the coherent receiver, but also there is no need to consider the delays between I and Q components in the PDM signal. Therefore, the four output ports of the optical hybrid can be also halved accordingly. However, this heterodyne technique can possibly be restricted by the bandwidth of the PDs. Furthermore, the external down conversion of the IF signals may enhance the complexity of the reception system. Currently, the tremendous progress in increasing the sampling rate and bandwidth of ADCs and PDs gives a high possibility to exploit a simplified heterodyne detection. With large-bandwidth PDs and ADCs, down conversion of the IF, I/Q separation of quadrature signal, and equalization for the PDM and nonlinear effect can all be realized in the digital domain, the DSP followed the ADCs. A heterodyne detection in transmission system is a limited 5 Gb s^{-1} 4-QAM signal over 20 km in [28] and limited 20-Mbaud 64- and 128-QAM over 525 km in [29], and then reaching Tbps by Dong *et al* [30]. High-order modulation formats, such as PDM-16-QAM and PMD-64-QAM, taking advantage of multi-level higher capacity, can offer spectral effectiveness for 100G or beyond by adopting heterodyne detection and digital signal processing.

For the 100G or beyond coherent system with required transmission distance shorter than 1000 km, the inferiority of the SNR sensitivity in heterodyne detection is not so obvious. Conversely, a smaller number of ADCs and easy implementation of the DSP for IF down conversion, make the heterodyne detection a potential candidate for a 100G or beyond transmission system. Figure 3.46(a) shows a schematic diagram of the heterodyne reception with digital processing. The quadrature amplitude modulated signals are transmitted and imposed onto the $\pi/2$ hybrid coupler which is now simplified and there is no $\pi/2$ phase shifter as compared to the hybrid coupler. The coherent mixing of the LO and the modulated channel would result in the time domain signals which are an RF envelope covering the lightwaves and limited within the symbol period. Thus, both the real and imaginary parts appear in the electronic signals produced after the balanced detection in the PDPs in which the beating happens (refer to figure 3.46(b)). The electronic currents produced after the PDPs are then amplified via the TIA to produce voltage-level signals which are conditioned to appropriate levels so that they can be sampled by the ADCs. Thence doing the FFT will produce the two-sided spectrum which exhibits the frequency shifting of the baseband to the RF or IF frequency. Both the in-phase and quadrature components are embedded in the two-sided spectrum. One sideband of the spectrum can be used to extract the I- and Q-parts of the QAM signals for further processing. Ideally the compensation of the CD should be done in the first stage then followed by the carrier phase recovery and then resampling with correct

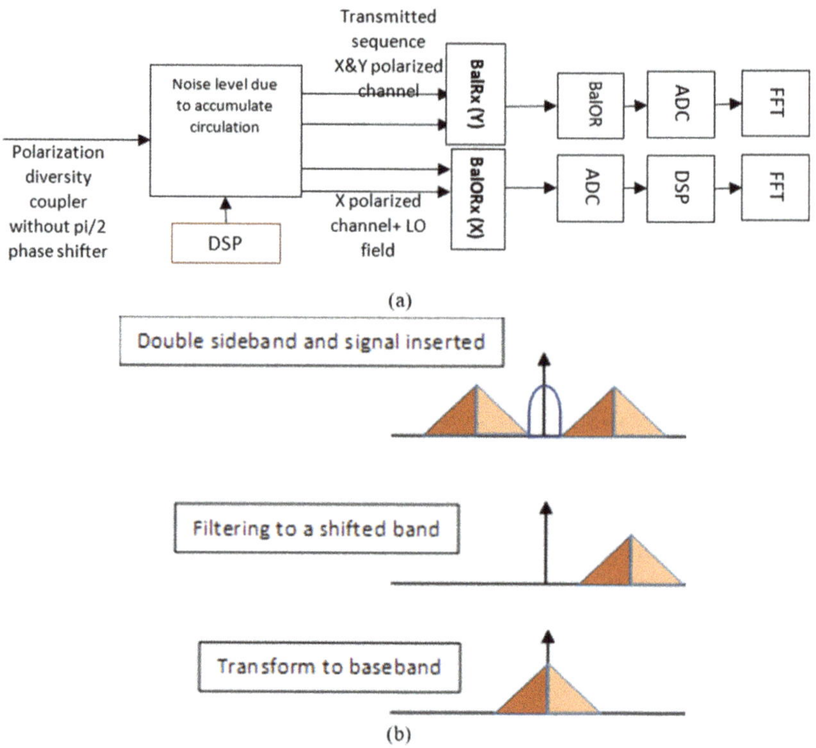

Figure 3.46. Principles of heterodyne coherent reception (a) principal blocks (b) spectra to recover baseband signals (adapted from reference [20]).

timing. Appendix B depicts the schematic of a balanced optical receiver and its principles of operation.

The following sequence of processing in the digital domain was conducted: (i) sampling in ADC and conversion from analog voltage level to digital sampled states; (2) FFT to obtain frequency domain spectrum; then (iii) extract one-sided spectrum and do a frequency shifting to obtain the baseband samples of the spectrum; (iv) resampling with twice the sampling rate; hence (v) compensation of CD and (vi) carrier phase and clock recovery; (vii) using the recovered clock to resample the data sequence; then (viii) conducting normal CMA and three-stage CMA to obtain the initial constellation; then (ix) frequency equalization of frequency difference of the LO; (x) feed-forward phase equalization, LMS equalization for PMD compensation; then finally (xi) differential decoding of the samples and symbol to determine the transmission performance BER with respect to certain OSNR measured in the optical domain and launched power at the input of the transmission line.

It is further noted that for the equalization based on DSP, a T/2-spaced time-domain finite impulse response (FIR) filter is used for the compensation of chromatic dispersion (CD). The two complex-valued, 13-tap, T/2-spaced adaptive

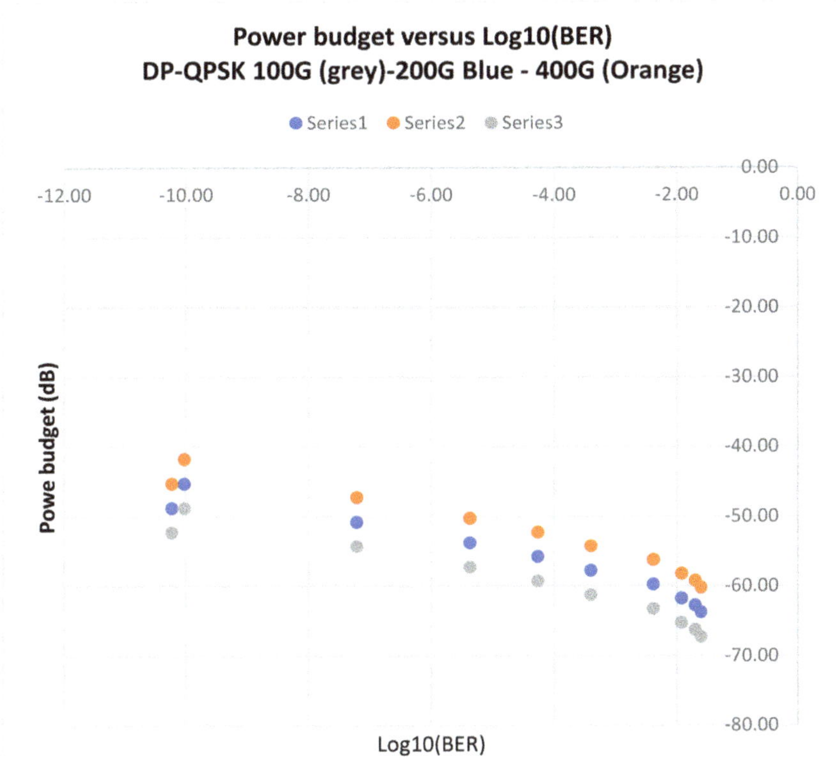

Figure 3.47. B2B BER versus OSNR of the heterodyne reception system of PDM-16-QAM 112 Gb s^{-1} ×8 superchannel with different IF frequency (reproduced from [20] with permission).

FIR filters are based on the classic CMA followed by three-stage CMA, to realize multi-modulus recovery and polarization de-multiplexing.

The B2B BER versus the OSNR of the heterodyne reception optical transmission system is shown in figure 3.47 for PDM-16-QAM 128 Gb s^{-1} per sub-channel with different IF and compared with a homodyne detection scheme. It is noted that there is a 3 dB penalty of heterodyne as compared with homodyne reception due to the fact that only one-sided spectrum is employed. Thus, at BER of 4e−3 the required OSNR is about 23 dB. The transmission distance is 720 km of SSMF incorporating EDFA and non-DCF.

3.11 Remarks

The total information capacity transmitted over an SSMF can be increased by spectral packing of sub-channels by pulse shaping techniques and DSP processing algorithms as well soft FEC and/or pre-emphasis/distortion of the transmitting signals so as to allow higher order of modulation QAM to be realized. Thus, digital processing in real time and digitally-based coherent optical transmission systems have been proven to be the most modern transmission systems for the global internet in the near future. The principal challenges are now lying on the realization of ASIC (application specific integrated circuits using microelectronics technology) or ultra-fast FPGA (field programmable gate array) based systems. The concepts of processing in the digital domain have been proven in offline systems and thus more efficient algorithms are required for real-time systems. These remain topical research topics and engineering issues. Higher symbol rates may be possible when wider bandwidth optical modulators and DAC and ADC systems are available, e.g. grapheme plasmonic silicon modulators [31] and their integration with micro-electronic DAC and ADC, DSP systems.

The processing algorithms for QAM vary from level to level but essentially, they can be based on the number of circles existing in multi-level QAM as compared to a mono-cycle constellation of the QPSK scheme. The algorithms developed for QPSK can be extended and modified for higher level circular constellation.

3.12 PAM4-IM/DD systems

3.12.1 Generating PAM-4 signals

In generating the PAM-4 signals the most common method is to use the arbitrary generator (AGW) or DSP generated outputs and then feed them through a DAC so that the analog signal waveforms can be amplified in an RF amplifier and then fed to an optical modulator. Complementary channels can also be available for differential drive. The main problem for this technique is the amount of power consumption, especially when the sampling rate is increased as a cubic function of the sampling.

In order to avoid such power consumption photonic technique can be employed with two optical modulation paths. One modulator is driven with NRZ and the other also by NRZ but with 6 dB amplitude reduction. Thus PAM-4 format can be generated, as recently reported. Another method is to use a multi-section electrode of an MZIM. Each section is driven by a phase element of a phase pattern which is equivalent to the digital waveform. This section describes these advanced generation techniques for PAM-4 sequence so as to double the bit rate of the transmission systems.

3.12.2 DSP-based PAM-4 generation for low cost, broadband channels

Two main contemporary approaches have been reported [32, 33], one approach is from the optical domain and the other in the analog electronic domain. In the optical domain approach, two parallel optical path incorporating EA modulation paths whose amplitudes are 3 dB different in intensity generate PAM-4 signal channels. On the other hand, the electronic approach is by generating a multi-section electrode which is placed on one optical path followed by phase modulation of the sections of the electrode by electronic amplified signals whose amplitudes are digitally equivalent to a DAC generated bit.

3.12.2.1 DAC-free PAM-4 generation by dual optical path modulation

DAC-free optical transmitter is preferred in a low cost transmission link as the power consumption can be maximally reduced. One of the ways is to use two optical modulators which are driven by a pair of synchronized signals of the same RF broad bandwidth. Their amplitudes are 3 dB different in power, as shown in figure 3.48(a).

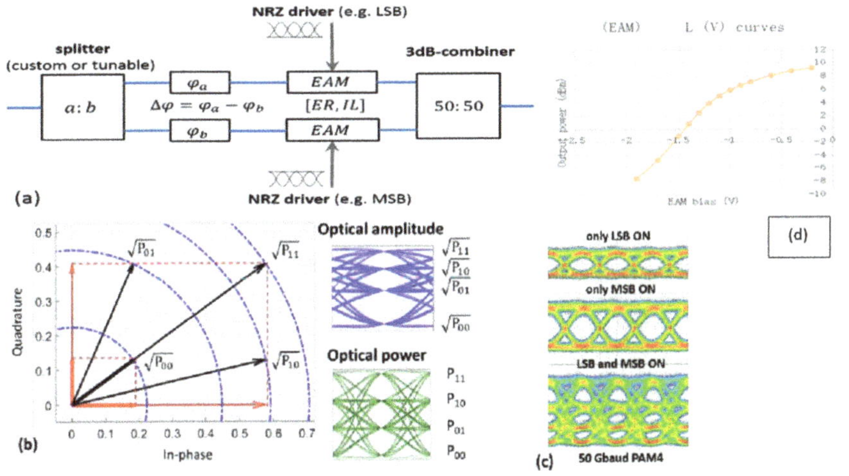

Figure 3.48. (a) Block diagram of the DAC-less Tx; (b) example of an equidistant PAM-4 generation using the first quadrant of the complex plane where the power split ratio a:b was chosen as 0.33:0.66, $\varphi = 90°$ and the EAMs exhibit an absorption factor which is limited to 10 dB or 10% in terms of power of the guided optical waves passing through the waveguide branch. In this special case, the optical power eye openings decrease with decreasing extinction ratio (ER) but they remain equidistant, without changing φ. (c) Optical eyes for the DAC-free transmitter at 50 GBaud [32]. (d) Transfer curve (power versus bias voltage) of EAM modulator—output power versus applied voltage across waveguide on absorption region.

Two optical amplitude modulators of electro-absorption (EAM) type are fabricated by deposition of Ge thin-film on top of Si waveguides. Thus, Ge can act as an absorber that is the same as a PD in the C-band window. Thence, this acts as an in-line EAM. The phase tuning sections placed in front of the EAM can be added so that the tuning of the phase synchronization can be made. The outputs of the two EAMs are combined to give PAM-4 signals. Therefore, QAM signals, for example 16-QAM, can be generated if a phase shifter is placed in a cascade of one EAM, i.e. *I*/2 phase shifting, thence the complex signals can be generated. The phasor diagram and the eye diagrams can be generated as depicted in figure 3.48(b). Much more complex signal channels can be generated if DACs are used to alter the patterns feeding into the EAM. It is observed that the PAM4 eye diagram is amplitude dependent because the delay time is amplitude dependent on the electro-absorption effects in the EAM. Hence a tuning in the phase shifters placed in front of the EAM branches would balance the PAM4 pattern. More details of the operation of the DAC-free PAM4 TX can be referred found in [32]. Figure 3.48(d) depicts the power transfer as a function of the applied voltage to the electrode of the EAM.

3.12.2.2 Multi-sectional DAC-free modulator

Multi-section electrodes can be used to drive the optical path of an MZIM [33] as shown in figures 3.49(a) and (b). This structure is different with the DAC-less EAM PAM-4 modulator shown in figure 3.48. The electrodes can be single-drive or deferentially driven in figure 3.49(b). The input impedance of these sectional electrodes must be impedance matched to the transmission line characteristic

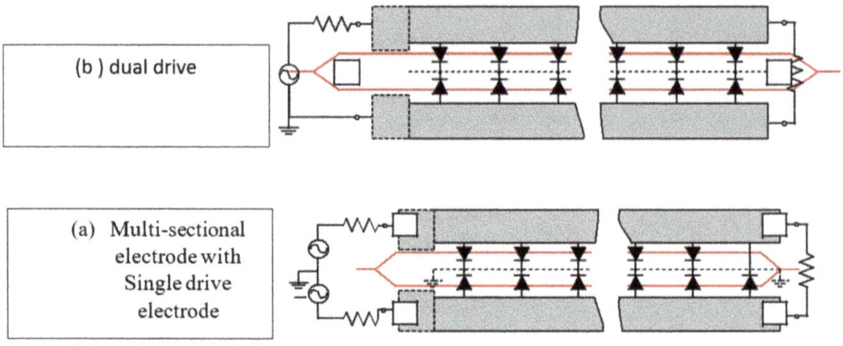

Figure 3.49. Schematic of (a) series push–pull modulator which can be characterized by two ports, and (b) a dual-drive modulator which requires four port characterization. The dual-drive modulator shown here also has a p–n junction connected **B2B** and has a virtual ground at the middle due to its longitudinal symmetry. The red lines represent optical waveguides forming the MZMs.

impedance. The main problem of this multi-sectional approach is the impedance matching. Thus, the delay can be different between the signal bits entering the electrodes so there would be distortion. Furthermore, the electronic delay would also be another additional difficulty. The analog electronics can also be more complicated compared to the DAC-free technique.

3.13 Beyond 100G low-cost PAM-4 systems for last mile access

3.13.1 112 Gbps PAM4-OM4 by band-limited 10G components

Low cost modules for DCN are critical for lowering the cost for DCN, especially in distributed DCN. One way to lower the cost of the high-speed module is to use components for 10 Gbps but for 100G (25G × 4 lanes) but equalized by MLSE algorithm implemented in the DSP processing. This sub-section presents the technique of this low-cost approach. The system set-up is shown in figure 3.50. The 10G components can be from Manufacturer 1 (Sumitomo), Manufacturer 2 EML (EA modulator integrated with DFB laser); Manufacturer 2 DML, which is a laser modulated by direct driving laser current, and another DML version from manufacturer 2. These components are frequently used in the 10G transmission ink and operating in the O-band (1310 nm spectral window). At the receiver side there is an optical receiver of bandwidth of 40 GHz 3 dB passband. This receiving subsystem is composed of a high-speed PD in tandem with a TIA of a spectral noise current density of 40 pA/sqrt (Hz) and a trans-impedance (TI) transfer power gain at mid band of 150 V W^{-1}. The transfer power characteristics of the EAM are shown in the inset of figure 3.50.

As observed, the effective bandwidth can be 27 GHz with 10 dB and higher SNR. The number of bits per symbol that can be used varies from 2 to 6 over the 27 GHz band. Thus, the order of higher modulation formats can be employed accordingly in the OFDM or DMT. Some impedance mismatch is causing some degradation of SNR as observed in figure 3.51.

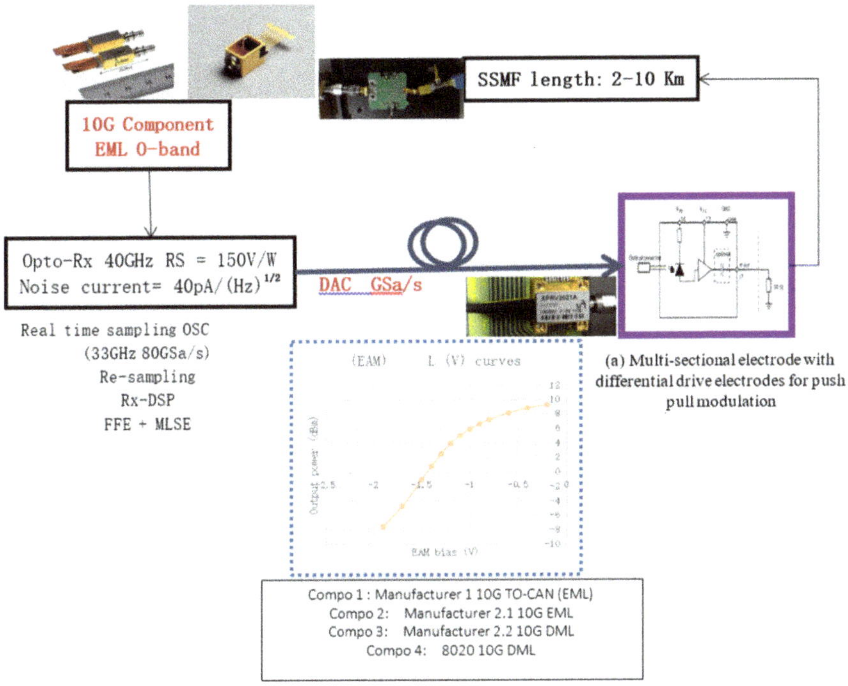

Figure 3.50. Low cost 100G transmission system by 10G components. Inset at bottom: EAM transfer curve—optical power output versus applied voltage and biasing of the EAM.

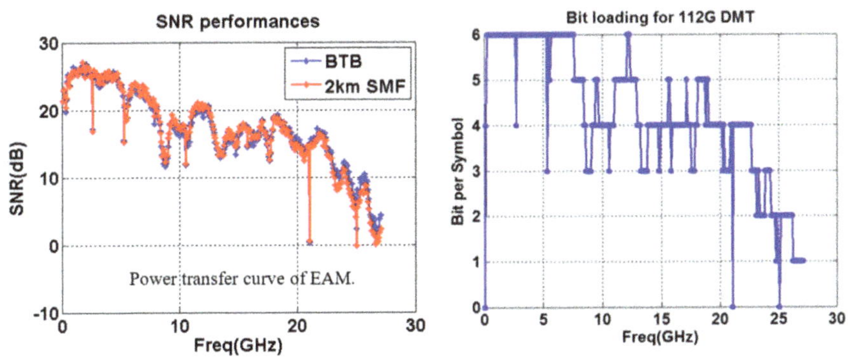

Figure 3.51. Electrical SNR and the bit per symbol versus frequency to 30 GHz.

The digitally modulated sequence is generated and output to drive the TOSA (transmitter optical assembly) via a DSP platform whose sampling rate is 100 GSa s^{-1}. 112 Gbps is successfully transmitted over 2 km SMF with a sensitivity of −2.5 dBm which is only 0.2 dB power penalty after 2 km SM. Improvement is required to reduce the error-floor and the highest bit rate is 120 Gbps for both B2B and over 2 km length

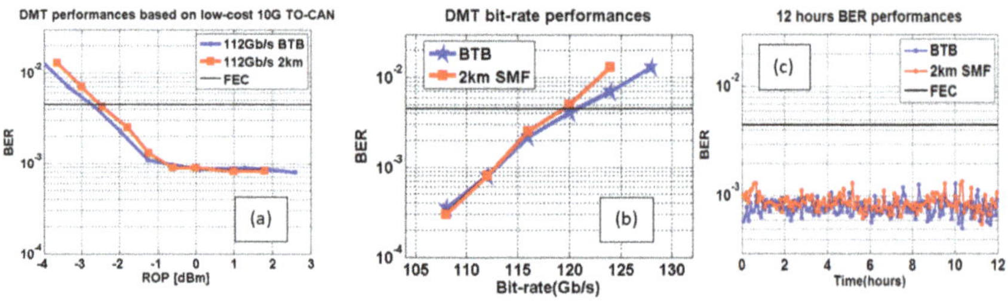

Notes: 112Gbps is successfully transmitted over 2km SMF with a sensitivity o f -2.5dBm; Only 0.2dB power penalty after 2km SMF; Improvement is required to reduce the error-floor; and the highest bit rate is 120Gbps for both BTB and 2km cases.

Notes: (i) The highest used frequency is around 27GHz; (ii) Power fading possibly caused by mismatch between the PCB and component?

Figure 3.52. Sumitomo TO-CAN TOSA: BER versus (a) receiver sensitivity (dBm) and (b) bit rate from 105 to 130 GBps. BTB = back to back. (c) Stability of BER with respect to time in hours. BTB = back-to-back.

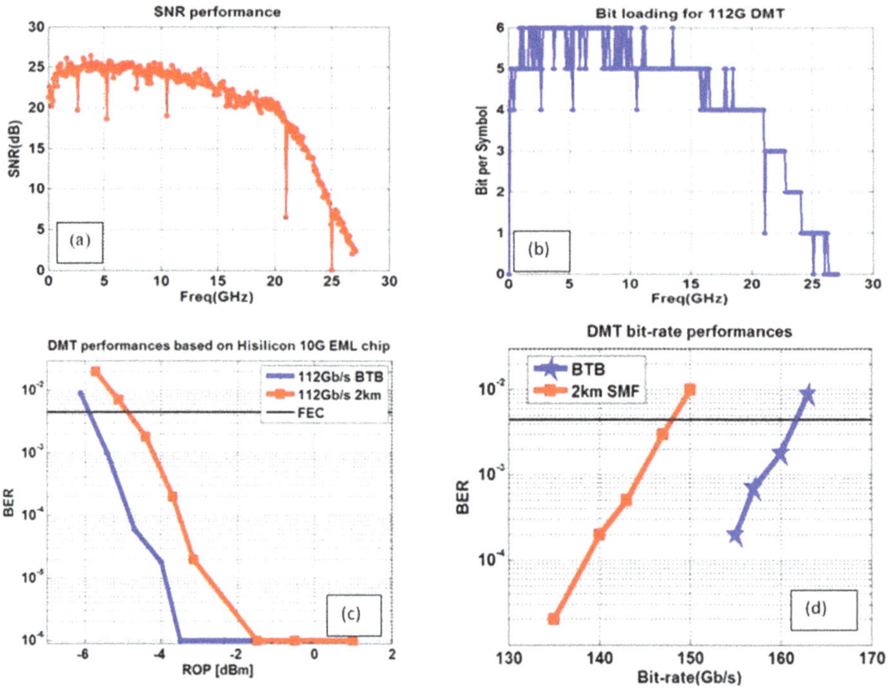

Figure 3.53. Manufacturer 2 EML: (a) SNR and (b) bit-loading of 10G EML chip. The highest used frequency is around 27 GHz; narrow power fading is caused by clock leakage; and Manufacturer 2 EML: (c) ROP and (d) capacity performances for 2 km SMF.

of SSMF. Figure 3.52 shows the ROP and capacity performances for the Sumitomo TO-CAN 10G component transmitted over 2 km of SSMF (figures 3.53–3.55). Physical components are shown in the insets in figure 3.50. For the TOSA of different manufacturers, their performances are listed in table 3.3.

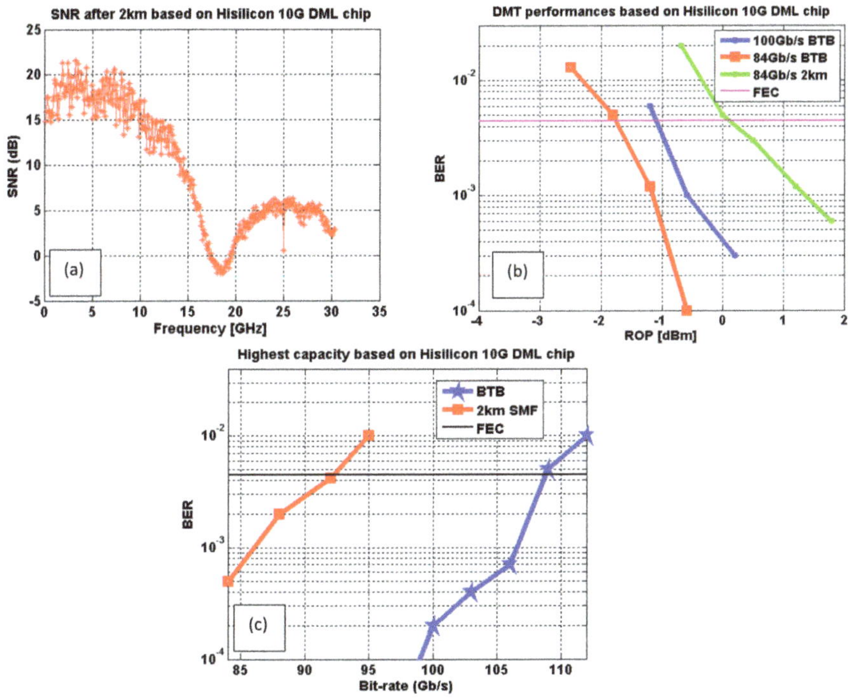

Figure 3.54. Manufacturer 2 10 GHz DML TOSA: performances for 2 km transmission. (a) Frequency response of SNR; (b) BER versus receiver sensitivity; and (c) BER versus bit rate for modulation DMT.

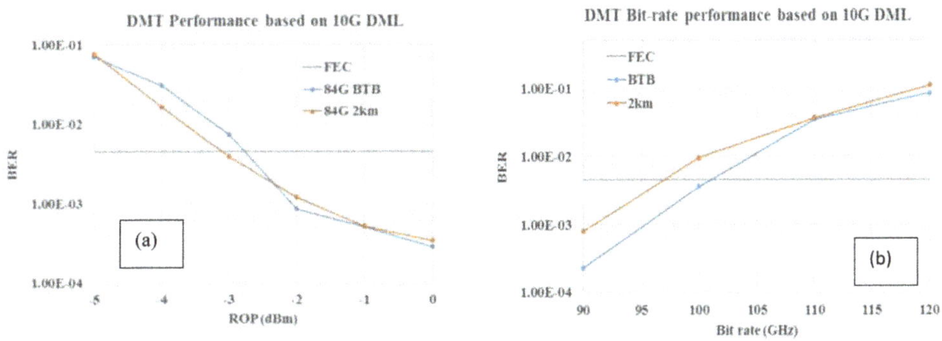

Figure 3.55. Manufacturer2 DML: BER versus (b) receiver sensitivity (optical power input to the receiver) and (b) versus bit rate → capacity performances for 2 km by Manufacturer2 DML.

3.13.2 PAM-4 150 Gbps by band-limited components

It is understood that the bandwidth of electronic and integrated optic devices, especially Si photonics is limited to around 30–40 GHz. So, the total effective bit rate can be limited to around 80 Gbps for NRZ modulation format. This section, however, attempts to describe an NRZ but with PAM-4 duobinary (PAM4 DuoB) using electrical filtering, not optical duobinary phase [34] in association with DSP to

Table 3.3. Summary of performances of different TOSA EML and DML components of different manufacturers (Sumitomo and Manufacturer (a) and (b)). BTB = back-to-back. ROP = received optical power.

	Sumitomo EML	Manufacturer 2 (a): EML	Manufacturer 2 (a): DML	Manufacturer 2 (b): DML
BTB ROP (dBm)	−2.7@112G	−7.6 @112G	−1.8 @ 84G	−2.8 @ 84G
2 km ROP (dBm)	−2.5 @ 112G	−7 @ 112G	0.2 @ (84G)	−3 @ (84G)
5 km ROP(dBm)		−6.5 @ (112G)		
10 km ROP(dBm)		−9.5 @ (84G)		
BTB capacity (Gbps)	120	162	108	101
2 km Capacity (GBps)	120	147	92	97
5 km Capacity	—	130	—	—
10 km Capacity	—	99	—	—

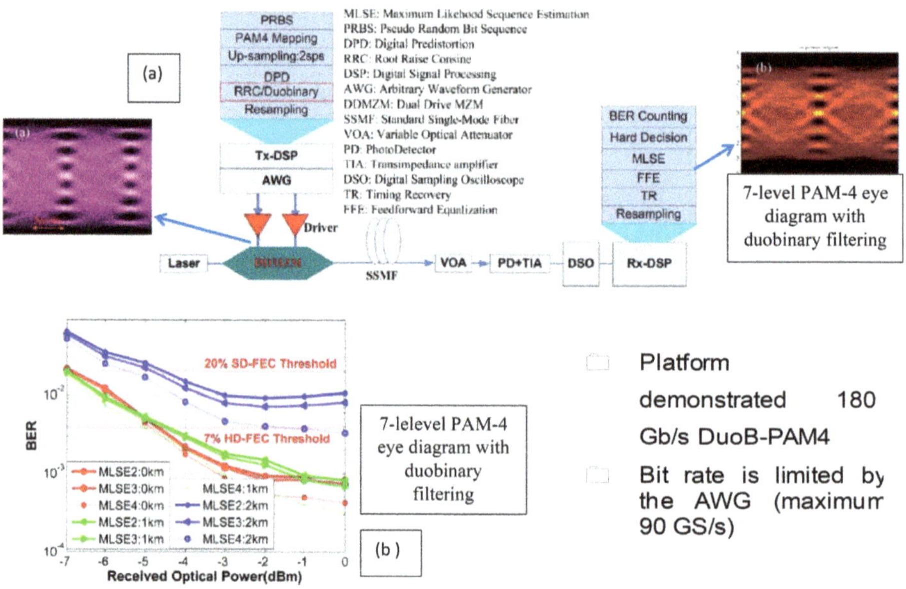

Figure 3.56. Demonstrated 180 Gb s^{-1} DB-PAM4—bit rate is limited by the AWG (maximum 90 GSs^{-1}), MLSE to recover the pulse sequence (a) schematic of transmission set-up and PAM-8 eye diagram; (b) BER versus receiver average optical power (dBm).

demonstrate the bit rate to more than 150 Gbps. The experimental transmission system is shown in figure 3.56(a). A laser is coupled to a dual electrode MZIM whose 3 dB bandwidth is about 32 GHz. The laser linewidth is about 2 MHz. The MAZIMM is driven by two RF amplifiers fed by the analog signals output of the DAC of an AWG. More details are given in the inset of figure 3.56(a). The transmission link is a 2 km SSMF. The transmitted signals are received by an optical receiver whose bandwidth is limited by the 3 dB bandwidth of 34 GHz of the TIA. The signals are sampled and processed in the DSP. The sampling rate is 90 GSa s^{-1}.

The PAM4 duoB eye diagram shown in figure 3.56 indicates a seven-level eye diagram and confirms the half band filtering of the PAM4 eye. These eyes at the transmitter input and at the front end of the DSP at the receiver. Both seem not affected very much after the transmission of the 2 km SSMF. The signal sequence is sampled and processed, recovered by MLSE and then BER is measured with respect to the variation of the receiver sensitivity by varying the attenuation coefficient of the VOA. An FEC is used and the BER (1e−3 to 1e−4) is set at the level at which 7% FEC and 20% soft FEC can be allowed, as indicated in figure 3.56(b). The variations of the transmission distance are from B2B to 2 km with a 1.0 km step. The sequence bit rate is 180 Gbps, that is, a baud rate of 90 GBaud by Nyquist sampling theorem.

In summary, the duobinary filtering in the electrical domain reduces the total effective bandwidth of the PAM-4 and thus permits the minimum distortion on the signals and the DSP processing to recover the original PAM signals with minimum reduction of the sensitivity of the receiver. Indeed, this bit rate can be increased over 200 Gbps if the bandwidth of the system components can be enhanced to the upper limit of about 40–45 GHz. Alternatively, the OFDM by direction or DMT technique can be employed to increase the transmission rate. This technique is demonstrated in the next few sub-sections. The disadvantage of OFDM is that the power consumption is a bit high due to its high sampling rate. It is noted that the power dissipation is proportional to the cubic function of the sampling rate.

3.13.3 Beyond 1.0 Tbps capacity by IM/DD systems

It is possible to Increase the bit rate to 150G or even 200G per channel so that with an aggregation of more optical carriers the total capacity per transmission line can reach more than 1.0 Tbps, as shown in figure 3.66. The system schematic of figure 3.56 and/or figure 3.50 needs to satisfy the following conditions: (i) DAC/ADC bandwidth should be increased from 15 GHz to 25 GHz+; (ii) optical components should be increased to 30 GHz+; (iii) driver and TIA should be increased to 30 GHz+; and (iv) the number of channels can be increased to 12 or even 16 from 8 optically modulated carriers, as shown in figure 3.57.

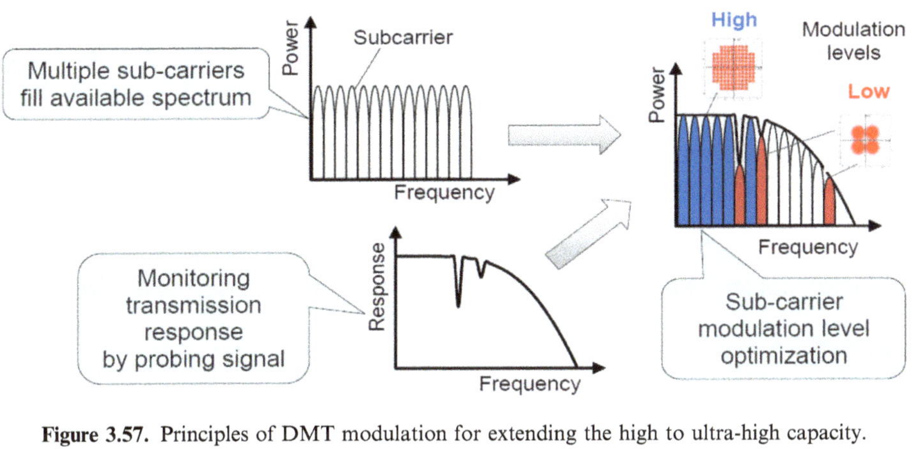

Figure 3.57. Principles of DMT modulation for extending the high to ultra-high capacity.

3.14 Discrete multi-tone transmission DD-OFDM systems

3.14.1 Beyond 200^{+}Gbps DMT transmission

The short-reach interface 400GE[4] has been standardized and has been commercialized towards the year 2023, which is based on 100 Gbps/carrier. Presently, the next-generation 800GE/1TE (gigabit ethernet/terabit ethernet) has been seriously discussed with three possible solutions: coherent optics, eight-lane by 100/125 Gbps intensity-modulated direct-detected (IM/DD), and four-lane by 200/250 Gbps IM/DD. Compared with the coherent solution as described in section 1.4 and in chapter 4 for multi-carrier comb source, simple IM/DD configuration is more attractive due to the advantages of complexity, cost and size. For IM/DD, systems with smaller number of optical lanes (higher per lane data rate) are strongly desired to reduce the complexity, size, power consumption and cost of the transceivers. Thus, four-lane solution with beyond 200 Gbit s^{-1} per channel carrier is a promising candidate for 800GbE/1Terabit-E. 200 Gb s^{-1} PAM4 transmission has been reported as described by using a 40 GHz DAC and in-house modulator with a bandwidth higher than 50 GHz [35]. Similarly, a 214 Gb s^{-1} PAM4 could be realized using an in-house EAM with a bandwidth of 56 GHz [36]. Currently, the next-generation commercial DAC/ADC, driver and optical components (TOSA and ROSA) would not be higher than 35 GHz and cannot reach 45 GHz due to electronic limitations of 65 nm analog fabrication with some analog peaking of the analog output stage of the amplifier followed by the digital stage. Thus, higher spectral-efficiency format is required to achieve 200+ Gbps per lane (carrier) based on 35 GHz bandwidth class components. DMT requires the narrowest bandwidth and possesses the highest spectral efficiency [37], which can be considered as an attractive candidate for 800GE/1TE solution. A DMT with a gross data rate of 200 Gb s^{-1} is realized, but without considering the FEC and other overheads [37]. A 250 Gb s^{-1} DMT is demonstrated by Yamzaki *et al* [38], however, two DACs are multiplexed to generate a high-bandwidth DAC, which increases the system cost and complexity.

In this section, it is experimentally demonstrated 200+ Gbps transmission using a platform consisting of a low-cost IM/DD DMT system using a single-wavelength, single-DAC/ADC, single-modulator and single-PD. We implement a new nonlinear equalization (NLE), which is based on Volterra filter upgraded by some absolute terms, so called ABS-based NLE. The mean SNR is improved by 2.4 dB and 1 dB, compared with linear equalization and conventional NLE. Besides, enabled by trellis coding modulation (TCM), the BER performance is improved by more than one decade. Concerning a net rate of 200 Gb s^{-1}, a 224 Gb s^{-1} (7% HD-FEC is assumed) and 214 Gb s^{-1} (KP4-FEC is assumed) data are successfully transmitted over 1 km single-mode fiber (SMF) with a BER of 2.4e−4 and 2e−5, respectively.

Furthermore, a low-cost IM/DD DMT transmission system using a single-wavelength, single-DAC/ADC, single-modulator and single-PD is described.

[4] IEEE P802.3bs 400 Gb/s ethernet task force. http://ieee802.org/3/bs/index.html

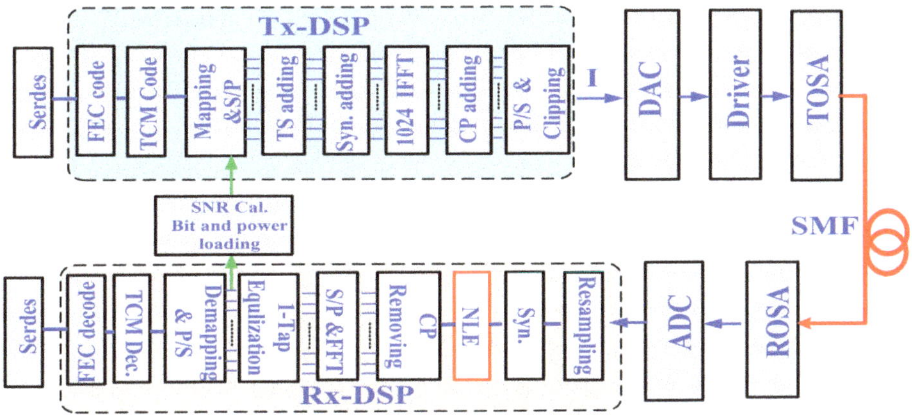

Figure 3.58. Algorithm processing architecture of DMT system at the transmitter section.

The operational principles of DMT are shown in figure 3.57. The optical carrier is modulated by electrical multi-carrier channels which are modulated in multi-level modulation formats and orthogonal with each other. These electrically digitally modulated signals are generated in a DSP-based generator. Only the real part of the complex signals is used, as in the IM/DD any complex parts also result in real intensity. The spectrum of the multiple sub-carriers is shown in figure 3.57. These sub-carrier channels are shown after passing through a transfer curve of the electrical or optical components, also shown in figure 3.57 with an insert of the multi-QAM constellation of high and lower order. Optimization can be designed to accommodate the sensitivity of the receiving sub-section. The software architecture and parameters of the DMT system at the transmitter and reception sub-systems are shown in figure 3.58.

3.14.2 Experimental set-up

The sectional sub-routines are also stated, including the FEC. The experimental platform is shown in figure 3.59 whose components' parameters of sub-systems are given in table 3.4. The DSP is loaded with the generation of the DMT signals as described in figure 3.58. We experimentally prove the possibility of transmission beyond 200 Gbps employing DMT operational principles with the set-up shown in figures 3.58 and 3.59. The transmitter DSP is similar to the traditional DMT DSP except that the 16-state TCM (trellis coded modulation) encoder is employed to improve the system performance. The IFFT size is 1024 and a cyclic prefix (CP) of 16 samples is added to alleviate the inter-symbol interference (ISI). The digital DMT data is converted to an analog signal by a Keysight AWG (AWG (8196A)) with a 3 dB bandwidth of 30 GHz and a sampling rate of 92 GSa s^{-1}. The TOSA side consists of an SHF amplifier driver (807S) which is used to boost the output of the AWG. The amplified electrical DMT signal with a V_{pp} of 2 V is employed to drive an MZIM (Oclaro SD40). The bandwidth of the MZIM is 30 GHz and the V_{pi} is

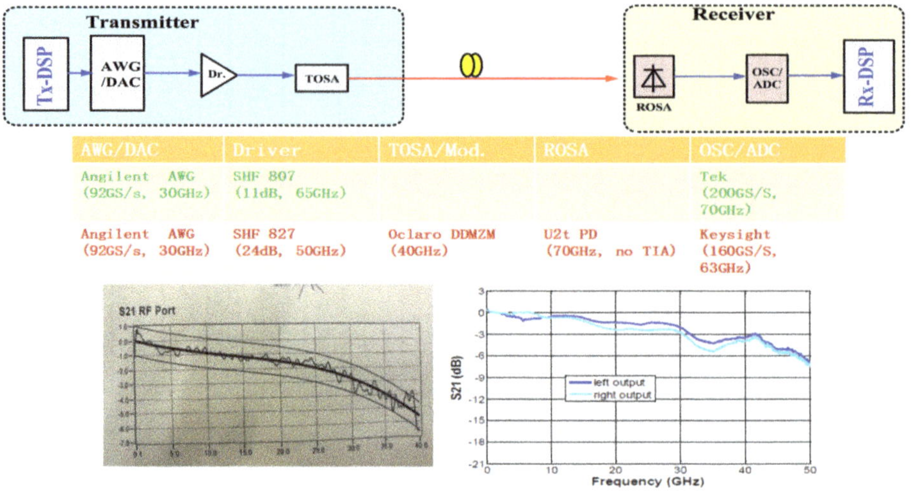

Figure 3.59. Beyond 200G DMT platform set-up whose parameters are shown in table 3.4.

Table 3.4. Parameters of DMT system at transmitter sub-system (a) FFT size and (b) DAC and ADC of different manufacturers.

(a)

FFT Size	1024
CP length	32
FEC (7%)	4.5E-3
Data rate	112Gbps
Subcarrier space	79.5MHz
Total bits	1408

(b)

DAC	TOSA	Driver	ROSA	ADC
Fujitsu 15GHz 85GS/s	10G EML (1577nm) (Sumitomo, Hisilicon)	SHF 40GHz	U2T 40GHz	Tek. OSC 33GHz 100GS/s

about 4.5 V. The frequency transfer functions of the optical modulator and electronic amplifier at the receiver are shown in the insets of figure 3.59 which are bandwidth limited to about 32–35 GHz.

In the receiver ROSA side, a 3 dB 50 GHz U^2t PD (XPDV 2120R) is employed to convert the optical signal to electrical. Since the PD has no TIA and the best input power is about 7 dBm, an optical pre-amplifier is employed to improve the sensitivity performance. A semi-conductor optical amplifier (SOA) is more suitable as an optical pre-amplifier with the advantages of low cost, low power consumption, and ease of integration. Due to the lack of SOA in our lab, an EDFA is used. A real-time oscilloscope (Keysight DSO X93304Q) is employed to capture the DMT signal at a sampling rate of 160 GSa s^{-1} and offline DSP processing is realized. In the Rx-DSP, the signal is firstly down-sampled from 160 GS s^{-1} to 92GS s^{-1} to match the DAC sampling rate. Conventional NLE and ABS-based NLE are used to mitigate system nonlinearities including signal-to-signal beating interference (SSBI) caused by direct detection process and modulator nonlinearity. After serial to parallel (S/P), CP removing and FFT, the signal is transformed to the frequency domain and the 1-tap equalization is used to compensate the linear distortions of the system. After de-mapping and P/S (parallel to serial) converting, a TCM decoder is used to demodulate the signal constellations to bit sequences by using Viterbi decoding algorithm.

Thus, in summary, the aggregate 200 Gbps bit rate can be achieved via the multi-level modulation format imposed on the sub-carriers of the DMT which are varied in the high frequency region so as to satisfy the receiver sensitivity. Furthermore, the equalization by MLSE enables the data recovery even in the high attenuation region of the transfer functions.

3.14.3 Performance

Figure 3.60(a) shows the SNR curves with a 3 dB bandwidth of about 23 GHz, which is mainly limited by the TOSA transmitter side. In the experiment, 508 sub-carriers are used, which corresponds to a bandwidth of 45.6 GHz. Without NLE, the mean SNR is 17.5 dB, as shown by the blue line in figure 3.60. An improvement of 1.4 dB is achieved by conventional second-order Volterra filter with terms of x_i, $x_i x_j$, and DC. When two new terms $|x_i|x_i$ and $|x_i|x_i^2$ are introduced (ABS-based NLE), the mean SNR is further increased from 18.9 dB to 19.9 dB. In order to simplify the equalizer, the equalizer uses only three samples, meaning that only 10

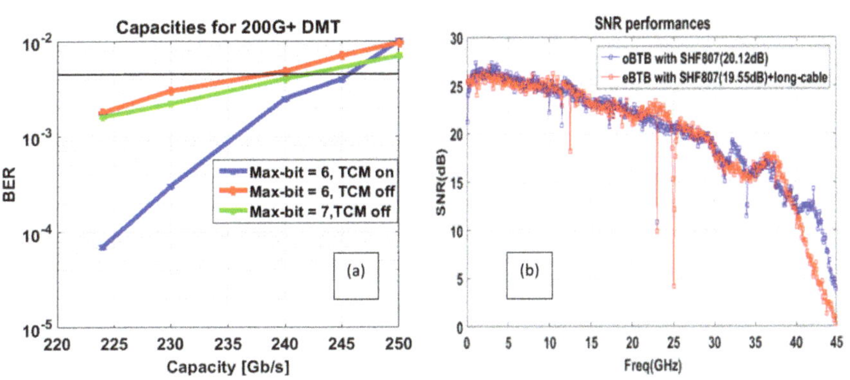

Figure 3.60. Optical BTB performance. BTB = back to back. (a) BER versus frequency with and without TCM; (b) SNR versus frequency with effects of variation of electrical coaxial cable.

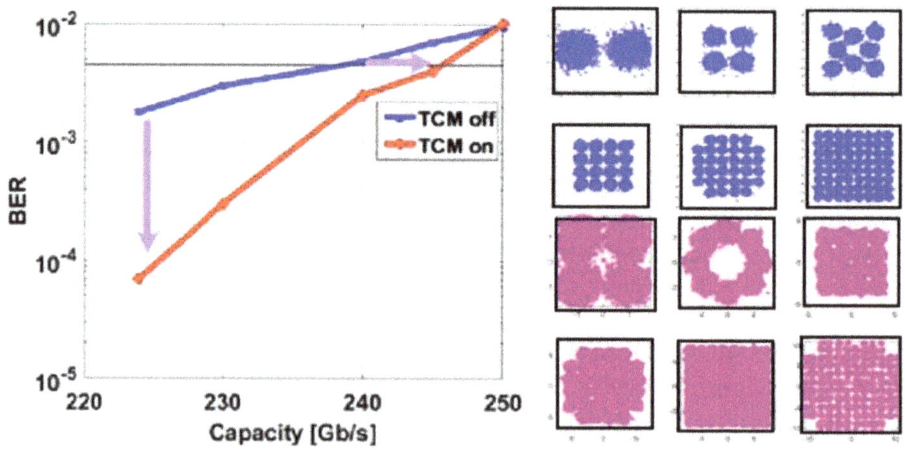

Figure 3.61. (a) B2B capacities with and without TCM; (b) constellations w/o (blue color) and w (pink color) TCM. All constellations are in the same two-dimensional scale.

and 16 taps are used in the conventional and ABS-based NLE modules, respectively.

We note the following: (i) TCM contributes a lot for the error-floor; (ii) for the TCM-on case, the best performance is 7e−5 for 224G (200G net rate); the maximum bit-rate is 245G@4e−3; (iii) for the TCM-off cases, the maximum 7 bits(128-QAM) bring some gain; thus, we can ask whether a 7 bits modulation would bring some gain for TCM-on case. The constellation of different modulation formats at different spectral regions of the transfer function of the sub-systems are shown in figure 3.61 operating under B2B condition and under TCM-on or TCM-off scenarios. The bit-loading maps of 224 Gb s^{-1} and 214 Gb s^{-1} DMT is shown in figure 3.62(a). Figure 3.62(b) shows the power sensitivities of 224 Gb s^{-1} DMT, then figure 3.62(c), the power sensitivities of 214 Gb s^{-1} DMT.

BER versus capacity (bit rate) for the B2B case with and without TCM is presented in figure 3.63(a). With trellis coding modulation (TCM), not only is the capacity improved by 5 Gb s^{-1} (@4.5×10^{-3}) but also the BER performance is significantly improved from 2×10^{-3} to 7×10^{-5} (@224 Gb s^{-1}). The constellations with and without TCM are shown in figure 3.61. The TCM encoder adds one bit more and doubles constellation points (e.g, increasing from QPSK to 8QAM). Although the constellation with TCM is a bit blurred, the TCM performance is improved due to higher Euclidean distance by the convolutional encoder. A more complex TCM encoder can potentially increase the performance; however, in the following experiments the number of TCM states was fixed to 16 to lower the system complexity.

We note the following: (i) the transfer function of electrical BTB is not consistent with the SNR in high-frequency; (ii) the noise is not white; the Tektronix digital sampling oscilloscope could have introduced this type of noise; and (iii) that the transfer function of B2B is well match with the SNR.

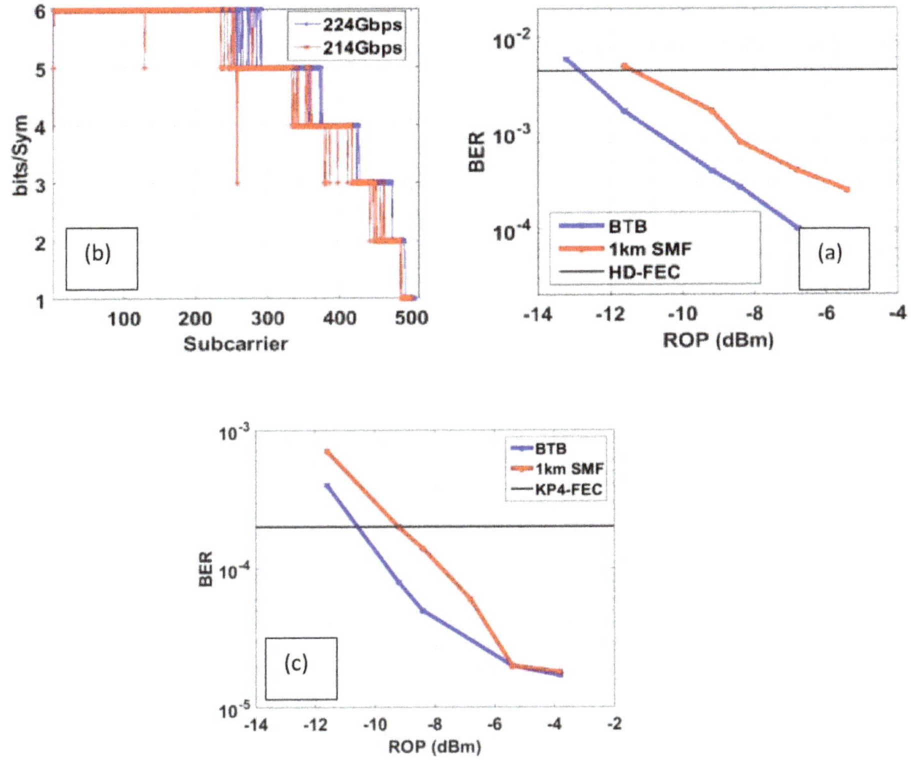

Figure 3.62. (a) Bit-loading maps of 224 Gb s^{-1} and 214 Gb s^{-1} DMT; (b) power sensitivities of 224 Gb s^{-1} DMT; and (c) power sensitivities of 214 Gb s^{-1} DMT.

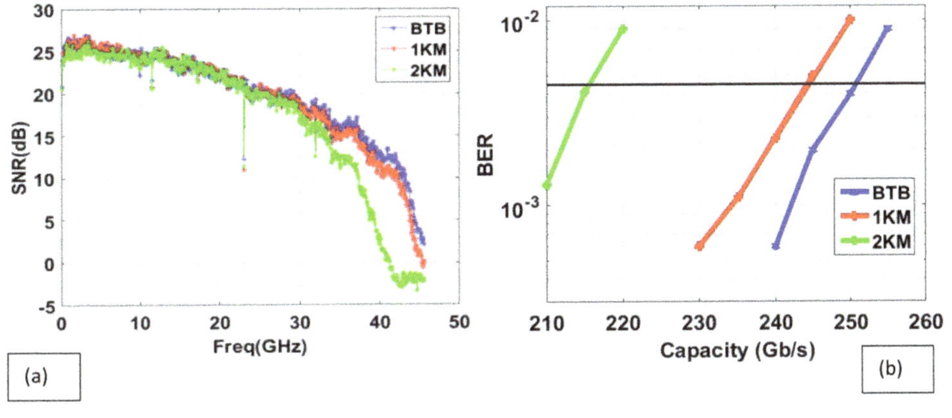

Figure 3.63. (a) SNR curves for BTB, 1 km, and 2 km links. (b) Capacities in BTB, 1 km and 2 km cases.

A net data rate of 200 Gb s^{-1} was investigated with flexible FEC redundancy in different scenarios. For a high-performance (better power sensitivity) scenario, HD-FEC is employed with a gross date rate of 224 Gb s^{-1} with the DMT bit-loading map shown in figure 3.62(a) (blue line). Although 7 bits/symbol can be loaded in the low-frequency spectral range, maximum 6 bits/symbol is mapped to enhance the system noise tolerance. Figure 3.62(b) illustrates the receiver optical power (ROP) performance for the 224 Gb s^{-1} scenario with a sensitivity of −12.8 dBm in the B2B case, which is measured before the pre-amplifier. About 1.2 dB power penalty is observed after 1 km C-band transmission, which is attributed to the 17 ps nm^{-1} km^{-1} chromatic dispersion (CD) parameter. The bit-loading map of 214 Gb s^{-1} is also shown in figure 3.62(a) (200 Gb s^{-1} net data rate and KP4-FEC), which is suitable for low-latency/power scenarios. The sensitivity of −10.4 dBm and −9.1 dBm is achieved for BTB and 1 km transmission, respectively. An error-floor at 1e −5 is observed, which is one decade bellow the FEC threshold and it can be sufficient for real applications.

Figure 3.63(a) shows the SNR curves of B2B, 1 km, and 2 km C-band transmissions, respectively. The SNR quickly decreases at frequencies above 30 GHz in the 2 km link, which is due to the accumulated CD (~34 ps nm^{-1} by 2 km SSMF). To study the maximum system capacity, we vary the data rate and record the BER performance. Transmissions of 250 Gb s^{-1}, 244 Gb s^{-1} and 216 Gb s^{-1} over 0 km, 1 km, and 2 km SMF with a BER bellow pre-FEC of 4.5e−3 (HD-FEC), as shown in figure 3.63(b), are successfully demonstrated, with a net data rate of 223 Gb s^{-1}, 218 Gb s^{-1},193 Gb s^{-1}. Since SSMF operating in the C-band exhibits much higher CD than that in the O-band, we believe that our solution can be used for 2 km or even 10 km O-band transmissions with negligible power and capacity penalties (figure 3.64).

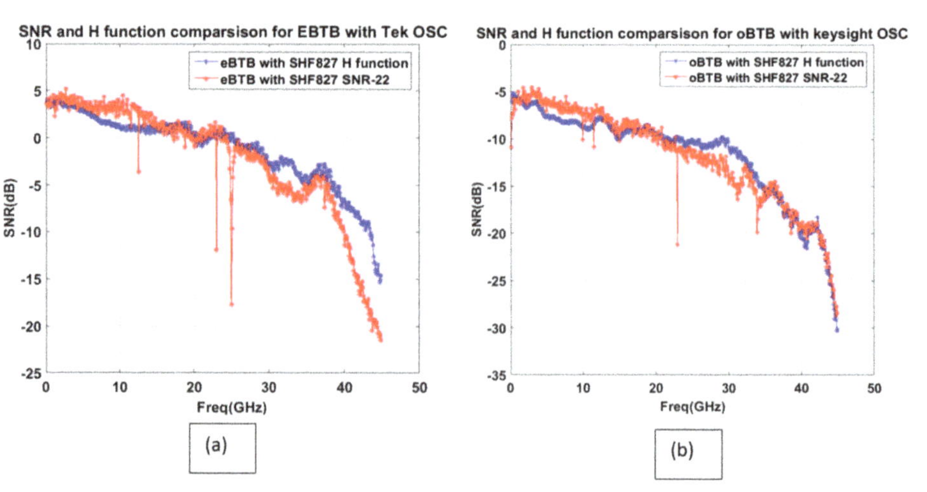

Figure 3.64. DMT transmission system: SNR versus frequency with H-function (a) electrical SNR versus signal frequency; (b) optical SNR versus signal frequency.

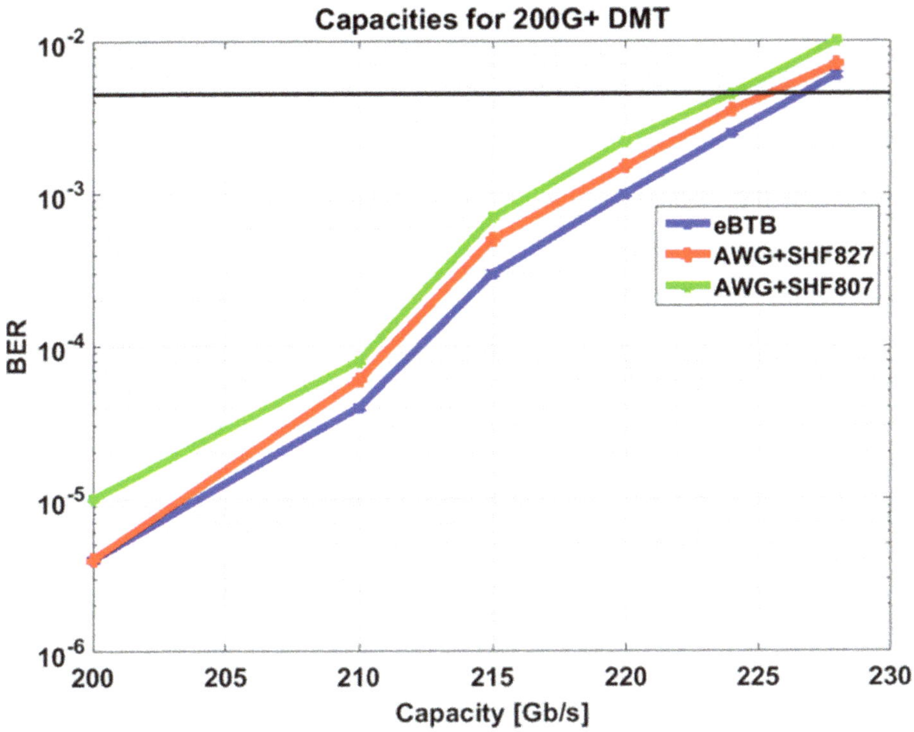

Figure 3.65. Capacities for 200G+ employing the components, as listed in table 3.4, and variation of the RF drivers.

Capacities for 200G+ per carrier can be seen as depicted in figure 3.65. Aggregation of many optical modulated channels can offer multi-Tbps total capacity, as shown in figure 3.66. The optics and signal processing optical receivers for such aggregation are shown.

3.14.4 Remarks

A beyond 200 Gbps DMT system is demonstrated with a simple architecture. Based on the proposed ABS-based NLE and TCM, the system sensitivity performance and error-floor are significantly improved. With the help of an optical pre-amplifier, receiver sensitivities of −12.8 dBm and −10.4 dBm are achieved for HD-FEC (hardware forward error coding) and KP4-FEC (class protcol-forward error coding), which currently is the best sensitivity performance for a 200 Gbps IM/DD system. The experimental results show that our 200 Gb s^{-1} DMT scheme is a very promising low-cost and high capacity candidate for the next-generation of 800GE/1TE systems by using four-lane optics. GE= gigabit ethernet.

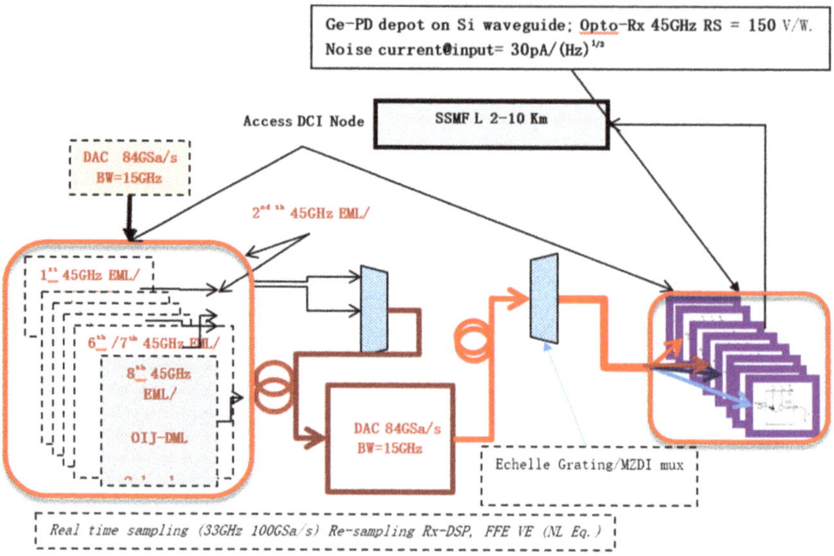

Figure 3.66. Multi-carrier multichannel to aggregate transmission capacity to beyond Tbps (8 × 150 Gbps giving a total 1.2 Tbps). Multiple-lane (8×) transceiver e-PIC system set-up: transceiver—analog e-PIC in association with DSP. RS = receiver sensitivity in V/W, that is for an optical power of one watt the receiver gives 150 V voltage output. So, for a typical average input power of 10 μW the output of the receiver is about 1.5 mV entering the ADC. OJ = optical injection, DML = direct modulated laser, RS = receiver sensitivity.

3.15 Higher-order modulation in IM/DD systems

Following the standardized short-reach 400 GE[5], the next-generation 800 GE or 1.6 terabit ethernet (TE) is needed [39] beyond 2020. For short-reach applications below 2 km, intensity modulation and direct detection (IMDD) solutions are more preferred than coherent solutions due to their lower complexity, cost and size and power consumption with some lower sensitivity. Compared with the 100G/lane approach, the 200G/lane solutions are widely viewed as more compact and cheaper [40–43]. Recently, Wei *et al* conducted an experimental demonstration for comparison of 200G per lane of different higher order modulation format under IM/DD techniques. The system is shown in figure 3.67 in which DSPs are used in both the transmitting and receiving sub-systems so as to generate via software flexible modulation format in these sub-systems. This flexibility allows users to increase or decrease the capacity in such links whenever desired. The feasibility of 200G/λ short-reach DCN based on 30 GHz components[1], [2] [5–8, 11] has been shown using advanced modulation formats and DSP. Demonstrations mainly considered single-carrier pulse amplitude modulation (PAM) [4, 5, 7–11] and multi-carrier DMT schemes [3, 6, 9]. TCM, which is a promising approach in the presence of bandwidth

[5] IEEE P802.3bs 400 Gb/s ethernet task force. (2018) http://ieee802.org/3/bs/

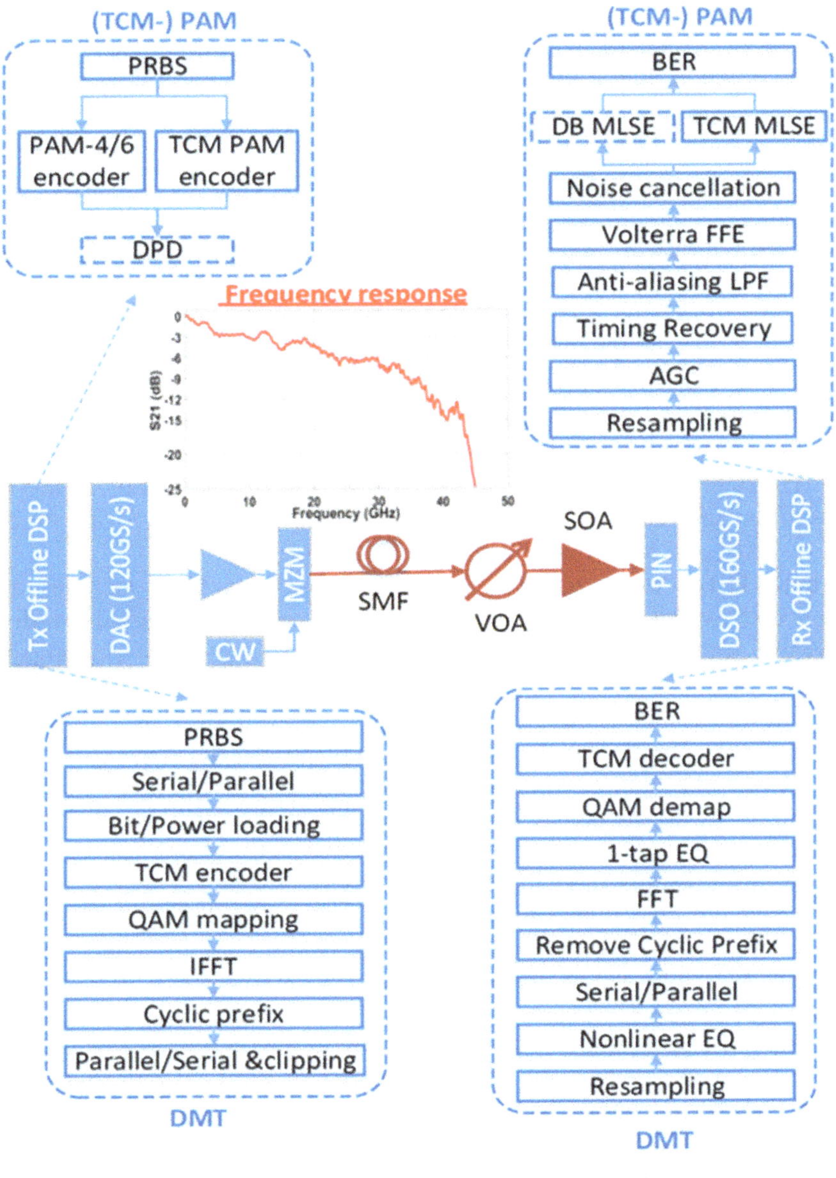

Figure 3.67. Schematic of the experimental set-up for 200G systems.

and noise limitations, has been used in both single-carrier and multi-carrier systems [6–8].

The choice of the modulation techniques influences the trade-off between transmission system performance and processing complexity due to cost and power factors. Experimental comparison of various advanced IMDD modulation formats including duobinary (DB) PAM-4, PAM-6, two-dimensional (2D)-PAM8 TCM and DMT for DCN applications has demonstrated the flexibility for capacity exchanges in these short haul transmissions, especially in DCI (data center interconnection) or short haul access. The DCI technology is based on band-limited components to 30 GHz to achieve 200 Gbits s^{-1} rate or an effective 7 bits Hz^{-1} IM/DD under DMT modulation format and PAM-4 or PAM-6.

This demonstration is performed in C-band with EDFA. The transmitter loads a soft DSP to generate the digital waveforms viaa 120 GSa s^{-1} 30 GHz bandwidth AWG to convert them into an analog electrical signal. A linear electronic driver with 50 GHz 3 dB bandwidth is used for a single-drive a 33 GHz 3 dB MZM coupled with a single-mode laser operating at 1550 nm, an SOA pre-amplified optical receiver plus a VOA, and a 60 GHz PD. The SOA gain is adjusted to optimize the input power of the PD. The detected signal is sampled and digitally processed in a 160 GSa s^{-1} 60 GHz 3 dB bandwidth real-time sampling oscilloscope. The end-to-end (from AWG to oscilloscope) frequency response of the optical B2B link is shown in the inset of figure 3.67, indicating an overall effective 3 dB bandwidth of only 19 GHz. The digital signal is processed by offline DSP. The detailed offline DSPs for the transmitter and the receiver are also presented in figure 3.67 for all modulation formats of PAM, TCM PAM, as well as DMT schemes.

For both DB PAM-4 and PAM-6, the transmitter DSP includes a PRBS generator, a PAM bit-mapper. Furthermore, a linear 16-tap digital pre-distortion (DPD) with resampling is employed at the transmitting sub-system so as to compensate for the limited overall bandwidth.

For PAM-6, the encoder maps the input bits to a 32-QAM constellation, whose I and Q components are transmitted over two consecutive symbols [43, 44]. The DPD aims to compensate the limited bandwidth of the transceiver[2] and can be optionally switched on or off. In the offline DSP at the receiver, the signal is resampled to two samples per symbol. Automatic gain control followed by timing recovery (TR) and an anti-aliasing low-pass filter are then applied. The signal is then equalized by a symbol-spaced Volterra filter, which contains linear, second and third order kernels. A DB or a non-DB FFE (feed-forward equalization) is applied, depending on the trade-off between noise enhancement and SNR requirement [43]. A following noise cancellation (NC) unit is used to suppress and whiten the additive noise [45]. If the FFE is trained to deliver a non-DB signal, the subsequent processing consists only of a decision device before the calculation of the BER. Otherwise, in the case of DB FFE, a DB MLSE [44] is used to decode the DB PAM signal.

2D-PAM8 employs a similar processing as PAM-4 and PAM-6 mentioned above with an incorporation of a convolutional TCM encoder and a TCM MLSE decoder in the transmitter and the receiver, respectively.

Figure 3.68 depicts the overall bandwidth of the channel launching into the transmission line and after the reception sub-systems.

The receiver sensitivity of the 200 Gbits s^{-1} (plus overhead rate) of PAM-4 and PAM-6 modulation format under different processing techniques is shown with BER in figure 3.69 for the B2B and 1.0 km SSMF transmission link. Table 3.5 tabulates the performance of the effectiveness of FEC/DSP on higher modulation formats on the system performance (figures 3.70–3.72).

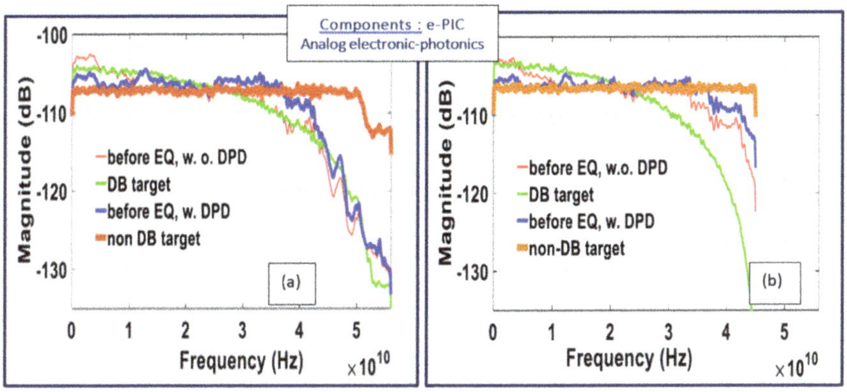

Figure 3.68. (a) 112 GBaud PAM-4 and (b) 90 GBaud PAM-6 signal spectra before and after receiver FFE. Labels: EQ = equalizer, DPD = detection photodetector, DB = duobinary.

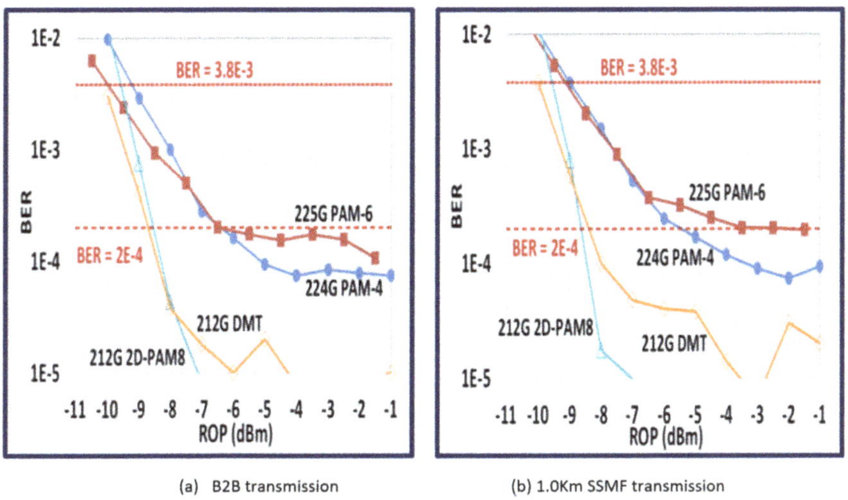

Figure 3.69. BER versus receiver sensitivity (minimum power at receiver) of each scheme for the (a) optical B2B, and (b) 1 km SSMF cases, under different higher order modulation as indicated.

In the above experimental comparison of the most promising 200 G/lane or (λ) DCN IM/DD techniques, employing band-limited 30 GHz components, DB-PAM4 and PAM6 are identified to offer very effective links with high sensitivity under PAM6 with simpler processing power. PAM4 would be an attractive solution if wider passband components are available. 2D-PAM8 allows the use of simpler FEC but requires relatively high power processing DSP.

3.16 Intra-DC networking and access transmission

The generic interconnection in an intra-DC is shown on the left side of figure 3.73 with relevant growth of the market related to this sector on its

Table 3.5. Performance and FEC/DSP algorithms for 200G under different modulation formats.

Format	224G DB-PAM-4	225G PAM-6	212G 2D-PAM8	212G DMT
1 km ROP sensitivity	−9 dBm	−9 dBm	−8.6 dBm	−8.4 dBm
FEC over head (%)	12%	12%	5.9%	5.9%
DSP Processing/equalization	DPD DB FFE, DB-MLSE	DPD, FFE	DPD, FFE, TCM	IFFT/FFT FFE, TCM
1 km ROP sensitivity	−9 dBm	−9 dBm	−8.6 dBm	−8.4 dBm
FEC over head (%)	12%	12%	5.9%	5.9%
DSP and equalization	DPD DB FFE, DB-MLSE	DPD, FFE	DPD, FFE, TCM	IFFT/FFT FFE, TCM IFFT/FFT

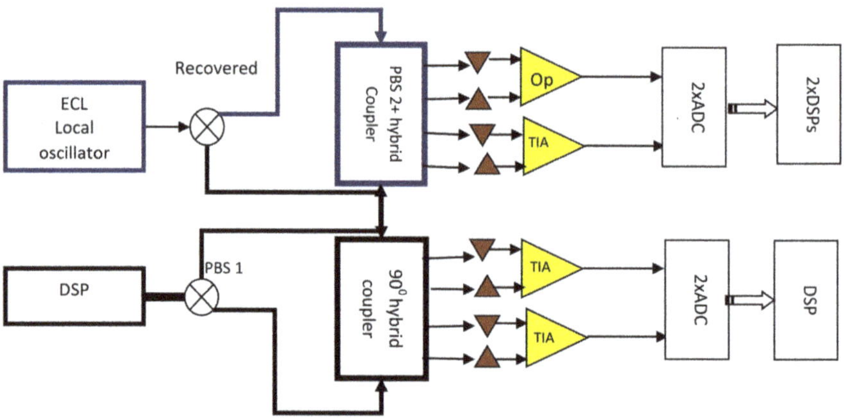

Figure 3.70. Block diagram of a digital coherent optical receiver (OFE) using PD pair and differential amplifier or two-port optical receiver. TIA = trans-impedance amplifier with a transfer resistance (current input giving output voltage) in the mid band or few Kilo-Ohms.

upper right-hand-side corresponding to its capacity growth. Table 3.6 summarizes the transmission technique and distance for 500 m to 2 km for intra-DC communications. Both incoherent and coherent receptions are outlined. Although coherent reception might be a bit more expensive, it is expected to overpass the direct detection due to its high sensitivity, thus no optical amplifier and DSP recovery can be used.

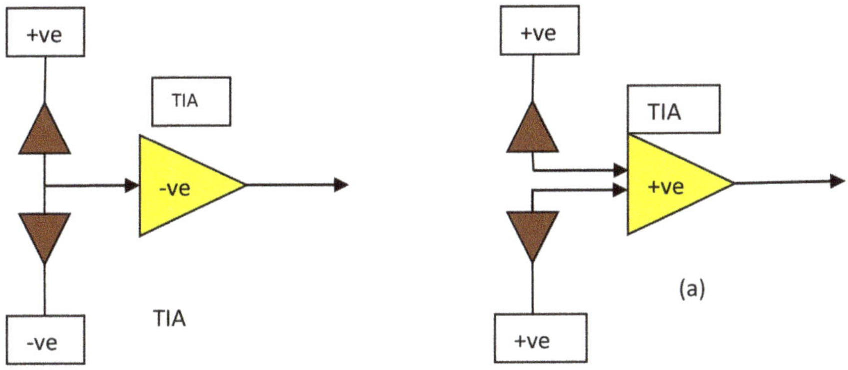

Figure 3.71. Arrangement of PD pair for balanced detection, (a) single end TIA, (b) differential balanced amplifier.

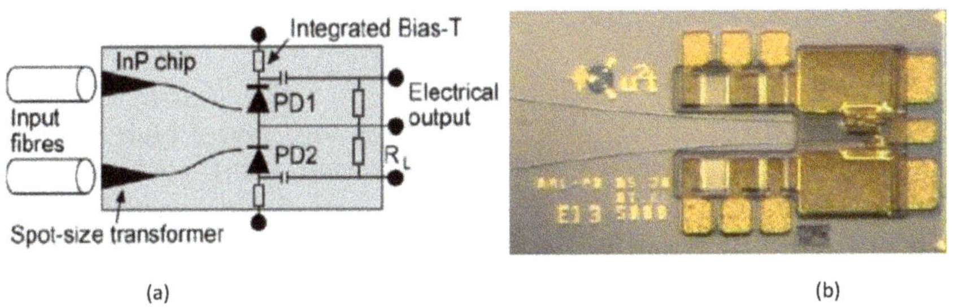

Figure 3.72. PDP differential amplifier optical receiver: (a) schematic diagram and (b) realized integrated circuit with two guided wave structures.

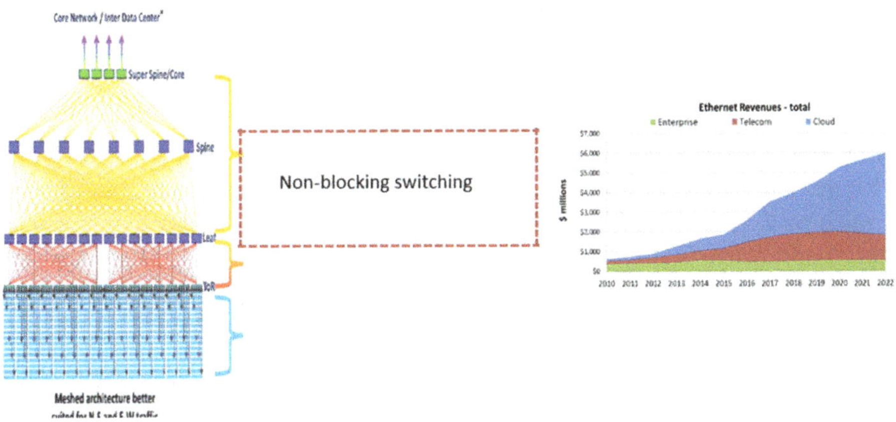

Figure 3.73. DC interconnections and growth strength to 2022 and trending beyond.

Table 3.6. 500 m to 2 km transmission link for intra-DCN, current and future technology. DCI = data center interconnection.

Application	Transmission distance	Port ratio	Current technology	Near future
DCI and	10 km	25%	Coherent 400ZR	SMF
Metro	80 km		80K0Km	Coh/DD
Spine↔	500 m–	25%	CWDM4,100G/ 400G	SMF
Core	2 km	TOR↔Leaf	SMF	DD
100–	50%		400G, MMF	
Leaf↔ Spine		300 m	8	
			parallel	
Server↔ TOR	⩽30 m	AOC	25G	

3.17 Concluding remarks

The explosive growth of traffic in DCNs drives the demand for high-capacity short-reach interconnections. Following the standardized short-reach 400 GE[4], the next-generation 800 GE or 1.6 terabit and then exabit ethernet (Tera-E/Exa-E) are essential and urgently needed [46] beyond 2020 when 5G network technologies are already deployed. For short-reach applications below 2 km, typically in DCN interconnection and ultra-high capacity access nodes, intensity modulation and direct detection (IMDD) solutions are preferred as well as coherent reception solutions. IM/DD are sometimes preferred because of their lower complexity, cost and size. However, unlike the coherent transmission technique in which polarized modes can be employed and then the quadrature phase components can increase the capacity four times, the IM/DD, the higher order modulation formats, can be used in association with the power of digital processing to meet the demands, especially in the effectiveness of the sensitivity and equalization. Compared with the 100G/lane approach, 200G/lane solutions are widely viewed as more compact and cheaper. However, currently there is a lack of commercial components that can satisfy the requirements on bandwidth and SNR. For this reason, a few demonstrations showing remarkable performance leveraged experimental high-bandwidth components such as a special high-speed DAC have been outlined and described in this chapter. Even with high-performance components, powerful DSP is required to mitigate linear and nonlinear distortions. This chapter has also outlined the coherent reception techniques that we expect for the coherent sub-systems with low cost Si-photonic components availability in the near future.

The high capacity transmission techniques described in this chapter allow us to understand the main principles and specific applications for DCI networking.

3.18 Appendix A: Principles of DSP-based coherent transmission

For DSP-based coherent transmission systems, both the transmitter and receiver require DACs, ADCs and DSPs so that digital signal processing can be employed in association with algorithms to recover the transmitted data sequences of both polarized channels. These digital signal processing systems are the distinct difference in contrast to the first coherent systems developed in the 1980s [47]. The optical channels are polarization multiplexed and modulated with the in-phase and quadrature signals under appropriate biasing points so that complex QAM signals can be generated. In the reception sub-system, the transmitted channels are mixed with an optical local laser via a hybrid coupler so that the polarized channels can be demultiplexed and even the in-phase and quadrature components can be split. These optical domain channels are then detected via balanced reception and electronically amplified and sampled by ADCs and thence processed in the DSP systems. More details of these reception systems are described in section 3.3.

3.19 Appendix B: Balanced detection in coherent receivers

The quality of balanced detection in a coherent receiver is very important due to the non-ideal condition of the two PDs connected B2B. The impact of this imbalance generates distortion which can then be defined by a term common mode rejection ratio (CMRR). Furthermore, the optical power split and polarization diversity in the $\pi/2$ hybrid coupler has significant impact on the optical receiver. Thus, the impact of the characteristics of the optics on the balanced detection is analyzed. Parameters that characterize the performance of the balanced detection suitable for the optical front end (OFE), are proposed.

3.19.1 Optical front end

The structure an OFE using polarization diversity and coherent detection is shown in figure 3.70. The structure consists of two hybrid couplers for the X- and Y-polarization channels of the transmitted data streams. The LO is also polarization split and fed into the coupler so that mixing between the oscillator and the signal polarized modes can occur.

3.19.2 Optical mixing and polarization diversity

Optical signals and LO are split in terms of polarization and power, and their mutual mixing happen in the $\pi/2$ hybrid coupler. The 900 optical section ensures that the distinction between the I and Q components of the optical signal can be distinguished.

Total insertion loss per polarization would be about 10 dB including the 3 dB power splitting to the two polarized channels. For each channel of I and Q there are two complementary outputs in the optical domain where the phase difference would be π. Thus, a balanced PDP (photodetection pair) would give an extra 3 dB gain in common mode and depleted in difference mode.

3.19.3 Differential amplification

Differential detection and amplification of optical signals (now photodiode currents) is normally done by using a single input port with the two PDs connected B2B and then fed into a TIA, as shown in figure 3.71(a). Alternatively, the photocurrents can be fed separately into two differential inputs of a differential TIA, as shown in figure 3.71(b).

The input fields fed into the two PDs can be written as:

$$E_s(t) = \sqrt{P_s(t)}\, e^{j\omega_s t} e^{j\phi_s(t)} \tag{3.11}$$

$$E_{\text{LO}}(t) = \sqrt{P_{\text{LO}}(t)}\, e^{j\omega_{\text{LO}} t} e^{j\phi_{\text{LO}}(t)} \tag{3.12}$$

Thus after mixing an polarization splitting, the detected currents at the outputs of the PDP are given by

$$i_1(t) = \frac{1}{2}\Re\left\{P_s(t) + P_{\text{LOs}}(t) + 2\sqrt{P_{\text{LO}}P_s(t)}\sin(\omega_s - \omega_{\text{LO}})t + \phi_s(t) - \phi_{\text{LO}}\right\} \tag{3.13}$$

$$i_2(t) = \frac{1}{2}\Re\left\{P_s(t) + P_{\text{LOs}}(t) - 2\sqrt{P_{\text{LO}}P_s(t)}\sin(\omega_s - \omega_{\text{LO}})t + \phi_s(t) - \phi_{\text{LO}}\right\} \tag{3.14}$$

In equations (3.13) and (3.14), there are three terms: (i) photocurrents by the signal itself; (ii) currents produced by the power of the LO; and (iii) useful signal current whose amplitude is boosted by the local oscillator power. The first current due to signal happens at a frequency far away from the spectral sensitivity of the PD material, the second term produces a DC photocurrent and then turns into shot noise current, which creates more noise in the signal output. The third term is a useful term that represents the phases of the signals in I and Q components.

3.19.4 Unmatched detector frequency responses

A further difficulty that may degrade the performance of a dual-detector receiver occurs when the two detectors have differing frequency responses. This problem could arise, for example, if the detectors have different bandwidths. For good cancellation, the LO noise photocurrents in both detectors must be well matched both in amplitude and phase over the entire frequency band of interest. If one detector begins to roll off before the other near the edge of the frequency band, poor cancellation could result. For 20 dB of cancellation with a 50:50 beam splitter and the two attenuators equal to 1, the detector responses must be matched within at least 1.6 dB (assuming matched phases) and 11%° (assuming matched amplitudes). Clearly, if one detector has rolled off by 3 dB and acquired a 45° phase shift relative to the other, poor cancellation will result. Unmatched detector frequency responses can also occur because of line resonances in the two signal paths when the detectors are physically connected into a circuit. If the noise in each detector is unequally amplified due to differing line resonances in the two signal paths, the balanced mixer action of the dual-detector receiver may be disrupted. In fact, an attenuator designed to maximize the SNDR (signal-to-noise differential ratio) of a particular receiver might change the line resonances of the signal path in which it is placed enough to lower the overall SNDR rather than increase it as predicted.

References

[1] Soref R A and Lorenzo J P 1986 All-silicon active and passive guided-wave components for lambda = 1.3 and 1.6 microns *IEEE J. Quantum Electron.* **22** 873–9
[2] Jalali B and Fathpour S 2002 Silicon photonics *IEEEJ J. Lightwave Technol.* **24** 4600–15
[3] Kikuchi K 2016 Fundamentals of coherent optical fiber communications *IEEE J. Lightwave Technol.* **34** 157
[4] Bell T E 1986 Communications: Coherent optical communication shows promise, the FCC continues on its path of deregulation, and satellite communications go high-frequency *IEEE Spectr.* **23** 49–52
[5] Savory S J 2010 Digital coherent optical receivers: algorithms and subsystems *J. Sel. Topics in Quant. Elec.* **16** 1164–78
[6] Pfau T, Hoffmann S and Noe R 2009 Hardware efficient digital receiver concept with feed forward carrier recovery for M-QAM constellation *IEEE J. Lightwave Technol.* **27** 989–99
[7] Mendinueta J M D, Bayvel P and Thomsen B C 2011 Digital lightwave receivers: an experimentally validated system model *IEEE Photon. Technol. Lett.* **23** 338–40
[8] Zhou X and Yu J 2009 200-Gb/s PDM-16QAM generation using a new synthesizing method ECOC2009, Paper 10.3.5
[9] Gnauck A H and Winzer P J 2009 10 × 112-Gb/s PDM 16-QAM transmission over 1022 km of SSMF with a spectral efficiency of 4.1 b/s/Hz and no optical filtering ECOC2009, paper 8.4.2
[10] Stojanovic N 2008 An algorithm for AGC optimization in MLSE dispersion compensation optical receivers *IEEE Trans. Circ. Syst.* I **55** 2841–7
[11] Mao B, Stojanovic N, Xie C, Chen M, Binh L and Yang N 2011 Impacts of ENOB on the performance of 112 Gbps PDM-QPSK ECOC2011, Paper We.10, P1.46
[12] Kester W 2006 ADC input noise: the good, the bad, and the ugly. Is no noise good noise? *Analog Dialogue* **40-01** 13–7
[13] Regenbogen L K 1980 Nonlinearity model with variable knee sharpness *IEEE Trans. Aerosp. Electron. Syst.* **16** 874–7
[14] Weiner J S *et al* 2003 SiGe differential transimpedance amplifier with 50-GHz bandwidth *IEEE J. Solid-State Circuits* **38** 1512–7
[15] Ip E and Kahn J M 2008 Compensation of dispersion and nonlinear impairments using digital back propagation *IEEE J. Lightwave Technol.* **26** 3416
[16] Fatadin I, Ives D and Savory S J 2010 Laser linewidth tolerance for 16-QAM coherent optical systems using QPSK partitioning *IEEE Photon. Technol. Lett.* **22** 631–3
[17] Louchet H, Kuzmin K and Richter A 2008 Improved DSP algorithms for coherent 16-QAM transmission *Proc. ECOC'08, Paper tu.1.E6 (Belgium)*
[18] Noe R, Pfau T, El-Darawy M and Hoffmann S 2010 Electronic polarization control algorithms for coherent optical transmission *IEEE J. Sel. Quant. Electron.* **16** 1193–9
[19] Technical specification of Corning fiber G.652 SSMF given in ed Binh L N 2010 *Digital Optical Communications* (Boca Raton, FL: CRC Press) ch 3, appendix
[20] Godard N 1978 Passand timing recovery in an all-digital modem receiver *IEEE Trans. Commun.* **26** 517–23

[21] Stojanovic N, Gonzalez N G, Xie C, Zhao Y, Mao B, Qi J and Binh L N 2012 Timing recovery in Nyquist coherent optical systems *Int. Conf. Telecommunications Systems (Serbia)*

[22] Fang T T and Liu C F 1993 Fourth-power law clock recovery with pre-filtering *Proc. ICC* 2 *(Geneva)* 811–5

[23] Dong Z, Li X, Yu J and Chi N 2012 128-Gb/s Nyquist-WDM PDM-16QAM generation and transmission over 1200-km SMF-28 with SE of 7.47 b/s/Hz *IEEE J. Lightwave Technol.* **30** 4000–6

[24] Don Z *et al* 2012 128-Gb/s Nyquist-WDM PDM-16QAM generation and transmission over 1200-km SMF-28 with SE of 7.47 b/s/Hz *IEEE J. Lightwave Technol* **30** 4000–6

[25] Zhou X, Nelson L E, Magill P, Isaac R, Zhu B Y, Peckham D W, Borel P I and Carlson K 2012 PDM-Nyquist-32QAM for 450-Gb/s per-channel WDM transmission on the 50 GHz ITU-T grid *IEEE J. Lightwave Technol.* **30** 553

[26] Zhou X, Nelson1 L E, Isaac R, Magill P, Zhu B, Peckham D W, Borel P and Carlson K 2012 1200 km Transmission of 50 GHz spaced 504-Gb/s PDM-32–64 hybrid QAM using electrical and optical spectral shaping *OFC/NFOEC (Los Angeles, CA, 4–8 March 2012)*

[27] Bosco G, Curri V, Carena A, Poggiolini P and Forghieri F 2011 On the performance of Nyquist-WDM Terabit superchannels based on PM-BPSK, PM-QPSK, PM-8QAM or PM-16QAM Subcarriers *IEEE J. Lightwave Technol.* **29** 53

[28] Zhu R, Xu K, Zhang Y, Li Y, Wu J, Hong X and Lin J 2008 QAM coherent subcarrier multiplexing system based on heterodyne detection using intermediate frequency carrier modulation *2008 Microwave Photonics* pp 165–8

[29] Nakazawa M, Yoshida M, Kasai K and Hongou J 2006 20 Msymbol/s, 64 and 128 QAM coherent optical transmission over 525 km using heterodyne detection with frequency-stabilized laser *Electron. Lett.* **42** 710–2

[30] Dong Z, Li X, Yu J and Yu J 2013 Generation and transmission of 8 × 112-Gb/s WDM PDM-16QAM on a 25-GHz grid with simplified heterodyne detection *Opt. Express* **21** 1773

[31] Liu M and Zhang X 2012 Graphene-based optical modulators *Proc. OFC (CA USA)* paper OTu1I.7

[32] Verbist J *et al* 2017 DAC-less and DSP-free PAM-4 transmitter at 112 Gb/s with two parallel GeSi electro-absorption modulators *Proc. ECOC 2017 (Gottenburg)*

[33] Patel D, Pavizi M, Ben-Hamida N, Rolland C and Plant D V 2018 Frequency response of dual-drive silicon photonic modulators with coupling between electrodes *Opt. Express* **26** 8409–15

[34] Binh L N 2009 *Digital Optical Communications* (Boca Raton, FL: CRC Press, Taylor and Francis) ch 9

[35] Kanazawa S *et al* 2016 Transmission of 214-Gbit/s 4-PAM signal using an ultra-broadband lumped-electrode EADFB laser module *Proc. OFC, Th5B.3*

[36] Lange S *et al* 2018 100 GBd intensity modulation and direct detection with an InP-based monolithic DFB laser Mach–Zehnder modulator *J. Lightw. Technol.* **36** 97–102

[37] Zhang L *et al* 2018 Nonlinearity-aware 200 Gbit/s DMT transmission for C-band short-reach optical interconnects with a single packaged electro-absorption modulated laser *Opt. Lett.* **43** 182–5

[38] Yamazaki H *et al* 2016 300-Gbps discrete multi-tone transmission using digital-preprocessed analog-multiplexed DAC with halved clock frequency and suppressed image *Proc. ECOC, PDP Th3B.4*

[39] Wei J, Cheng Q and Penty R V *et al* 2015 400 Gigabit Ethernet using advanced modulation formats: performance, complexity, and power dissipation *IEEE Commun. Mag.* **52** 182–9
[40] Yamazaki H, Nagatani M and Wakita H *et al* 2018 Transmission of 400-Gbps discrete multi-tone signal using >100-GHz-bandwidth analog multiplexer and InP Mach–Zehnder modulator *Proc. ECOC (Rome, Italy)* pp 1–3
[41] Estarán J M, Mardoyan H and Jorge F *et al* 2019 140/180/204-Gbaud OOK transceiver for inter- and intra-data center connectivity *J. Lightwave Technolnol.* **37** 178–87
[42] Zhang F, Zhu Y and Yang F *et al* 2019 Up to single lane 200G optical interconnects with silicon photonic modulator *Proc. OFC (San Diego, CA)* pp 1–3 Th4A.6
[43] Wei J, Zhang L and Prodaniuc C *et al* 2018 Linear pre-equalization techniques for short reach single lambda 225 Gb/s PAM IMDD systems *Proc. ECOC (Rome, Italy)* pp 1–3
[44] Stojanović N, Prodaniuc C and Zhang L *et al* 2019 4D PAM-7 trellis coded modulation for data centers *Photon. Technol. Lett.* **31** 369–72
[45] Wei J, Stojanović N and Xie C 2018 Nonlinearity mitigation of intensity modulation and coherent detection systems *Opt. Lett.* **43** 3148–51
[46] Wei J, Cheng Q and Penty R V *et al* 2015 400 Gigabit ethernet using advanced modulation formats: performance, complexity, and power dissipation *IEEE Commun. Mag.* **52** 182–9
[47] Li G 2009 Recent advances in coherent optical communication *Adv. Opt. Photon.* **1** 279–307

IOP Publishing

Transmission, Processing, and All-Optical Routing for Ultra-High Capacity Data Center Networking (Second Edition)

Le Nguyen Binh

Chapter 4

Super-channel transmission by multi-carrier sources

In data center (DC) networking the reduction of space and the energy consumption of communication and switching systems are critical to reduce the power or energy supplies to DCs. Parallel transmission of several optical channels can save the use of several optical and optoelectronic components at the transmitter and reception systems of the networks. Comb generation of laser sources can offer the solution of several carriers from the same laser, such as the resonant cavity of wideband stimulated emission medium in the same lasing waveguide.

This chapter gives an introduction and related technologies to such generation of comb line lasers whose linewidth is sufficiently narrow for polarization division multiplexed complex modulation and coherent reception.

Transmission with ultra-high capacity for interconnection of DCs, in networking cores and networking within DCs are critical for the Zetta-bps[1] global Internet. The low-cost aspect is so important so as to lower the total cost per user. Thus, for multi-channels carrying 100G and beyond, the employment of a multi-carrier source would drastically reduce the cost if several single sources are used. This chapter presents the stable generation of comb lasers by recirculating fibre loop incorporating integrated synchronous optical modulators (ISOMs). The ISOM is composed of a master and slave configuration of Mach–Zehnder waveguide interferometers operating in such mode that the optical phase paths are in synchronization. The generated sub-carrier spectrum can be frequency shifted to the left or to the right side or to both sides of the main optical carrier without suffering from the interference of the other sideband shifted sub-carriers by applying Hilbert transformed electric oscillation driving signals to the dual electrodes of the ISOM, hence suppression of

[1] Zetta = 10^{+18} in Greek numerals

doi:10.1088/978-0-7503-5849-1ch4

one shifted band. The generations of sub-carriers in the shifting left-side or in the right-side and to the dual-band spectrum are experimentally demonstrated. The single sideband (SSB) comb generators are then incorporated in tera-bits s^{-1} (Tbps) optical transmission systems and demonstrated in installed fibre network allowing the transmission of Tb s^{-1} super-channels (SCs) over long-haul non-compensating optically amplified lines and optical coherent receivers incorporated digital signal processors (DSPs). Nyquist pulse shaping of data sequences is implemented by using a digital-to-analog converter (DAC) with 56–64 GSa s^{-1} so as to achieve effective spectrum packing of the modulated sub-carrier channels. The obtained transmission performance (BER versus OSNR) of 10^{-3} with OSNR of 16 dB, shows that the sub-carriers of the SSB comb generator can provide a stable source for multi-Tbps transmission systems. It is noted here that with 50 GHz channel spacing the baud rate can reach 112 GBaud with Nyquist or raised-cosine pulse shaping. Thus, for PAM4 or PAM8 the aggregate bit rate can reach 200 Gbps or 300 Gbps with the shorter distance at which the OSNR can be satisfied. Such a short distance and ultra-high rate can be most useful for intra-C networking or the client side, i.e. access super-nodes.

We note here also that such a comb source can be employed in testing the spectra of multi-channel systems such as EDFA operating in C^+ spectral window with the total time span of less than 1.0 milli-second, but this is not within the scope of this chapter/book. In this chapter the comb generation source, described in detail, is a recirculating type. There are many types of comb generators and the number of comb lines can vary from four to several hundreds. This chapter focuses on the application for transmission of ultra-high capacity and naturally the comb lines are in the C^+-band of the 1550 nm spectral window.

4.1 Introduction

SCs reaching a total capacity of a few to tens of Tbps have attracted significant attention for the emerging optical transport networks. Comb generation of optical narrow line optical sources of high stability and minimum jitters, from a single primary source offers significant advantage in the SC optical transmitters [1–3]. This section presents the generation and modulation of SCs defined as multi-channels generated by the lightwaves coming from the same reference laser source or the primary laser line, by generation of multi-carrier laser (MCL) sources from a primary lightwave carrier recirculating sideband (SSB) frequency shifting techniques. The uniqueness of the SSB generation presented here is the synchronous modulation of the lightwave field recirculating around the loop which incorporates an ISOM composed of a master–slave interferometric intensity or phase modulation. Their electrodes are controlled in such a way that the synchronization of the modulated lightwaves can be in constructive interference, hence maximizing the power of the generated sub-carriers. The formation of Tbps SCs can be 10 × 100G (1 Tbps) or 20 × 100G (2 Tbps) with the amplitudes of the sub-channels equalized using wavelength selective switches. In order to achieve efficient spectral packing of the modulated channels, pulse shaping of the data sequence is also implemented using a DAC at a sampling rate of 56–64 GSa s^{-1}, so that the modulated spectra of sub-channels exhibit 'brick-wall-like' shapes, hence minimizing the channel crosstalk. By employing the sub-carriers of the multi-sub-carrier laser (MSCL) as laser sources, modulation of individual sub-carriers is implemented, with and without pulse shaping, to generate an SC in which each individual channel can be modulated using polarization multiplexing technique and QPSK modulation format. Thus, the two polarized channels at 28 GBaud and QPSK (2 bits/symbol) give 112 Gbits s^{-1} (Gbps). So, if 10 or 20 sub-carriers of the comb generator are employed then the resulting aggregate bit rate reaches 1.0 Tbps and 2.0 Tbps. These SCs can be generated for transmission over a long-haul multi-span optically amplified fibre line which consists of several 80 km spans of non-dispersion compensating standard single-mode optical fibres (SSMFs).

So, if combining both spectra of the shifting left and shifting right with respect to the central original carrier frequency position, then a dual-sideband shifted sub-carrier comb generation can be achieved. This type of comb generator is demonstrated by using two recirculating frequency shifting ring comb generators integrated with an optical switching and coupling sub-structure. This is possible provided that there is no interference between the previously generated and the current generation, hence allowing the generation of modulated channels whose total capacity reaching the limit and beyond of 25 Tbps. To the best of the author's knowledge, the dual-band comb generator is experimentally demonstrated here for the first time.

The rest of this chapter is thus organized as follows: section 4.2 outlines the principles of operation of the comb generation with the frequency shifted to either the left or right side of the original lightwave carrier. Experimental results obtained for such comb-generated lines are given, especially the measured linewidth of these comb lines. Section 4.3 presents our experimental observation of the generation of dual sideband sub-carriers using a double ring configuration. Tbps SC optical transmitters implemented using these SSB comb generators are reported in section 4.4. Their integration into optically-amplified fibre transmission systems is also described and the transmission performance of these transmission systems is then given. Finally, some concluding remarks and potential future works are stated.

4.2 Comb generation of multi-sub-carriers

4.2.1 General structure

The schematic of a constructed recirculating frequency shifting (RCFS) comb generator is shown in figure 4.2 in which a recirculating loop is the basic configuration incorporating an optical modulator, an optical amplifier to compensate the insertion losses of various inline devices and an optical attenuator whose coefficient can be adjusted so that the overall loop gain is slightly less than unity so as to avoid self-oscillation or lasing. Such a recirculating comb source was demonstrated in [4, 5], whose schematic and generated spectra are shown in figure 4.1. A 50:50 optical coupler or 3 dB split is employed for coupling into and out of the loop the continuous lightwaves and combed sub-carrier lines, respectively. This fibre coupler is to be replaced by an optical switch and associate couplers for generation of shifted carriers to both sides of the main lightwave carrier, the dual

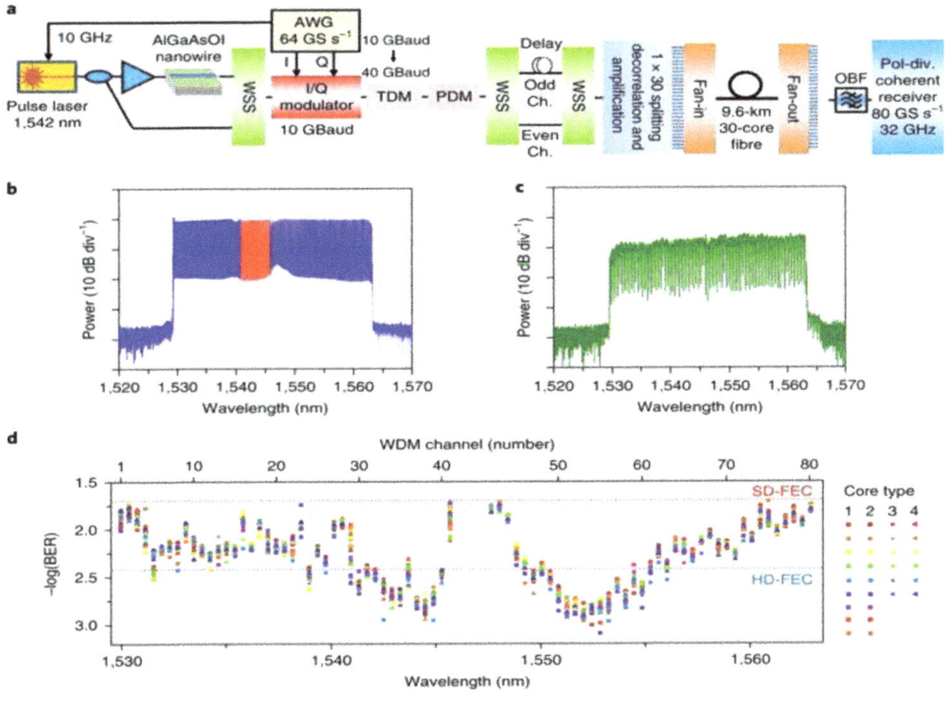

Figure 4.1. Schematic and generated spectra of a high-density comb lines laser.

shift recirculating comb generator which will be described in section 4.3. An electrical synthesizer generating sinusoidal waves whose frequency determines the frequency shifting of the optical lines of the generated comb, is used to apply into the electrodes of the modulator, hence via the electro-optic effect to modulate the phase of the guided lightwaves and thus optical modulation.

4.2.2 Synchronous optical modulator

The ISOM incorporated into the loop is a synchronous type whose structure can be seen by the inset of figure 4.2 composed of a master Mach–Zehnder interferometric modulator (MZIM) whose two arms are connected by two slave electro-optic interferometers (EOIs). The lightwave going through the two slave interferometers can be biased and activated such that constructive rather than destructive interference can be achieved at the combined output. The optical waveguides of the ISOM are formed by diffusion of thin Ti film deposited on top of the *Z*-cut or *X*-cut $LiNbO_3$ substrate. Optical waveguides support only single-mode and both TE and TM polarizations. However, only one polarized mode is excited in the circulating ring. A thin SiO_2 film is deposited on top prior to the deposition of electrode film to match the velocity of the RF travelling wave to that of the guided optical waves. Travelling wave electrodes are independent with respect to the DC electrodes (i.e. unconnected) so that independent biasing to create the initial phase states of the guided optical waves can be done. Besides the two DC electrodes deposited in one light path of the slave EOI, another DC electrode is available in one path of the master MZIM and ideally at the output of one of the slave EOIs so that tuning of the phase of the interfered lightwaves emerging from the outputs of the two slave EOIs can be achieved in synchronization. Thus, this modulator is termed the interferometric synchronous optical modulator or in short, ISOM.

As depicted in the inset of figure 4.2, the two arms of ISOM are driven with a $\pi/2$ phase shift with respect to each other so that SSB shifting of the lightwave can be achieved at the output of the modulator. Implementing the $\pi/2$ phase shift is indeed doing a Hilbert transformation. The principles of the shifting of the optical carrier frequency to the right or to the left of the original lightwave carrier are shown in figure 4.3. Applying sinusoidal or co-sinusoidal RF signals to the electrodes of the slave EOIs will shift the lightwave carrier to the right or left, respectively, as the summation of the two modulated lightwaves emerging at the output of the slave EOI are constructively or destructively combined in either band of the spectrum, thus resulting in an SSB frequency shifting. The SSB shift is critical so that multiple lines or interference of the other sideband sub-carriers with the shifted frequency sub-carriers generated in the precedent modulation can be avoided. It is noted here that the SSB principle was first demonstrated by Smith *et al* [6] to transport RF signals over a long length fibre. In our recirculating comb generator, the optical wave is shifted and then circulating through the loop and then shifted again and again to generate other multiple shifted sub-carriers.

Thus, the SSB modulation serves two purposes: firstly, to shift the original carrier only to either the right or the left side of the original carrier, and secondly to minimize the interference of the other sideband generated carriers if double sideband (DSB) modulation is employed. This SSB modulation technique was also employed by several research groups for transmission over long-haul fibre to carry radio channels [7–9] or digital channels using orthogonal frequency division multiplexing (OFDM) [10–12] and in the comb generation for Tbps transmission but not in synchronization [13–16].

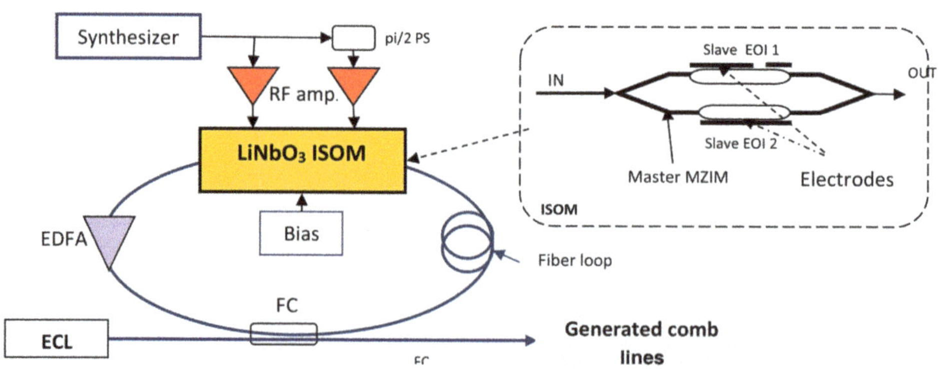

Figure 4.2. Schematic structure of the RCFS comb generator. ECL: external cavity laser; EDFA: Er:doped fibre amplifiers. FC = fibre coupler; PS = phase shifter; EDFA = Er:doped fibre amplifier; ECL = external cavity laser, ISOM = interferometric synchronous optical modulator. EOI = electro-optical intensity manupulator; MZIM = Mach–Zehnder intensity modulator.

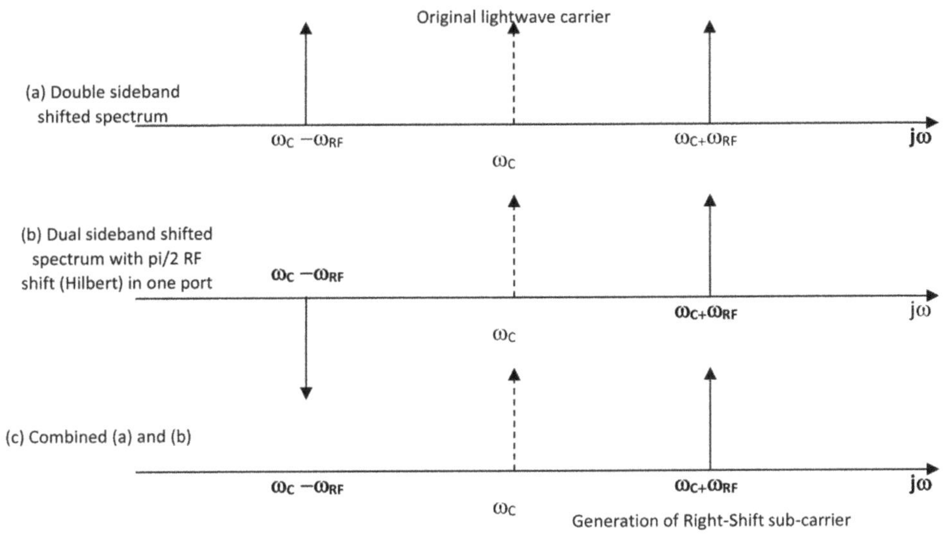

Figure 4.3. Spectra of (a) the RF modulated carrier and (b) $\pi/2$ phase shift or Hilbert transformed RF wave driven signal modulated lightwave, (c) the constructive and synchronous spectrum, the generated right-shift sub-carrier.

Therefore, for the case of dual-band shifting of the lightwave main carrier, described later in section 4.3, it would require two comb generators combined with an optical switch and optical couplers to generate the shift-left and shift-right sub-carriers with respect to the central original lightwave following the principles illustrated in figures 4.3(b) and (c) with both the normal the reverse $\pi/2$ phase shift.

4.2.3 Implementation of the comb generator

The comb generator is implemented using fibre-based devices. A ring of fibre of length about 25 meters is formed with one input and one output port of a 3 dB optical fibre coupler (FC). This means that there is always a 50% intensity of RF shifted lightwaves coupled back into the ring and 50% coupled to the output port. A $LiNbO_3$ ISOM is incorporated in the fibre ring. The ring length is accounted for with all the pig-tailing of optical components associated within the ring. The overall gain of the ring must be less than unity so as to avoid any lasing. Er-doped fibre amplifiers are included so as to compensate for the optical loss of the fibre, the modulator and the coupler FC. The open loop is constructed first and the modulation of the lightwave carrier by a purely sinusoidal wave from a reference synthesizer is applied. The shifting of the lightwave carrier to the left or to the right is observed, as shown in figures 4.4(a) and (b), respectively. Note that this frequency shifting can be either shifting left or right depending on the relative phase shift applied to the two slave EOIs embedded in the two arms of the master MZIM. As observed, the suppression of the residual lightwaves can reach higher than 25 dB by the tuning of the biasing voltage applied to the arms of the EOI, especially the DC bias of the master MZIM. This is why synchronization of the optical modulator is essential for efficient and

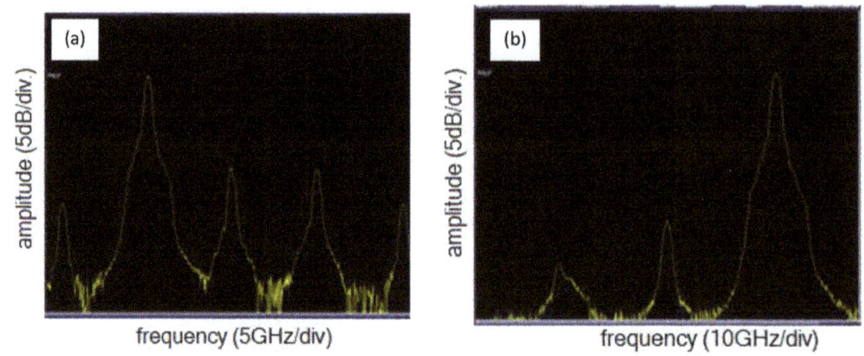

Figure 4.4. Spectrum of the SSB lightwave at output of the synchronous optical modulator (a) shifting right; and (b) shifting left in frequency scale. Resolution of horizontal scale (for (a) 40 pm/div. and for (b) 80 pm/div.) and 5 dB/div. in vertical scale for power.

high carrier-to-noise ratio (CNR) to be achieved. Under the synchronization scenario a minimum phase noise can be obtained. This is very important for these comb lines to be employed as the low noise sources for long haul and coherent reception of the modulated sub-carriers over many thousand kilometers of dispersion non-compensated fibre. These residual sub-carriers must be at this level so that noise due to the interference with other generated sub-carriers can be considered negligible.

The control of the stability of the generated comb lines is implemented by a control feedback mechanism with feedback signals coming from the synchronization of the jittering of the referenced RF synthesizer, the injection optical power level into the recirculating ring, and the pumping level of pump laser into the optical amplifying devices incorporated in the ring. To set spectral spacing between the comb lines, a synthesizer launches the RF waves to the two arms of the EOIs with RF amplifiers to boost the electrical signals to appropriate level so that maximum modulation depth can be achieved with minimum nonlinear effects. We note that although the modulator 3 dB bandwidth is specified at 23 GHz, it is still possible to operate the modulator at 33 GHz although some RF signal loss, hence demanding higher RF signal power. This is because the driving signal is purely sinusoidal and has a very narrow band. We have successfully generated comb lines from 28 GHz to 33 GHz for channel spacing between sub-carriers of 28G, 30 G and 33 Gb s^{-1} as shown in figure 4.5. Note also in figure 4.5(b) the residual original carrier in the far right of the spectrum. This appears because of the use of moderate gain optical amplifiers (EDFAs) and it takes some time for the optical signals to reach saturation state in the amplifier. Depending on the loop and the gain of the EDFA with respect to those components embedded in the recirculating ring, there may be one or two residual frequency shifted lines in the spectrum, as observable in figure 4.4. The spectrum is flat over some considerable spectral range, about 5.5 nm which is sufficient to support 10 sub-carriers whose equal spacing is in the order of 50 GHz (or equivalently 0.4 nm) in the 1550 nm region. With these 10 sub-carriers and the

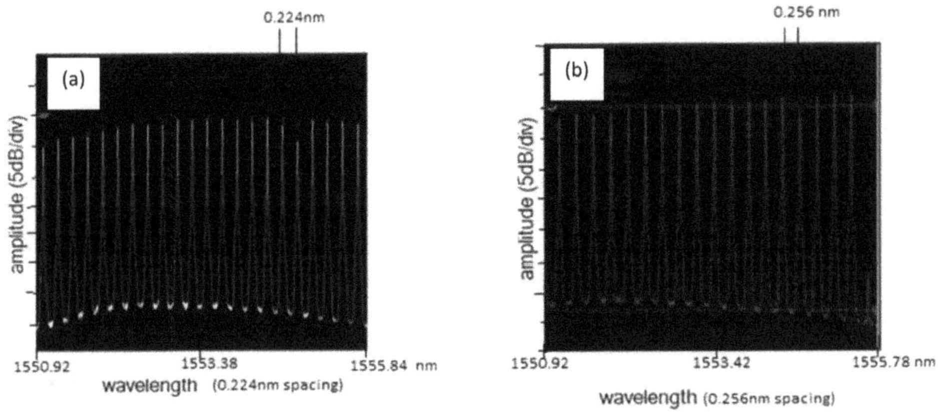

Figure 4.5. Spectrum of multi-sub-carriers, comb lines generated by the RCFS loop. Vertical scale: power amplitude of arbitrary units and horizontal scale: frequency of 50 GHz/div.

symbol rate of 28 GBauds s^{-1} per sub-carrier channel, the aggregate bit rate is 100 Gbps per sub-carrier if using two polarized channels and QPSK or two bits per symbol. Therefore, a total capacity of 1 Tbps can be obtained for 10 sub-carriers generated by this comb generator. For 2.0 Tbps we must design the conditions for the comb generator such that 20 sub-carriers can be generated.

The comb lines can also be equalized by using a WSS (wavelength selective switch) to obtain amplitude equalized sub-carriers, the ripple level of the comb lines depends on the flatness of the optical amplifier incorporated in the ring, thus a flat spectral EDFA would normally be selected for this type of recirculating RF shifted comb generator. On both the left- and right-shift spectra, the CNR is achieved in order of 27 dB. This is quite remarkable as compared with other comb-generated lines reported so far [14, 17]. We believe that this is contributed from the synchronization of the modulated lightwaves passing through the master MZIM and slave EOIs.

A low-phase noise external cavity laser (ECL) is used as the primary lightwave which is fed into the SSB frequency-shifted fibre ring under non-resonant condition to generate comb sub-channels. By non-resonance we mean that the resonance condition for lasing is not permitted by ensuring that the total gain and loss of the lightwaves circulating in the ring are less than unity. The ECL output power is set at its highest level of 13 dBm. Higher power level can be used if an EDFA booster amplifier is employed.

The spectra of the comb lines generated from the recirculating comb generator are shown in figure 4.5(a) and (b). Thus, we note the power level of the SBS line at −10 dBm and −4 dBm for the comb lines which has also been passing through an EDFA at saturation power of 16 dBm which is distributed to all sub-carrier lines.

Both sides of the primary laser lines can also be generated by employing a dual-ring structure with the SSB shift to the right or the left of the primary line as described in the next section. This ultra-wide spectral comb line source has also been generated in our laboratory.

The optical amplifier incorporated in the recirculating ring is operating in the saturation mode only when the power level of the lines is sufficiently high to boost it into the saturated region, otherwise it would operate in the linear region. For this reason, we observe that the frequency shifting lines would not be in the saturated level for the first few sub-carrier lines, the far most left side of the spectrum (figure 4.5(a)). The tuning of the DC bias supply voltage to the electrodes is quite sensitive and would be in the tens of mV range. One most important point, we must point out, is that the output fibre port must be angled connector type so that no reflection of sub-carrier combed lines can feed back into the recirculating loop to avoid the noise oscillation of the generated comb lines. Note that the suppression of the other harmonic lines is more than 25 dB.

The comb generator described in this section is then employed in a multi-Tbps transmission system which will be described in section 4.4.

4.3 Dual-band frequency shifting recirculating comb generator

Further to the possibility for generation of comb sub-carriers over the whole C-band in the region 1520–1565 nm for optical communication systems with the referenced central lightwave carrier located at the centre of the band, the frequency shifting of the original carrier must be into both the left and right sides of the carrier. We thus propose a dual-band comb generator structure, as shown, in which two recirculating frequency shifting ring comb generators share 'a common optical switching structure', as shown in figure 4.6.

Further detailed principles of the experimental generation of the dual-band comb lines can be described as follows: firstly, the continuous laser line is fed into a 2 × 2 FC (one port left unused) whose outputs can then be fed to 2 × 2 optical switches, as shown in figure 4.6. The integrated optical guided-wave switch is shown as an inset

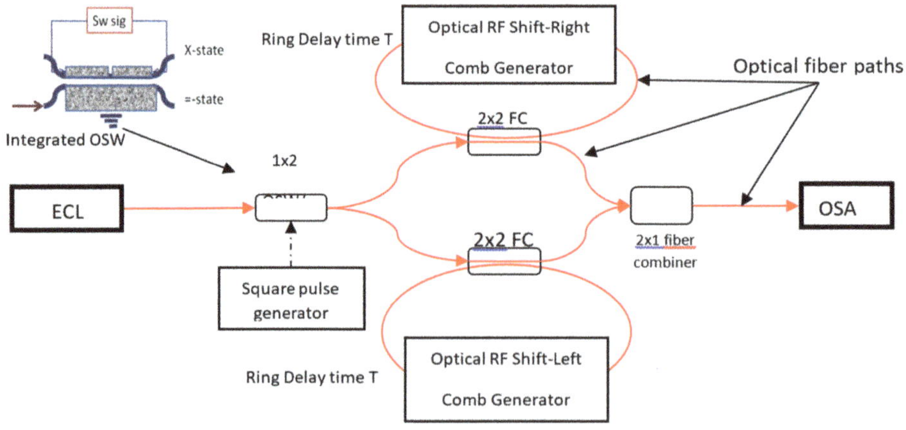

Figure 4.6. Schematic of the experimental dual-band comb generation. Labels: OSW = optical switching device with switch time much shorter than that of the ring propagation delay time *T*; FC = fibre coupler; OSW = optical switch; OSA = optical spectrum analyser. All continuous lines are optical paths. Dotted lines are electrical lines. Inset is the structure of an integrated OSW.

of this figure. The optical switch is of $LiNbO_3$ type and fabricated by Ti in-diffusion technique [18, 19]. The electrical switching signals are fed into the electrical port of the OSWs, as shown in the figure by a dotted line. All optical paths are shown as continuous lines. The bandwidth of the OSW is 18 GHz to allow the switching of 100 ns to be achieved without any difficulty. The switching electrical signals are complementary to each other indicating that one switch is on and the other is off. The switching period of the electrical switching signal is just less than that of the circulating delay time which is estimated as 100 ns for a 20-meter-long fibre loop of each comb generator. Each output of the 1 × 2 optical switch OSW is fed into a comb generator whose RF signals are fed into an ISOM with the $\pi/2$ shift in reverse with each other in the RF signals applied to electrical ports so that the shift left and shift right of the sidebands can be obtained as described in section 4.2.

The lightwaves are injected at a power level such that the optical amplifiers incorporated in the ring, similarly to that given in figure 4.2, can operate in the saturation mode. The outputs of the two ring comb generators are then combined via another passive FC. The output of the dual-band comb generator is monitored by an optical spectrum analyser (OSA). The switching time of the optical switch must be much less than that of the ring delay time so that no interference will occur. It is obvious that the delay time of the lightwaves circulating through a loop is sufficiently long, about 75 ns for 25 m length ring. Thus, a moderate bandwidth optical switch can be employed. Such electro-optic switches of a 3 dB bandwidth of 20–30 GHz in $LiNbO_3$ are commercially available. These switches can thus provide switching time between shifting left and right within 50–33 ps[2] [20, 21].

The observed dual-band spectral comb lines are shown in figure 4.6. We note here that as the switching efficiency of the OSW is not 100% but about 95%, there are some residual optical carriers which are embedded in the shifted optical sub-carriers that could cause some interference noise in subsequent comb lines. Thus, we expect some degradation to the performance of BER versus OSNR as compared with those depicted in figure 4.17 (described later in section 4.4). We also observed some undulation of the amplitudes of the comb lines shifted to the left side due to unequal transmission loss in the recirculating loop. Indeed, we have also set up the dual-ring comb generator using a passive FC instead of the OSW shown in figure 4.6. The optical spectrum of the comb sub-carriers can also be observed in figure 4.7. We note here that the optical amplifiers embedded in both comb rings of the dual-ring comb generator are implemented with high output power of about 20 dBm and they are set such that they are always operating in the saturated region under the optical launched power. Hence the comb lines reach the saturated power level in the first circulation. The difference between the switched configuration using OSW and the passive FC is that some residual lines at switching speed would appear in the spectrum when it is monitored by a high-resolution spectrum analyser. However, under the spectrum monitored by the OSA of minimum resolution of 0.1 nm these residual lines cannot be seen. The continuous original laser lightwaves are fed into

[2] See reference for 18 GHz optical switches at http://eospace.com/pdf/EOSPACE-custom-optical-switch.pdf.

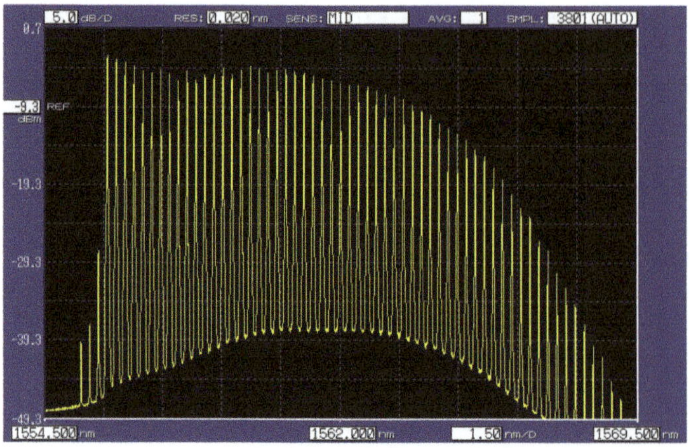

(a) Shift right comb generation

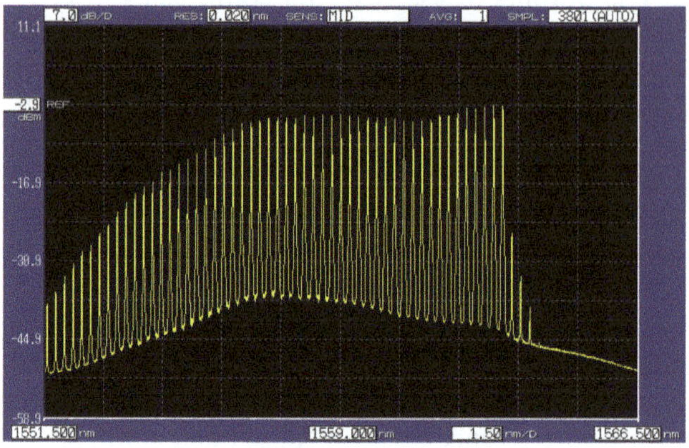

(b) Shift left comb generation

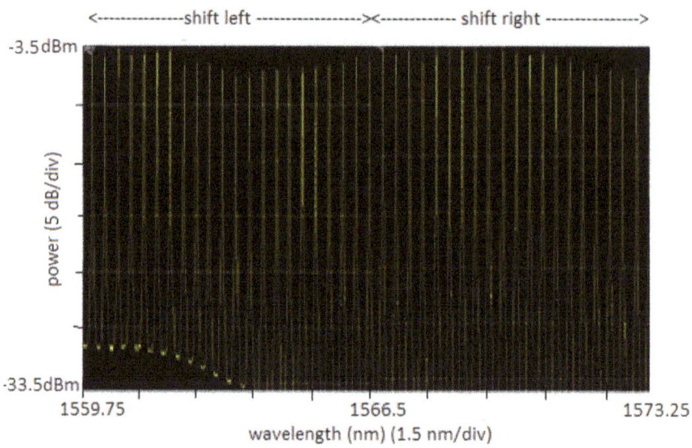

(c) Combined left and right shift spectrum of comb generation, part of the complete spectrum.

Figure 4.7. Observed optical spectra of dual-ring comb generator with double sideband shifting, sub-carriers shown as combined RF frequency shifted left and shifted right sub-carriers, (a) shift right, (b) shifted left, (c) combined right and left comb generation, a section of the spectrum only.

the two rings alternatingly so that the first order shifting of the optical sub-carrier can be generated at all times and thence the higher order shifted lines are generated due to the recirculating and frequency shifting via the ISOM. The two ISOMs are modulated with the $\pi/2$ phase shift in opposite ways, as described in section 4.2. In figure 4.7 the shifting left and shifting right are referred to in frequency terms, so if specified in terms of wavelength then the remarks must be in reverse.

A transmission experiment is yet to be conducted for these dual-band comb sub-carriers.

4.4 Comb generator in multi-Tbps optical transmission system

The comb generator can be employed to generate the referenced laser comb lines which are then modulated with Nyquist pulse shaping implemented in a Fujitsu 56 GSa s^{-1} DAC [22]. The principal purposes of Nyquist shaping are to achieve more effective bandwidth and minimizing the inter-symbol interference (ISI) between adjacent symbols (figure 4.8). The modulated spectra by Nyquist shaping and non-shaping obtained in the experiment, are shown later in this section, figures 4.9(a) and (b). However, it is well known that a direct relationship between a rectangular spectrum and the $\sin x/x$ impulse response indicating that for a sequence of Nyquist pulse sequence, the zero crossing points occur at the sampling point. In practice it is not that simple to generate a Nyquist impulse or pulse sequence due to the nature of the sharp rectangular shape of the spectrum, but their approximately equivalent cosine functions can substitute [23]. Furthermore, the

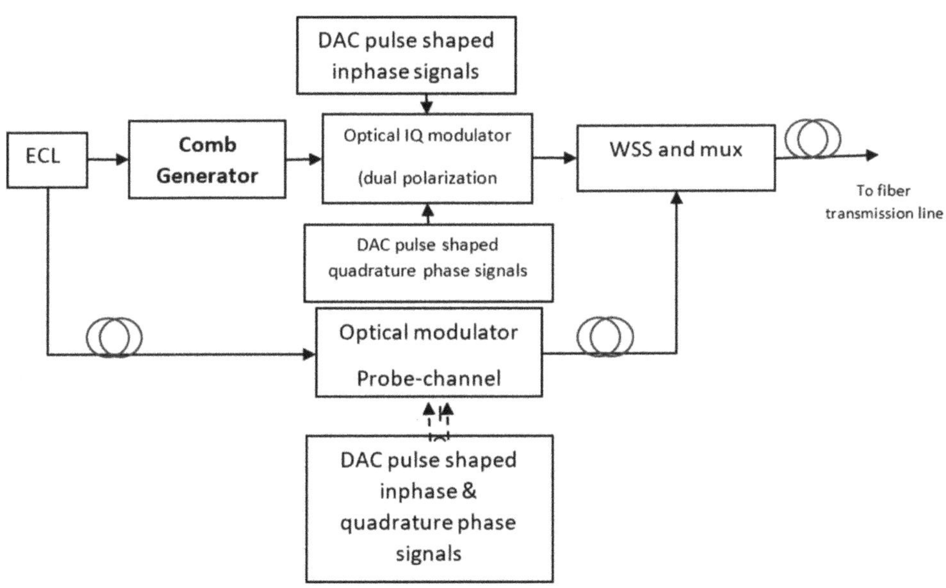

Figure 4.8. Optical transmitter employing comb-generated sources and pulse shaping modulation. DAC = digital-to-analogue converter; ECL = external cavity laser; WSS = wavelength selective switch; Mux = multiplexer; continuous line = optical path; dotted lines = electrical paths.

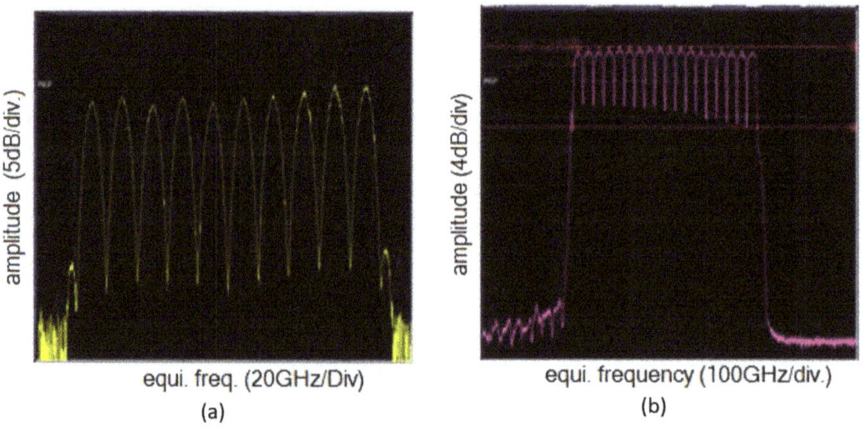

Figure 4.9. Spectrum of QPSK SC modulation with narrow spacing, 18 channels under Nyquist pulse shaping after transmission over 2000 km optically amplified SSMF spans; (a) non-shaping QPSK modulation 10 sub-channels of 28 GS s^{-1} polarization multiplexed QPSK for 1 Tbps SCs, (b) Nyquist shaping SCs, 20 at 28 Gb s^{-1} for 2 Tbps.

bandwidth of the DAC and that of the ISOM are not infinitely wide, so it is more realistic that the raised-cosine pulse shape is employed.

Experimentally, the comb generator described above with 10–20 sub-carriers can be used for creating modulated SCs of aggregate bit rate of 1 and 2 Tbps, respectively, with spacing as closed as possible. We have demonstrated these Tbps SC optical transmitters in transmission systems over a 1500 or 3500 km multi-span optically amplified non-compensating line consisting of cascade spans of SSMFs incorporating optical amplifiers and no dispersion compensating fibres (DCFs) depending on the modulation formats of either QPSK (quadrature phase shift keying) or 16-QAM (quadrature amplitude modulation of 16 points square constellation points). The pulse shaping is essential and implemented as described in the next section, for close packing the channels so that more information can be transmitted over a limited spectral window. These SCs are now attracting lots of interest in flexible grid and bandwidth optical transport networks [24, 25].

4.4.1 Packing modulated comb channels in the frequency domain using Nyquist shaping

4.4.1.1 Analytical representation for pulse shaping

In ultra-high capacity optical transmission systems, the higher the compact frequency spectrum of the signals, the better the spectral efficiency one can achieve. As the time-domain pulse is well related to its spectrum, the shaping of the pulse sequence can be implemented using a high-speed sampling DAC and a DSP. A rectangular spectrum is the ideal case as it is 'brick-wall'-like, so if such an ideal spectrum of the system response is desired then its corresponding equivalent impulse response must take the well-known shape of a sin $c(t)$ function [26]. At the sampling instants $t = kT(k = 1, 2 \ldots . N)$, the amplitudes reach zero. That implies that at the ideal sampling instants, the ISI from neighbouring symbols is negligible, or free of ISI. It is well known that a Nyquist pulse corresponds to a brick-wall-like frequency spectrum; hence one can pack channels whose spectra are close to each other and their zero crossing of the time-domain response passing through one another. Note that the maximum of the next pulse raise is the minimum of the previous impulse of the consecutive Nyquist channel.

Considering one sub-channel carrier 25 GBaud polarization division multiplexing (PDM)-DQPSK signal, hence the aggregate capacity of 100 Gbps per sub-channel with two multiplexed polarizations, hence to reach 1 Tbps, 10 sub-channels would be required. To increase the spectral efficiency, the bandwidth of these 10 sub-channels must be densely packed together. The most likely technique for packing the channel as close as possible in the frequency with minimum ISI is the Nyquist pulse shaping which is described earlier in this section. However, in practice such a 'brick-wall'-like spectrum, cannot be realizable, thus an equivalent and very closely approximated function can be employed, the raised-cosine shape [15]. Mathematically speaking, the raised-cosine filter can be implemented using a low-pass Nyquist filter whose vestigial symmetry property can be the main feature, indicating that its spectrum exhibits odd symmetry about $1/2T_s$, where T_s is the symbol-period. Its frequency-domain representation is a 'brick-wall-like' function. This frequency response is characterized by two values: β, the *roll-off factor*, and T_s, the reciprocal of the symbol rate in Sym s^{-1}, that is $1/2T_s$ is the half bandwidth of the filter. The impulse response of such a filter can be obtained by analytically taking the inverse Fourier transformation of equation (4.1), in terms of the normalized sinc function, as

$$H(f) = \begin{cases} T_s & |f| \leqslant \dfrac{1-\beta}{2T_s} \\ \dfrac{T_s}{2}\left[1 + \cos\left(\dfrac{\pi T_s}{\beta}\left\{|f| - \dfrac{1-\beta}{2T_s}\right\}\right)\right] & \dfrac{1-\beta}{2T_s} < |f| \leqslant \dfrac{1+\beta}{2T_s} \\ 0 & \text{otherwise} \end{cases} \longleftrightarrow h(t) = \mathrm{sinc}\left(\frac{t}{T_s}\right)\frac{\cos\left(\frac{\pi\beta t}{T_s}\right)}{1 - \left(2\frac{\pi\beta t}{T_s}\right)^2} \quad (4.1)$$

with $0 \leqslant \beta \leqslant 1$

where the roll-off factor, β, is a measure of the *excess bandwidth* of the filter, i.e. the bandwidth occupied beyond the Nyquist bandwidth as from the amplitude at $1/2T$. It is also well known that the frequency spectra of raised-cosine pulse with various roll-off factors. In our transmission systems the pulse shaping is conducted with the roll of factor less than 0.3.

For a sequence of symbols, the Nyquist filter offers the property of eliminating ISI, as its impulse response passing through the zero crossing points at all nT (where n is an integer), except when $n = 0$. Therefore, if the transmitted waveform is correctly sampled at the receiver, the original symbol values can be recovered completely. However, in many practical communications systems, a matched filter must be used at the receiver, so as to minimize the effects of noises. For zero ISI, the net response of the product of the transmitting and receiving filters must equate to $H(f)$, thus we can write:

$$H_R(f)H_T(f) = H(f) \tag{4.2}$$

where $H_R(f)$ and $H_T(f)$ are the transfer functions of the receiving and transmitting filters, respectively.

Alternatively, we can rewrite equation (4.2) as

$$|H_R(f)| = |H_T(f)| = \sqrt{|H(f)|} \tag{4.3}$$

The filters which can satisfy the conditions of equation (4.3) are the root-raised-cosine type. The main problem with root-raised-cosine filters is that they occupy larger frequency band than that of the Nyquist sinc pulse sequence. Thus, for the transmission system we can split the overall raised-cosine filter with one root-raised-cosine filter each at both the transmitting and receiving ends, provided the system is linear. This linearity is to be specified accordingly. An optical fibre transmission system can be considered to be linear if the total average power of all channels under simultaneous transmission is under the nonlinear self-phase modulation (SPM) threshold limit [27] which is about 10 dBm for SSMF. When it is over this threshold a weakly linear approximation can be used and some nonlinear distortions would occur. Therefore, the adjustment of the power level of each sub-carrier generated by the comb source is critical to avoid this nonlinear distortion effect.

The design of the Nyquist filter influences the performance of the overall transmission system [28]. Oversampling factor, selection of roll-off factor for different modulation formats, and the design of a finite impulse response (FIR) Nyquist filter and its key parameters are determined to optimize for the case when sub-carriers of the comb generator are employed. A wavelength selective switch (WSS) is used to equalize the power level of all sub-carriers for this purpose (see later in figure 4.11). If considering the transfer functions of the overall transmission channel including fibre, the WSS and the cascade of the transfer functions of all optical-to-electrical (O/E) components, the total channel transfer function is more Gaussian-like. To compensate this effect of the DSP-based transmitter (DSP-Tx), a special Nyquist filter acting as pre-emphasis filtering is cascaded to achieve the overall frequency response which is equivalent to that of the rectangular or raised-cosine with a roll-off factor of 0.9.

4.4.1.2 Implementation

The experimental implementation of the comb generator in a multi-channel optical transmitter for Tbps transmission is shown in figure 4.10. The comb lines can be split in the optical domain and individually modulated and then combined to launch into the optical fibre transmission line. Alternatively, as a matter for proving the principles, the complete set of the comb lines of sub-carriers can be modulated using one optical modulator. One channel of these modulated channels is then filtered by a WSS and then independently replaced by a probe modulated channel with respect to the combed channels. The observed Nyquist pulse shaped spectrum of the modulated comb sub-carriers is shown in figure 4.9(b) for 18 generated sub-carriers. Figures 4.10(a) and (b) show, respectively, the experimentally observed

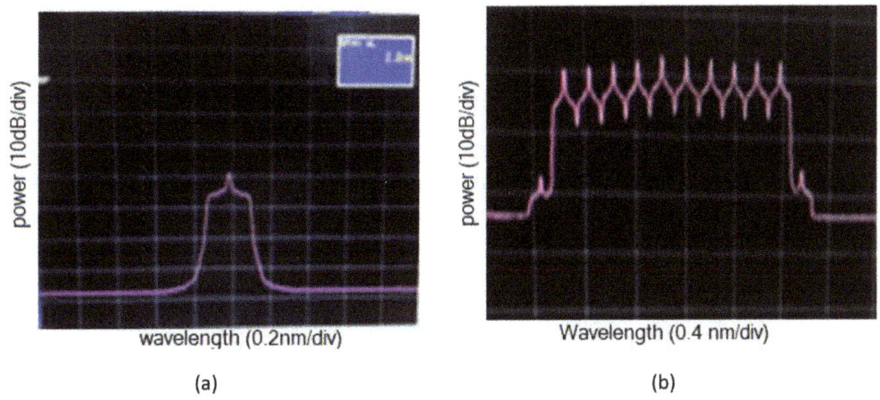

Figure 4.10. Spectrum of (a) extracted spectrum of a modulated channel using Nyquist shaping monitored after the optical filter (see also later in figure 4.11); and (b) packing of 10 comb-generated sub-carriers. Note that at wavelength 1550 nm and neighbouring spectral region 0.8 nm is approximately equal to 100 GHz.

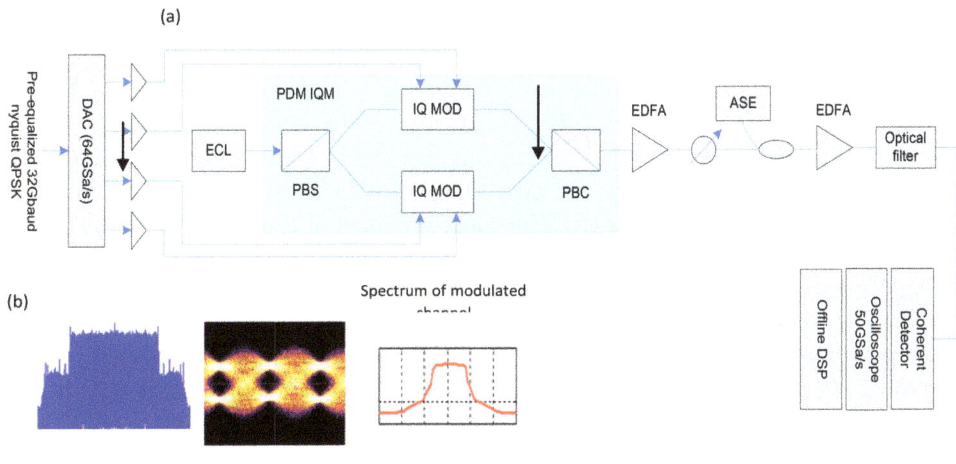

Figure 4.11. (a) Functional block diagram of an integrated 56–64 GSa s^{-1} (courtesy of Fujitsu Co. Ltd). (b) Experimental results as observed on eye diagram, received signal spectrum an modulated signals spectrum of 118 Gb s^{-1} Nyquist PDM-QPSK transmitter and back-to-back (B2B) performance evaluation; (b) ECL = external cavity laser; PBC = polarization beam combiner, PBS = polarization beam splitter; PDM IQM = polarization division multiplexed in-phase-quadrature phase modulator; MOD = modulator; EDFA = Erbium doped fibre amplifier; ASE = amplification stimulated emission noises.

spectrum of a raised-cosine modulated sub-carrier obtained in an optical spectrum analyser (OSA) and ten modulated channels for 1.0 Tbps field-trial transmission. The dual polarization multiplexed IQ modulator used in this transmitter is a Fujitsu type[3] in which two parallel IQ modulators are formed, one each for the transverse electric (TE) and transverse magnetic (TM) polarized modes. Thus, in order to

[3] Fujitsu Semiconductor Europe GmbH, 8-Bit, 55–65 GS/s, 0.9 V/1.8 V Digital-to-Analog Converter FMLY50FEAMEDJ8A0, Issued 2011

provide modulation of both polarizations, two DACs must be employed. The modulation and bias electrodes of the modulators are independent of each other.

4.4.1.3 Pulse shaping using ultra-high sampling rate DAC

This sub-section gives a brief insight into the pulse shaping for packing SCs whose carriers are generated from the SSB comb generators described in section 4.3. The principal sub-system of the SC transmission system is the DSP-based optical transmitter in which the DAC plays the central role in the pulse shaping, channel equalization and pattern generation. A schematic structure of the DAC and functional blocks is shown in figures 4.11(a) and (b), respectively. An external sinusoidal signal is fed into the DAC so that N times multiplying clock source can be generated for sampling at 56–64 GSa s^{-1}. Thus, the jittering and clock accuracy depends on the stability and noise of this synthesizer. A stable low phase noise signal generator Agilent model N9310A is employed to provide the referenced oscillation source for the DAC[3] injected into the port REFCLKP and REFCLKN given in figure 4.11(a). Four DAC sub-modules are integrated in one unit to provide four pairs of eight outputs of $(V_I^+, V_Q^+)(H_I^+, H_Q^+)and(V_I^-, V_Q^-)(H_I^-, H_Q^-)$ which are then employed to modulate the horizontally and vertically polarized optical channels of an IQ optical modulators. Thus with a baud rate of 25 GBaud and 2 bits/symbol with polarization multiplexing, the combined optical channel will carry 100 Gbits s^{-1}. Thence when 20% error coding is used, 25 GBaud becomes 28 GBaud and hence the aggregate bit rates become 112 Gbits s^{-1}. The Fujitsu DAC is specifically designed for this type of IQ modulator, as shown in figure 4.11(a), there are two DACs with output sequences dedicated to positive (P) and negative (N) logics. The functionalities of the other input and output ports are dedicated to memory loading and necessary signals, and then the biasing voltages required for the DAC operations.

The generation of I and Q components of the modulated lightwaves can be implemented as follows: the electrical outputs from the quad DACs are in a pair of positive and negative and complementary with each other. Thus, we would be able to form two sets of four output ports from the DAC development board, that is $(V_I^+, V_Q^+)(H_I^+, H_Q^+)and(V_I^-, V_Q^-)(H_I^-, H_Q^-)$. The positive superscripts indicate the normal pulse sequence while the negative sign superscripts indicate the complementary signals. The subscripts I and Q are used for the in-phase and quadrature phase components. Each output can be independently generated with offline uploading of pattern scripts into the memory of the DAC for execution. The arrangement of the DAC and PDM-IQ optical modulator is depicted in figure 4.11(b). Note that we require two PDM IQ modulators for generation of odd and even optical channels. As Nyquist pulse shaped sequences are required, a number of pressing steps are conducted to ensure the functionalities of the DAC: (i) characterization of the DAC transfer functions to see whether it can handle the signal bandwidth, if not then (ii) pre-equalization in the RF domain by implementing a pre-emphasis on the DAC generated sequence to achieve equalized spectrum in the optical domain, that is at the output of the PDM IQ modulator.

The characterization of the DAC is conducted by launching to the DAC a sinusoidal wave at different frequencies and measuring the waveforms at all eight output ports. As observed in the inserts of figure 4.11, the electrical spectrum of the DAC is quite flat provided that pre-equalization is implemented. The spectrum of the DAC output without equalization is shown in figures 4.12(a) and (b). Their corresponding eye diagrams are shown in figures 4.11(e) and (f). The transfer characteristics are not flat due to the non-flatness of the transfer function of the DAC, as shown in figure 4.11, which is obtained by driving the DAC with sinusoidal waves of different frequencies and recording the output magnitudes. This shows that the DAC acts as a low pass filter with the amplitude of its passband gradually decreasing when the frequency is increased. This effect can originate from the number of samples per period of the sinusoidal wave being reduced when its frequency is increased because the sampling rate can only be set in the range of 56–64 GSa s^{-1}. Thus, a pre-emphasis is required for equalization this DAC response.

The equalized RF spectra are depicted in figures 4.12(c) and (d). The time-domain waveforms corresponding to the RF spectra are shown in figures 4.12(g) and (h), respectively, as observed by a real-time sampling oscilloscope Tektronix DPO 73304A or DSA 720004B. Furthermore, the noise distribution of the DAC indicates that the sideband spectra have been shifted from the noise floor by an amount which accumulates from the circulating paths through the optical amplifier in each turn. It is noted by observing the inset eye diagram of figure 4.11(b) and those given in figures 4.12(g) and (h) that the zero crossings of the pulses are converging to a focal point. So, if one can do the sampling of the received pulse sequence at these zero crossings then the probability of error would be minimal (figure 4.13).

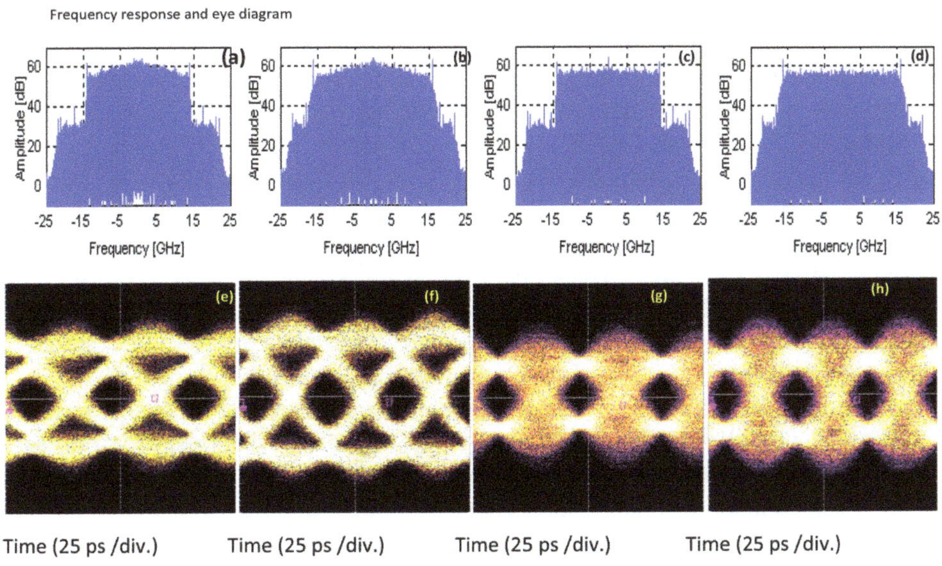

Figure 4.12. Generated spectra (a) and corresponding eye diagrams (e) of 28 GBuad electrical signals using DAC without pre-equalization, (b) and (f) for 32 Gbaud; spectrum (c) and eye diagram (g) of 28 Gbuad electrical signals after DAC with pre-equalization, (d) and (h) for 32 Gbaud. All observed using input to the electrical port of a sampling oscilloscope.

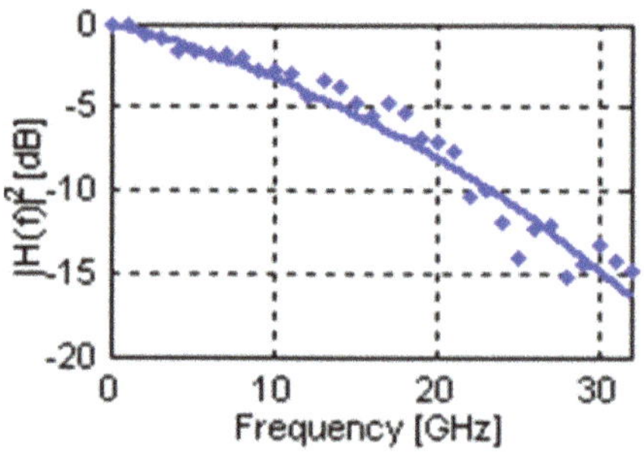

Figure 4.13. Frequency transfer characteristics of the DAC. Note the near linear variation of the magnitude as a function of the frequency.

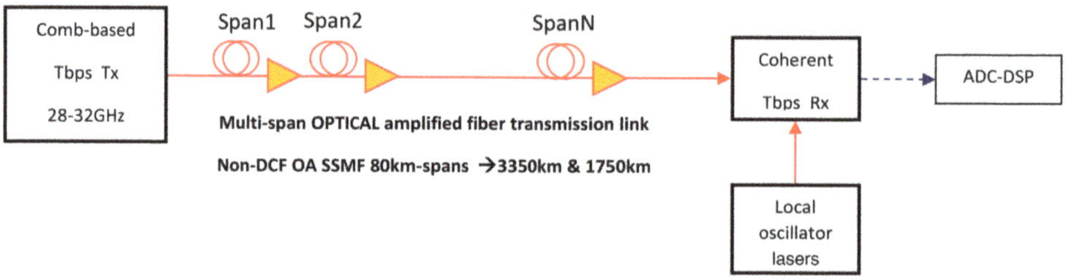

Figure 4.14. Optical transmission systems employing comb-generated sources and coherent reception techniques in association with digital signal processing. DAC = digital-to-analogue converter; ECL = external cavity laser; Tx = transmitter. Dotted line = electrical line; continuous line = optical path.

4.4.2 Transmission of super-channels formed by modulated MCL

The optical transmitter employing the comb generator is shown in figure 4.14. The combed sub-carriers are generated with a spacing varying from 28 to 32 GHz to accommodate 28–32 Gbps QPSK modulation formats, respectively. The receiver is a coherent type, thus a local oscillator (LO) of the same frequency of the probe channel sub-carrier, must be employed to mix with the incoming probe channel to bring the data sequence back to the baseband. The roll-off factor of the raised-cosine filter for all sub-channels is set at 0.9. The optical spectrum is very close to the raised-cosine shape. Due to the resolution of the OSA the exact roll-off factor cannot be determined accurately but only to a close approximate. Figure 4.15 shows four sections of the DSP-based coherent receiver. Firstly, the SCs are fed into a tuneable optical filter through which a particular channel is selected for the measurement of its transmission performance. The filtered channel is then going through a 90°optical hybrid coupler in which the two polarized channels are separated. Furthermore, via the use of a $\pi/2$ optical phase tuning both the in-phase and quadrature-phase components are also extracted and then coherently detected by mixing with an LO laser.

At the coherent receiver, a comb generator can be employed to mix with all sub-channels. However, due to the limited bandwidth of the optical detector where the mixing and beating between the LO and the channel may occur, our current receiver is demonstrated by demultiplexing all sub-channels and one by one, the probe

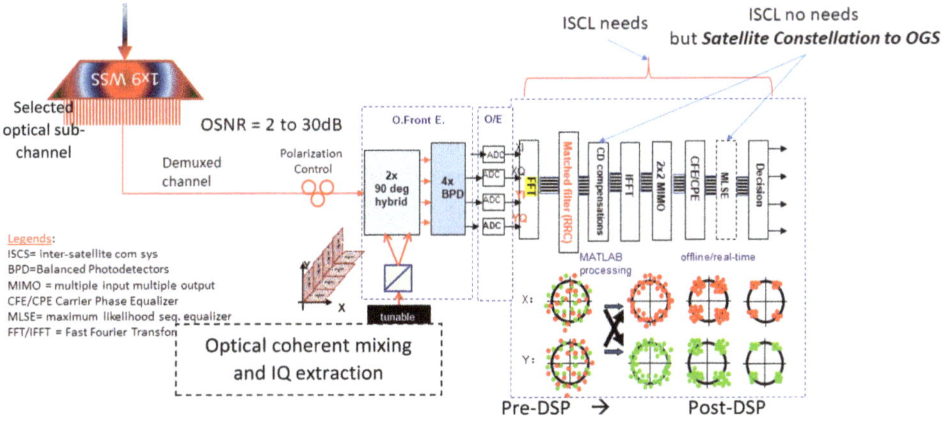

Figure 4.15. Generic schematic of the coherent receiver. ADC = analogue-to-digital convertor; MIMO = multi-input multi-output; IFFT = inverse fast Fourier transformer; CFE = carrier phase estimation; MLSE = maximum likelihood sequence estimator; CD = chromatic dispersion; BPD = photodetector pairs and electronic pre-amplifier; WSS = wavelength selective switch.

channels are recovered in the electronic domain for processing. The outputs of the hybrid coupler are detected by a pair of PDs connected B2B to maximize the openings of the eye diagram of the sequence. These electronic signals are then pre-amplified by a TIA balanced receiver. In our coherent receiver the LO is tuned to the frequency of one of the sub-carriers for which homodyne reception can be achieved. However, in practice there is a mismatching in frequency of the sub-carrier of the modulated channel and that of the local laser source. This mismatching can be compensated in the digital domain using an appropriate algorithm, for example the constant modulus amplitude type [30]. Thus, this type of coherent reception is commonly termed as serrodyne detection.

The electronic signals at the outputs of the balanced receivers are then sampled by a 64 GSa s^{-1} analogue- to-digital converter (ADC) to give the received sequences in digital domain for further processing by the DSP. The ADCs sampling rate can be performed at 56–64 GSa s^{-1} for 28–32 GBaud, respectively. Thus, we obtain two samples per symbol/baud period that satisfy the Nyquist sampling criteria. These digital data are then processed offline by a number of DSPs in which the clock information is recovered and the bit-error-rate (BER) can be determined. The third block in the far right of figure 4.15 depicts a schematic of the DSP section in which individual blocks indicate the main principles of processing. First the received samples are transformed into the frequency domain by the fast Fourier transform (FFT) block, the compensation of the distortion due to chromatic dispersion is performed and then the carrier phase recovery. The samples are then processed in the discrete time domain using certain algorithms such as the maximum likelihood sequence estimation (MSLE) or decision feedback estimation (DFE), to recover the phase constellation as shown by the inset in the post-DSP section of figure 4.15. Finally, the BER can be obtained. The cross lines between the X (horizontal polarized channel) and Y (vertical polarized channel) indicate mutual cross-processing between these two polarized channels to maximizing the recovery of the transmitted pulse sequence.

The equalization in the power amplitude of the SCs is implemented using a WSS. Individual comb modulated channels are coherently detected via the use of a $\pi/2$ polarized hybrid coupler and then digital processing of lines with an achieved BER in the region of 1e−3 which is acceptable for forward-error-coded transmission systems. The power sensitivity is also obtained with launched power in the normal acceptable range so that no significant nonlinear impairments occur. The transmission has also been used for field trials in Deutsch Telecom and Vodafone installed fibre transmission lines located in Germany, as shown in figure 4.16. The optical transmission path of the two field-trials using the optical transmission systems employing SCs generated by modulating the sub-carriers of the comb generator are: (a) Path 1: from Darmstadt to Nurnberg with additional several hundred kilometres of SSMFs inserted in the laboratories at Stuttgart and Nurnberg. (b) Path 2: from Esborn to Regensburg of Germany with additional kms of fibres inserted in different locations in the link. The number of sub-carriers generated from the comb generator employed in these field trials is 10 and 20 sub-carriers, thus 1.0 Tbps and 2.0 Tbps SCs can be transmitted, respectively. The probing and processing of individual channels and processing of the received signals show that BER of 1e−3 or better can be achieved without any difficulty. A typical plot of the BER versus the OSNR is depicted in figure 4.17. It is thus proven that the comb generator is very useful in multi-Tbps optical transmission systems. In the following we describe in more detail the transmission systems.

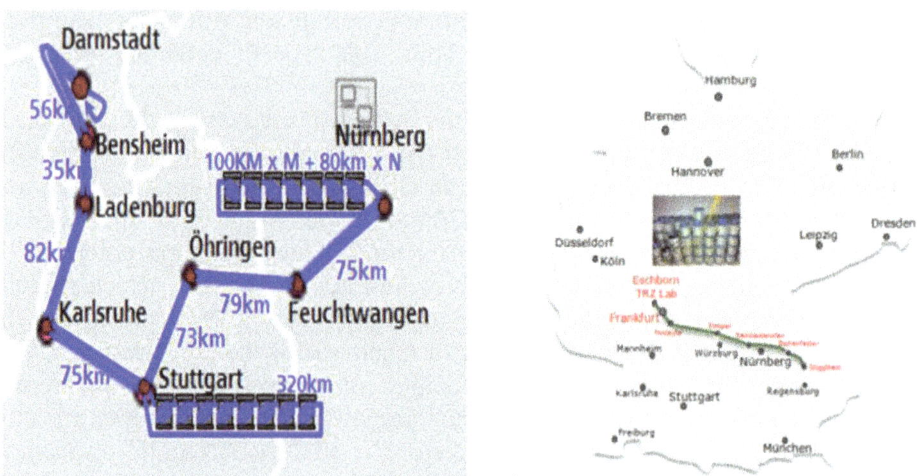

Figure 4.16. Field trial optical link path for comb-incorporated optical transmitter and coherent reception systems. (a) Darmstadt to Nurnberg with an additional several hundred kms in the laboratory inserted in Stuttgart and Nurnberg in Germany. (b) From Esborn to Regensburg in Germany with additional kms of fibres inserted in different locations as shown along the link.

It is noted that the optical transmission lines consist of standard single-mode optical fibres. No dispersion compensation of the fibre chromatic dispersion in the optical domain is used. All compensations or equalization of distortions are implemented in the digital domain, as shown in figure 4.15. Each of the spans of the fibre transmission line is established at 80 km, so the loss is about 19 dB over one whole span and the gain of the optical amplifiers is set at this level. The optical amplifier noise figure is about 5 dB. The field transmission distance was demonstrated at 3350 km and 1730 km for 100G and 200G bit rate per sub-channel, respectively. For 200G the modulation scheme was 16-QAM (4 bits/symbol).

To demonstrate that the comb line sub-carriers can be stable and employable in tera-bit s^{-1} optical transmission systems, we have measured the BER versus the OSNR of a probe channel, as shown in figure 4.17. The measured BERs for individual in-phase and quadrature phase components are also included in this figure. In order to save the cost of the field trial, we employ one IQ modulator to modulate all sub-carriers at the same time, then one selected channel is optically filtered out and a probe channel is injected. The source used in the probe channel is also derived from a sub-carrier of the comb generator. The total number of channels is 20 with 100 Gb s^{-1} per channel which is modulated by QPSK modulation format and two multiplexed polarizations with a baud rate of 28 GBauds. The noises are injected into the receiving sub-system via an optical coupler not shown in the diagram of figure 4.14 but shown clearly figure 4.11 with the unit ASE. The triangular points are measured under no error coding and the circles are obtained when error coding is employed. We observe no degradation of these performances for all sub-carriers of the generated comb. The linewidth of these comb sub-carriers was also measured and they match with that of the original external cavity laser (ECL). The phases of these comb sub-carrier lines are indirectly interpreted in the electronic digital domain when processing the modulated optical signals of each of the modulated channels shown in figure 4.10(b). The OSNR of 16 dB is about 3 dB

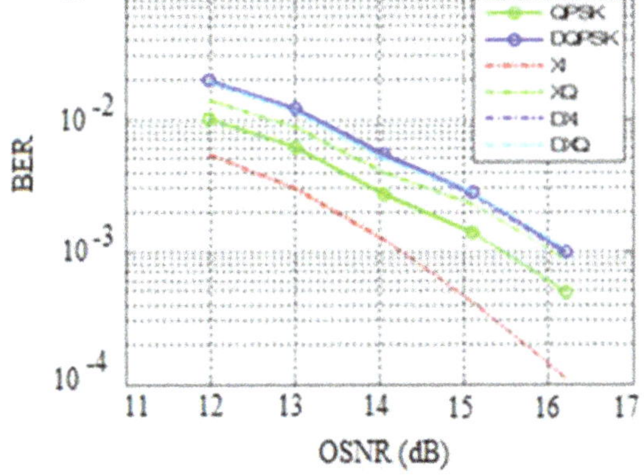

Figure 4.17. 2 Tbps channel measured performance by the variation of the BER versus OSNR, modulation scheme QPSK, 20 channels comb generator and 100 Gb s^{-1} per channel.

higher than the level required for single channel. This is expected as there may be some crosstalk from the left and right adjacent channels of the probe channel.

4.4.3 Remarks

This section has presented the generation of comb source whose spectra can be controlled by modulating the amplitude of the lightwave circulating in a recirculating fibre loop. Synchronization of the optical fields of the comb lines is achieved via the use of ISOM such that constructive interference of the lightwaves is achieved. Shift left, shift right or dual-sideband shift have been proposed and experimentally demonstrated. Only SSB-shift comb sub-carriers have been employed in the optical transmitters to generate the optical SCs for Tbps transmission systems. The SCs have been experimentally demonstrated over installed fibre transmission lines operating at aggregate bit rates of 1.0 Tbps and 2.0 Tbps. By modulating these comb lines and transmitted over long non-dispersion compensating optical fibre transmission lines, we have achieved acceptable error rates. A BER of 10^{-3} is considered to be error free when error coding is used. Thus, the linewidth of the comb generator can be considered to remain the same as that of the original external cavity laser. Currently we are conducting extensive characterization of the sub-carriers in terms of linewidth and tenability of their spectral positions. Our current works are expecting to employ these comb generators with enhanced bit rate at 400 Gb s^{-1} so that 10 or 20 × 400G can also be demonstrated by using the high-level modulation format e.g. 32-QAM or M-ary-QAM instead of QPSK so that 10 Tbps can be generated and transmitted and received with acceptable BER. Furthermore, we have reported a dual-band frequency shifting comb generator employing a dual recirculating frequency-shift ring in which two ISOMs are incorporated and inline optical amplifiers operating in saturation mode. However, no modulation and transmission of these sub-carriers are yet done. It is planned to demonstrate optical transmission systems employing these ultra-wide spectrum sub-carrier SCs in our future works. An optical phase locking may be needed for this application so that all local oscillation sub-carriers can be simultaneously locked onto a referenced phase as that of the transmitted SC. Finally, the reported comb generator given in this chapter can be potentially used as a multi-carrier LO for mixing with modulated optical sub-carriers in multi-channel coherent reception sub-systems for Tbps optical transmission systems.

4.5 Multi Tera-bits s^{-1} optical access transport technology

Tremendous efforts have been developed for multi-Tbps over ultra-long distance and metro and access optical networks. With the exponential increasing demand on data transmission, storage and serving, especially the 5G wireless access scenarios, optical internet networking has evolved for DC-based optical networks pressuring on novel and economical access transmission systems. This section describes: (1) experimental platforms and transmission techniques employing band-limited optical components operating at 10G for 100G based at 28G baud. Advanced modulation formats such as PAM4, DMT, Duo-binary etc are reported and their advantages and disadvantages are analyzed so as to achieve multi-Tbps optical transmission systems for access inter- and intra-DC-based networks; (2) integrated multi-Tbps combining comb laser sources and micro-ring modulators meeting the required performance for access systems are reported. Ten sub-carrier quantum dot com lasers are employed in association with wideband optical intensity modulators to demonstrate the feasibility of such sources and integrated micro-ring modulators acting as a combined function of demultiplexing/multiplexing and modulation, hence compactness and economy of scale. Under the use of multi-level modulation and direct detection at 56 GBd an aggregate of higher than 2 Tbps and even 3 Tbps can be achieved by interleaved two comb lasers of 16 sub-carrier lines; and finally (3) the fundamental designs of ultra-compact flexible filters and switching integrated components based on Si photonics for multi-Tbps active interconnection are presented. Experimental results on multi-channel transmissions and performance of optical switching matrices and effects on that of data channels are proposed.

4.5.1 Introductory remarks

The emergence of DC-centric networks have exerted tremendous pressure on traditional optical networks to be flattened and simplified as well as increasing the network capacity and the contents to be delivered to end users. The capacity of the access networks would reach several Tera-bits s^{-1} when 4K video for high definition and ultra-fast (faster than fast) delivery over fixed lines or wireless mobile 5G networks expected to be coming toward the end of this decade. The evolutionary network is schematically shown in figure 4.18 in which several Peta-b s^{-1} capacities would be transported in the long-haul core and metropolitan core networks via several clouds at the edge as well as at local distributed nodes. The exchanges of

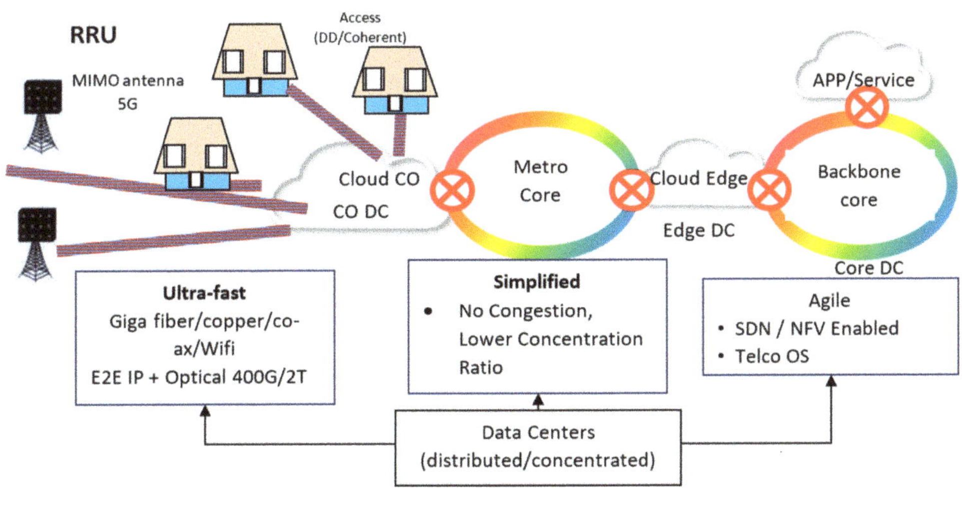

Figure 4.18. Evolutionary network modernization strategy. Cloud CO = central office cloud, DC = data center. 100G access transmission systems.

traditional telecom networks are expected to locate access clouds in their transformed flattened networks.

It is thus expected that multi-Tbps would be required in the new evolutionary networks to be deployed in the near future. The economy scale as well as the flexibility of such access transport networks or systems is a very important factor so that bandwidth on demand can be offered to end users. This section thus addresses the following: firstly, the transmission and direct detection of four optical lanes of different and packed wavelength channels each carrying 25–28 Gbps depending whether forward error coding (FEC) is required. Furthermore, several of such 100G modules would result in accumulated capacity reaching Tbps. Secondly, a baud rate of 28 GBd can be employed to produce 56 Gbps via the modulation format PAM4 producing 56 Gbps per channel. Then lastly the employment of comb sources and special modulators such as micro-ring types, to achieve multi-Tbps is given in both coherent and non-coherent reception. Lastly, we propose some integrated optical switching system to simplify the routing of ultra-fast optical channels to end user locations.

4.5.2 100G access systems

A typical access transmission system is shown in figure 4.19 [31], in which a low cost vertical-cavity surface-emitting laser (VCSEL) with limited bandwidth of 18 GHz is used under a baud rate of 56 GB and the data sequence is fed to directly modulate the VCSEL. Shown also in this configuration is the T-bias and the PAM4 data sequence obtained SHF-10 001 attenuator 6 dB and a high-speed combiner. The PAM4 eye diagram in the electrical domain is shown as the far left oscilloscope image in figure 4.19 followed by the optical output eye in back-to-back (B2B), transmitted through a length of SSMF, before feeding into the processor so as to evaluate the BER after passing through the minimum least square estimation algorithm. The performance of such a system is shown in figure 4.20. Under an

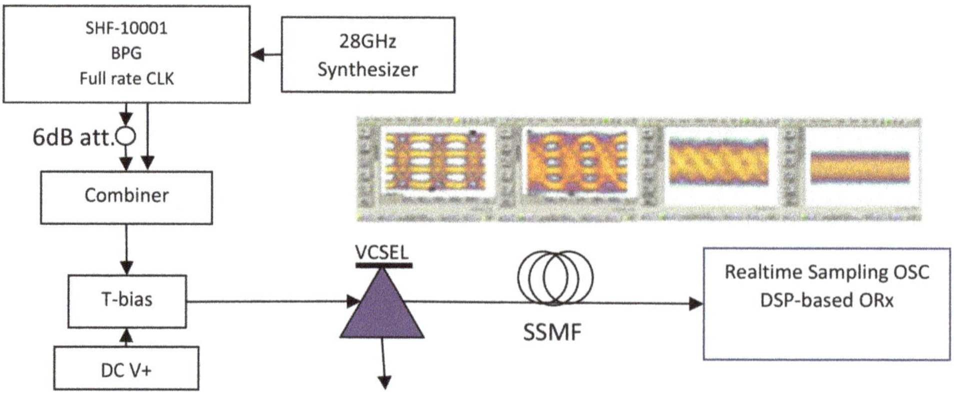

Figure 4.19. Typical access transmission system employing PAM4 Direct detection and associated DSP. Labels: SSMF = standard single-mode fibre; VCSEL = vertical-cavity stimulated emission laser; DSP = digital signal processor; att = attenuator; CLK = clock.

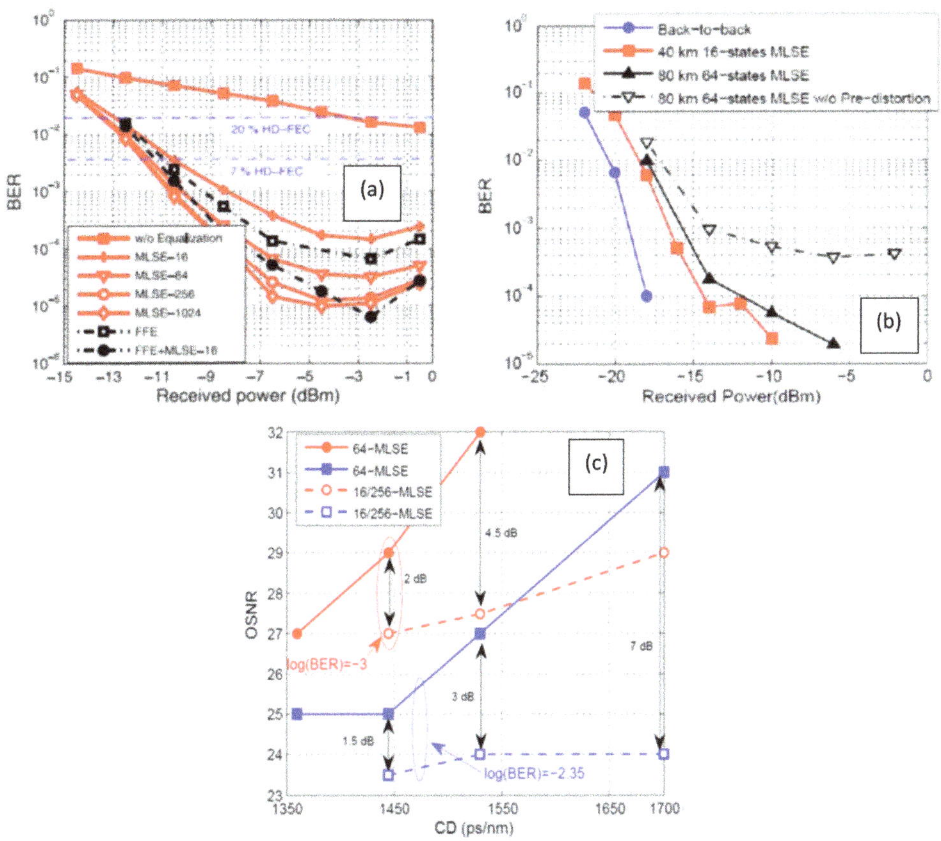

Figure 4.20. Performance of low-cost transmission system using band-limited optical components for metro access. (a) BER versus received optical power for PAM4 28 GBd; (b) BER versus received optical power for NRZ 28G using 10G components; and (c) OSNR versus residual chromatic dispersion in SSMF for NRZ 28G.

FEC of 20% and 7% we can expect that such a 56G PAM4 can be transmitted over a few to some tens of kilometers of SSMF.

Alternatively, another low-cost solution for access networks can be achieved by using components developed for 10G and operating at 28G with DSP processing and pre-emphasis at the transmitter so as to produce systems of 100 Gbps via four optical wavelength transmitters employing EML (external modulation laser) in which four lasers are integrated with electro-absorption modulators. The transmission performance for such optical transmission is shown in figures 4.20(b) and (c). It is believed that the footprint of such transponders employing 56 Gbps (PAM4) or 4 × 28 Gbps (NRZ) are the least costly to deploy into the access networks for DC-centric networks or flattened traditional telecom networks.

4.5.3 TBPS access transmission and routing technology

It is well known that the limitation of the electronic components is around 56 Gbps with bandwidth to around 35 GHz, especially when digital signals are generated such as in high speed DAC and ADC. Noise contribution in these signal generators must also be considered as it grows as a cubic function of the frequency. This noise can then reduce the allowable OSNR, and thus the amplitude levels when multi-level modulation format is employed. Multi-level level modulation is required in order to increase the bit rate or the number of bits per symbol so that 100G or 400G can be obtained. In this section the 112 Gbps via 56 GBd PAM4 modulation format signals is described. Furthermore, in order to achieve the Tbps SCs required, parallel channels are simultaneously generated and transmitting. We propose here the use of comb laser sources and equivalent modulators and multiplexers via the device known as the micro-ring modulator (MRM).

4.5.3.1 Multi-carrier comb source direct detection systems

The MRM can act as an optical filter as well as a modulation device. In comb generation one can generate multiple sub-carriers from one original source in the frequency domain or in the time domain with a very short pulse repetitive sequence followed by a periodic Fabry–Perot filter whose free spectral range meets the required frequency spacing in a DWDM grid spectrum. In the case of flex grid networks their periodic spaced grid can be occupied differently in a multiple equally spaced wavelength grid. Figure 4.22 shows the paths to generation of comb lines in these two domains. Furthermore, there are a number of techniques for generating comb lines in the frequency domain via different optical processing structures including: (i) frequency shifting recirculating optical loop with SSB shift of the guided lightwaves in the circulating cavity; (ii) phase modulation and carrier suppression return zero for shifting and then generating; (iii) modulation integrated in a laser cavity; and (iv) short pulse generation and then super-continuum via crystal waveguide then filtering of determined frequency spacing by a Fabry–Perot and then optical amplification to the desire level of the comb lines. Figure 4.23 shows the spectra of comb sources generated by frequency shift recirculating loop and combined phase and amplitude modulation in the nonlinear region. Furthermore, the modulated spectrum of the comb lines of figure 4.23(a) is given in (c). If the five-sub-carrier comb source is modulated to produce 400 Gb s^{-1} per line then the aggregate bit rate would be 2.0 Tbps. We have demonstrated that by shifting left and shifting right the comb lines we can cover the whole C-band with frequency shifting of 50 GHz, which is one frequency shift applied by the frequency synthesizer to the optical modulator incorporated inside the recirculating loop, as shown in figure 4.21.

In access networks the 1300 nm can be employed in order to avoid the dispersion effects of the SSMF. One such comb laser in the 1310 nm is shown in figure 4.24(a). The comb laser is fabricated using quantum dot structures in InGaAs with flat broadband mirror at the two ends of the laser cavity[4]. A comb line can be filtered, as shown in figure 4.24(b). Such a comb laser in the 550 nm region can be generated by an embedded modulator in a quantum dot laser cavity, as reported in [32, 33]. These

[4] Innolume GmbH, Dortmund Germany http://innolume.de/_pdfs/Comb/LD-1310-COMB-8.pdf.

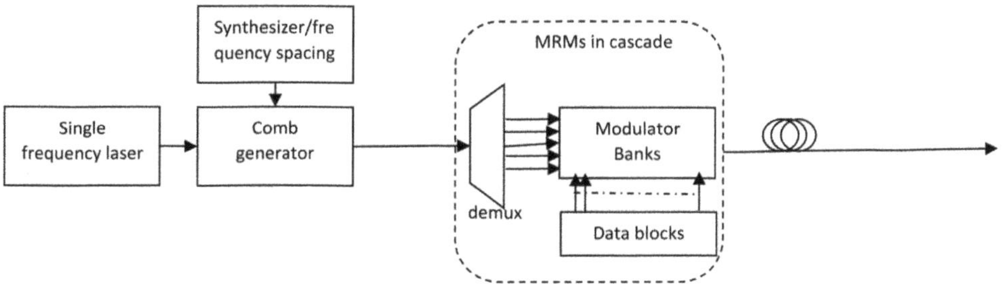

Figure 4.21. Schematic of multi-Tbps optical transmitter.

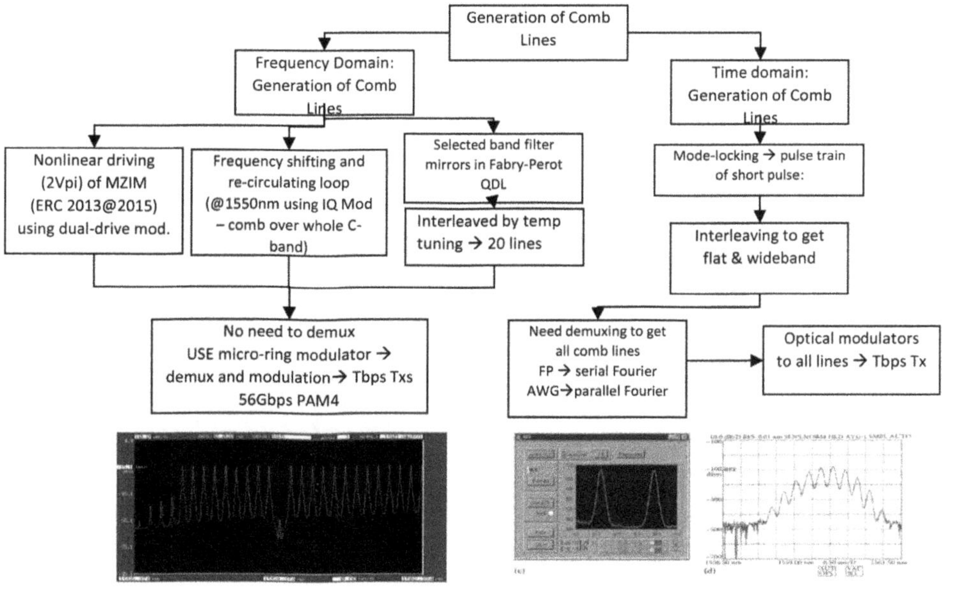

Figure 4.22. Methods of generation of comb sources in the frequency and time domain.

comb lasers would offer the generation of modulated channels reaching total capacity of 9.4 Tbps over the entire C-band with 100 Gb s^{-1} per channel and 94 sub-carriers of 50 GHz spacing. This band has now been extended to 94 wavelength carriers as optical amplifiers have been available and optimized over this C-band.

Employing the 1310 nm comb laser produced by Innolume GmbH and modulating using an external modulator with a 3 dB bandwidth of 35 GHz with the modulation format PAM4 we obtain 112 Gb s^{-1} per carrier, so the total aggregate capacity is 13 × 112 Gb s^{-1} or 1.45 Tbps can be produced by this comb laser. The 1310 nm region ensures that the transmission is attenuation limited and not dispersion limited. Fsigure 4.25 shows the spectra of the modulated comb line under matching of the filter centre wavelength with that of the carrier comb line. The PAM4 eye diagram under B2B are shown in figure 4.26. The BER is estimated at 1e−5. After 2 km transmission this is about 1e−3 under equalization using MLSE.

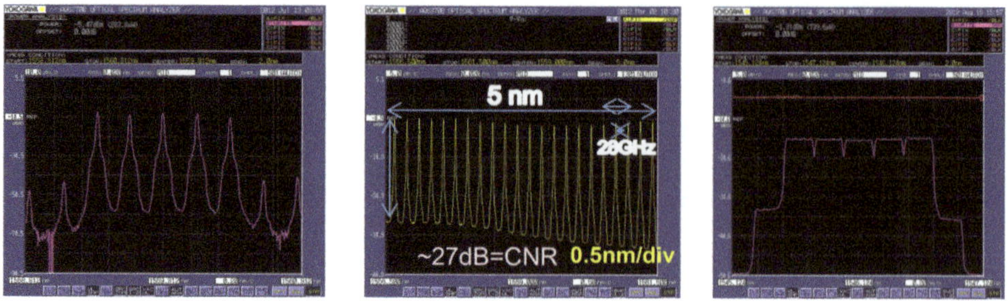

(a) Comb spectrum generated by phase and amplitude modulation.

(b) Comb spectrum by SSB frequency shifting recirculating loop.

(c) Modulated spectrum of (a).

Figure 4.23. Comb spectrum generated by (a) frequency shift recirculating loop; (b) phase and amplitude modulation in nonlinear overdrive region and (c) modulated spectrum of comb laser of (a).

(a)

(b)

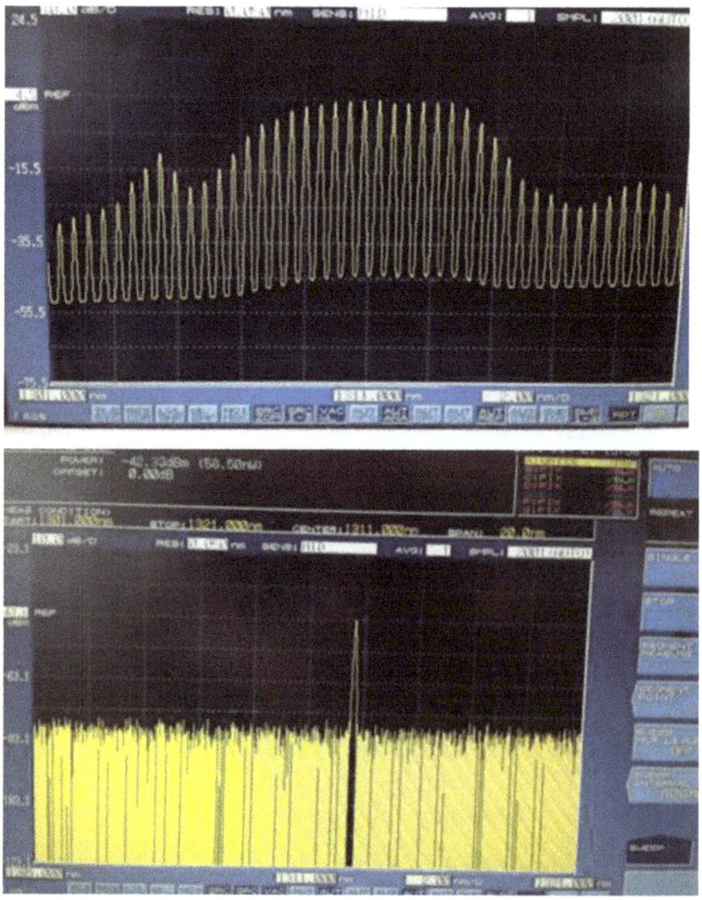

Figure 4.24. Spectrum of comb laser at 1310 nm region (a) all comb lines and (b) a filtered comb line.

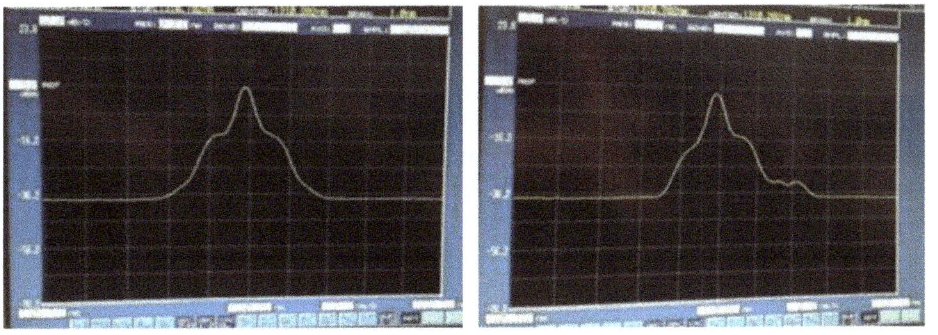

(a) (b)

Figure 4.25. Modulated spectrum of PAM4 56 GBd (a) filtered at center of spectrum and (b) misaligned filtered modulated spectrum at PAM6 56 GBd.

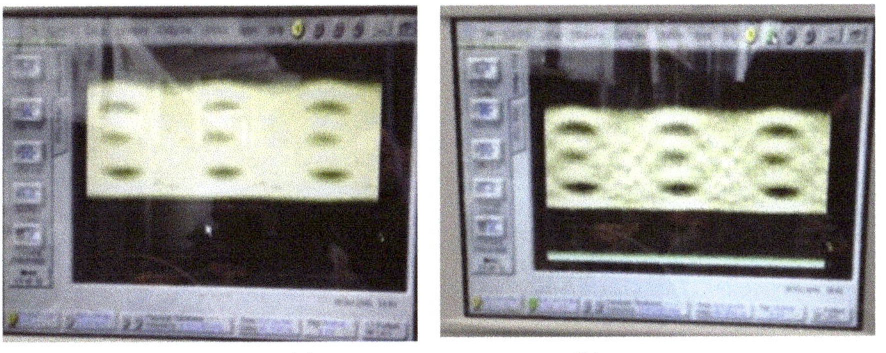

(a) (b)

Figure 4.26. Eye diagram at output of transmitter PAM4 56 GBd of a filtered channel of the comb modulated channels.

The BER is similar for all 13 comb lines indicating that the chromatic dispersion is not critical for this Tbps transmission system. Thus, they are most suitable, with the employment of FEC, for low cost Tbps access or DC environments.

4.5.3.2 Tbps coherent reception systems

Coherent transmission technology using polarization multiplexing (PDM) and M-ary quadrature amplitude modulation (M-QAM) has been deployed throughout the long haul and metropolitan core networks. The flattening of the telecom networks under multi-Tbps have put tremendous pressure on network carriers to transform several fibre rings topologies into mesh topology in the metro-core networks and close to the access users as much as possible. Although it is more complicated in terms of additional optical hybrid coupler and LOs in the coherent reception sub-systems but it offers much better operating margins and higher capacity of at least doubling by the PDM. It is also more expensive but this is expected to be lowered with the integrated Si photonic technology under current intense development and research [34–37]. Indeed, a number of significant integrated photonics based on Si CMOS technology are now available commercially.

This section provides an overview of our demonstration of a multi-Tbps transmission system based on coherent transmission and reception techniques. A general multi-Tbps optical experimental platform is shown in figure 4.27. The comb laser source is generated by using the frequency-shift circulating technique described above. The comb source is either demultiplexed into individual comb lines which are then fed into a pulse shape Nyquist transmitter through which the Nyquist sampling rate and roll-off is enforced on the optical modulator so that the modulated spectrum can reach close to the sharp-roll-off. In this way the optical channels can be packed close together to minimize the crosstalk as well as be efficient in the spectrum, hence providing higher capacity for all channels transmitted over the C-band. An arrangement of the PDM QAM transmitter is shown in figure 4.28. The spectrum of SC 2 Tbps is shown in figure 4.29. Alternatively, the mirroring

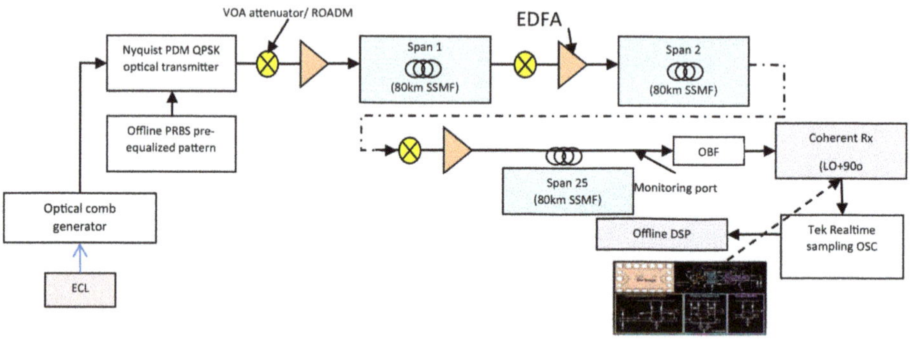

Figure 4.27. Schematic structure of the Nyquist PDM-QAM optically amplified multi-span transmission line. Inset is the advanced differential trans-impedance (TIA) optoelectronic receiver by CMOS 90 nm technology.

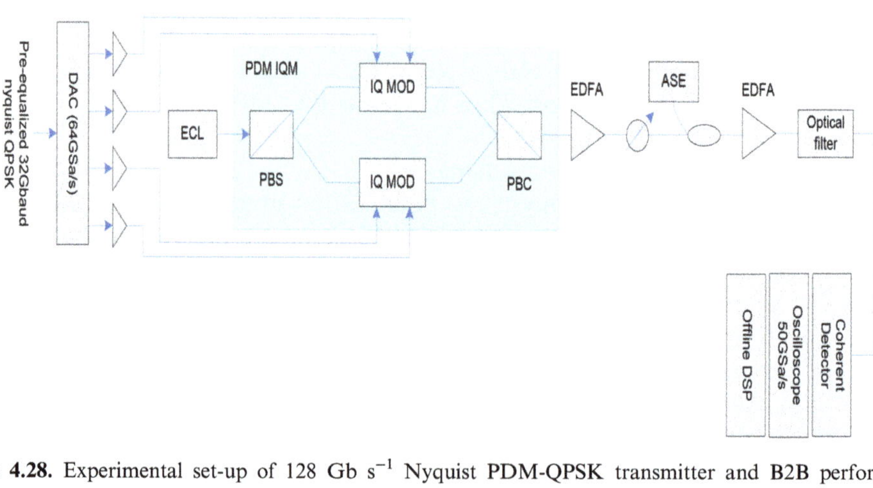

Figure 4.28. Experimental set-up of 128 Gb s^{-1} Nyquist PDM-QPSK transmitter and B2B performance evaluation (see also figure 4.14).

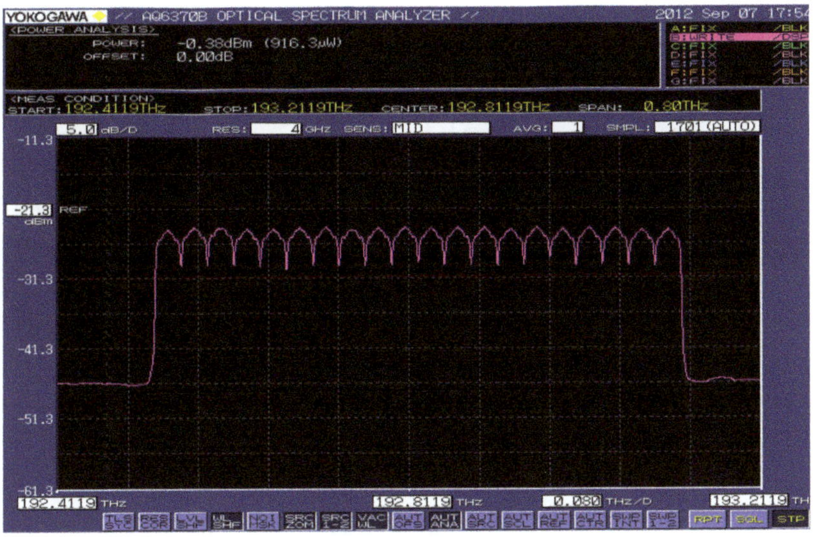

Figure 4.29. Spectrum of 2.0 Tbps optical channels using nonlinear comb generator with two probe channels and 18 dummy channels.

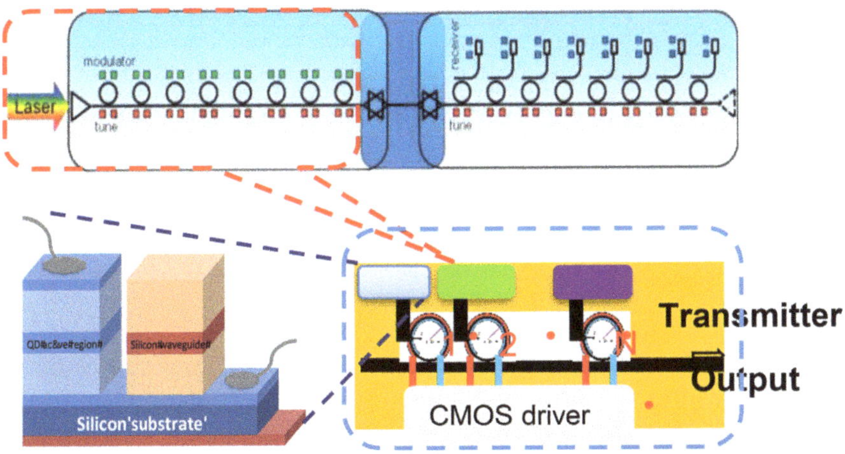

Figure 4.30. Tbps transmitter employing MRM as simultaneous mux/demux and modulation device.

modulator in Si integrated photonic platform, can be use as the Tbps transmitter, as shown in figure 4.30. The transmission performance of such a channel over 3500 km is shown in figure 4.29 and its transmission performance over 1600 km of optically amplified SSMF spans in figure 4.31(c). A typical spectrum and eye diagram of the Nyquist channel are shown in figure 4.31(a) and (b). This length of transmission the Tbps over this length allows the shorter transmission with a number of reconfigured add/drop mux (ROADMs) expected for metro networks, especially add/drop into access networks. It is noted that the coherent reception sub-system consists of an

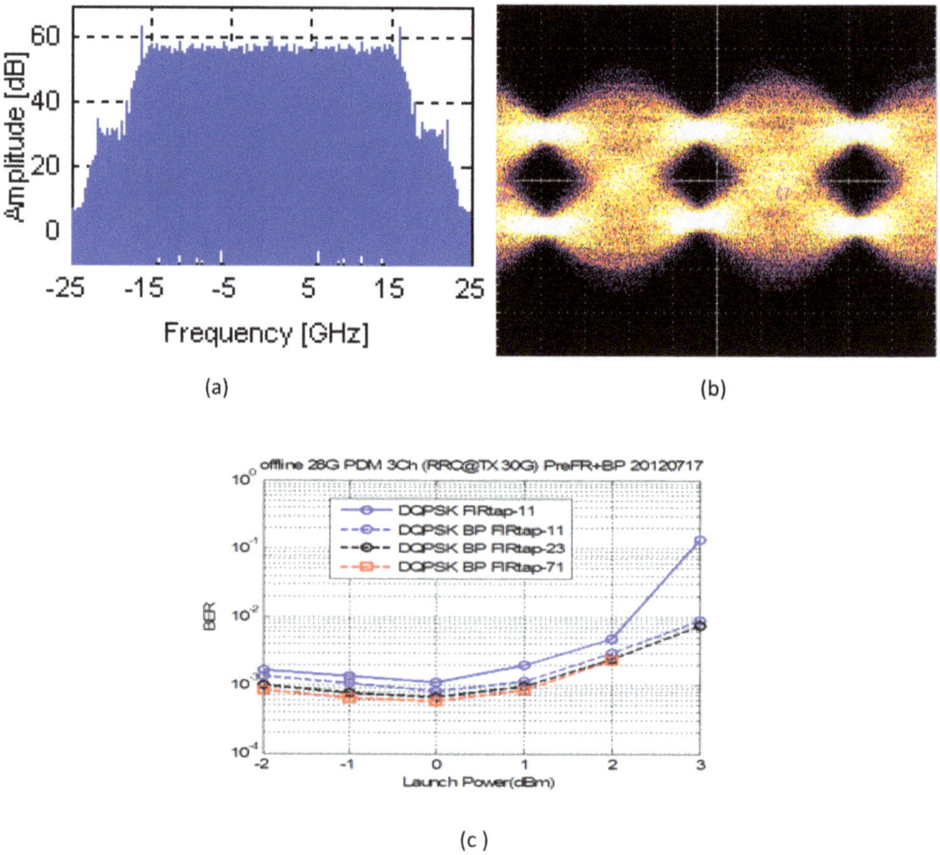

Figure 4.31. (a) Typical received spectrum (b) Nyquist eye diagram; and (c) BER versus launched power (dBm) of 20 channels with central channel as probe channel over transmission distance of 1600 km non-DCF, 20 optically amplified 80 km SSMF spans with VOA inserted in front of EDFA. 28 GBaud PDM-Nyquist QPSK (a) single channel (b) three channels.

optical hybrid coupler through which the mixing of a comb laser acts as the multiple line LO beating with all received channels in the PD after the optical demux. The balanced PDs are connected to the differential TIA so as to obtain the highest output voltage level signals which are compatible for feeding into the ADC and then to the sampling oscilloscope and digitally processed. The BER is sufficient for further FEC to achieve error free results. Such Tbps SCs can be integrated to transmit over multi-core fibres to give a total aggregate capacity reaching a few peta-bps.

4.5.4 Optical interconnect for multiple Tbps access networks

Optically added drop mux for access networks must be available from cloud to end users in the multi-Gbps per user or even several Tbps to MIMO antenna for the future 5G wireless access networks. We assume here that the multi-TBps access networks accept channels of at least 100 Gbps, to 400 Gbps and the SC 1 Tera, 2 T

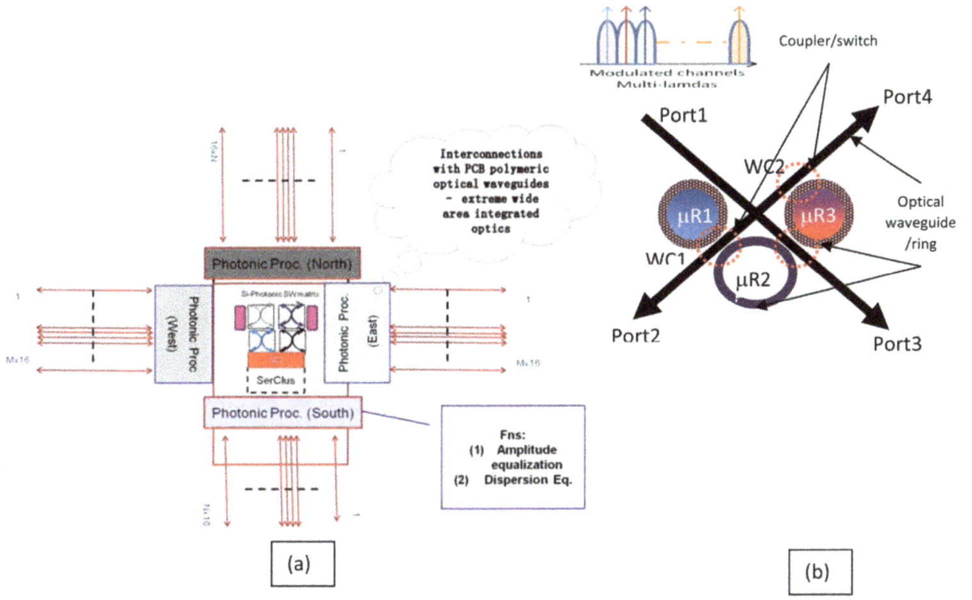

Figure 4.32. An optical interconnect consisting of an optical kernel and optical processor in north–south–east west with fibres carrying 25 T to 100 Tbps all in the C-band with 92 wavelength channels or SCs occupying multiple numbers of 50G grid spacing.

to 10 Tbps. Thus, the bandwidth of channels can be variable, hence the flexi grid WDM. This type of network will be installed in the next coming modernization of optical networking. Indeed, these networks are under deployment by carriers such as Telefonica, Deutsch Telecom etc. We propose an all optical interconnection for multi-Tbps capacity network between metro-core and metro access environment, as shown in figure 4.32(a). It consists of an optical kernel composed of a matrix of cross waveguides and micro-ring modulators/switches which can achieve not only demultiplexing or multiplexing but routing as well. Thus, lightwave paths can be routed and wavelength channels can be selected. Such wavelength switching and routing in three directions is shown in figure 4.32(b). This structure can be implemented in Si integrated photonic circuits with a density switching matrix of order 100 × 100 without many problems on optical waveguide paths but the electrical connections and heating tuning section.

4.5.5 Remarks

This section has addressed the issues of access technologies for network capacity reaching several Tbps. Such networks have already been deployed in 2023 for 5G mobile network and associated clouds, e.g. CRAN (cloud radio access networks for 5G networking) and 4K video delivery throughout the global internet. The transmission technologies at 100G per channel and Tbps have been presented and it is proven that they are appropriate for such multi-Tbps transport networks. Furthermore, we have proposed an optical interconnect structure for add/drop from the metro core to access networks. This optical interconnect can accept and operate with channels employing coherent or non-coherent reception. Tbps transmission systems operating under coherent or direct detection are described using comb generators and multiple wavalength mux/demux and modulation. There is no doubt that such multi-Tbps access networks and DC networking (DCN) would be extensively deployed in the near future.

4.6 Low cost 1.6 Tbps by un-cooled comb sources

4.6.1 DSP-assisted Tbps low-cost comb source system

For DCN and access networking with low cost and high capacity in the range of Tbps using coherent transmission technique, the cost of the lasers is important to reduce the total cost. However, the processing power of the DSP can give us an innovative approach to mitigate any difficulties due to the fluctuation of the comb laser lines with respect to those of the carriers. This section gives a brief presentation of this technique. The set-up of this platform can be referred to in figure 4.28, now depicted in figure 4.33, except that no Nyquist shaping is used so as to reduce the complexity of the DSP, hence the power consumption. The frequencies of the lasers of both transmitter and LO can be un-cooled and thus they can fluctuate slowly. Consider the channels whose spectral distribution is depicted in figure 4.35(a). The drift of the laser is assumed to be 0.1 nm °C^{-1}, total drift of the comb laser can be 3.6 nm over many hours, maximum drift of the central wavelength of comb lines 0.5 nm or about 5 °C during the processing time. There are different frequency spectra of the mixed channels after mixing with the LO comb lines (figure 4.35(b)), as shown in

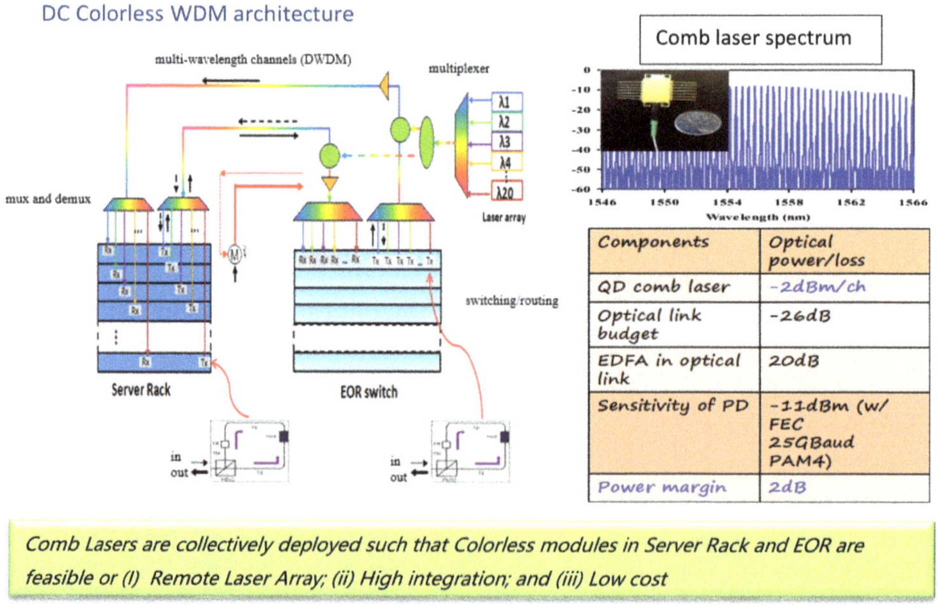

Figure 4.33. Multi-carrier single source (comb laser) and its use in both transmitter and coherent reception LO, inset is the chip size of the laser.

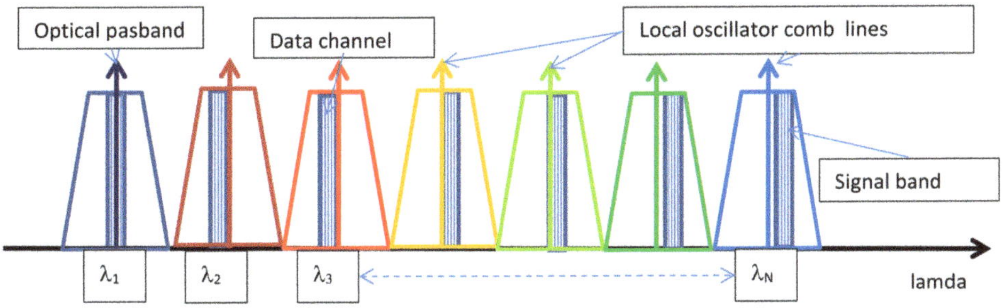

Figure 4.34. Spectral distribution of modulated channels, comb lines of multi-carrier laser, optical passband of optical components.

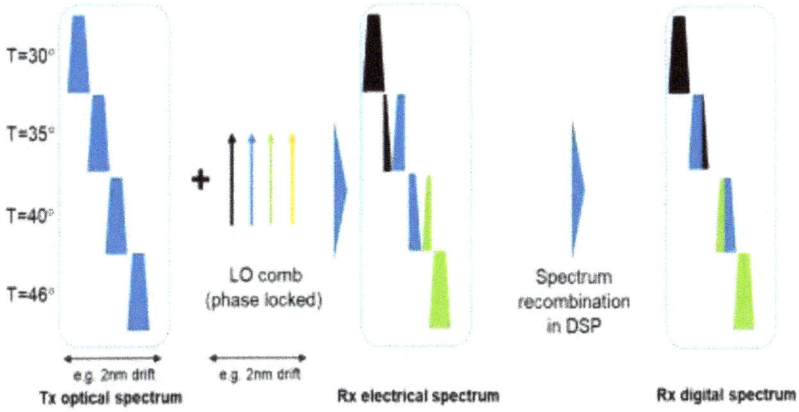

Figure 4.35. Spectra of channels (a) after transmission front of the optical receiver at different temperatures; (b) LO comb laser; (c) electrical domain after mixing with LO drifted spectra; and (d) after DSP processing or in digital domain.

figure 4.35(c), and then processed in the DSP to produce those given in figure 4.35(d). Therefore, the processing of different channels is specific for individual channels. The proposed technique offers significant simplification of the receiver structure, hence lowering costs.

4.6.2 A simple generation of comb lines via cascade modulators

A novel technique to generate an ultra-flat and power efficient optical frequency comb (OFC) by serial cascading of electro-absorption modulators (EAMs) and single drive Mach–zehnder modulator (MZM) has been shown in figure 4.36. In this proposed technique, the continuous wave (CW) light source is modulated and the spectrum broadened by one EAM and two MZMs, respectively. EAMs are in cascaded mode producing ultra-short pulses followed by MZM, introducing intensity modulation and tuning the power variation of even and odd order sidebands at the same level for flattening the optical spectrum. Here, the first

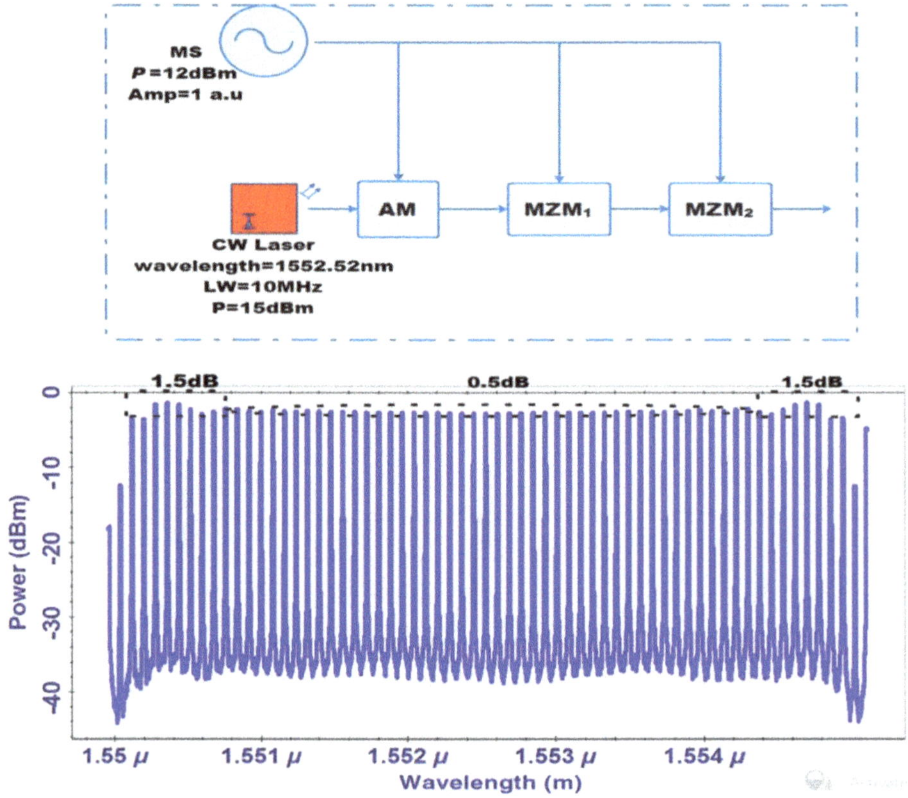

Figure 4.36. A simple comb generator employing three modulators in cascade.

EAM acts as sub-carrier generator, then the second EAM acts as sub-carrier enhancer and the MZM acts as sub-carrier flattening. By this technique, 63 sub-carriers with 10 GHz spacing and within 2 dB power fluctuation can be generated. The generated OFC has bandwidth of 630 GHz. Theoretical and principal analysis of cascaded EAMs and MZM can be referred to [38]. This technique can generate multi-carriers with uniform power level and reasonable high-power spectrum which can be advantageous for dense wavelength division multiplexing (DWDM) and short haul multi-Tbps transmission for data communications [38].

4.6.3 Optical injection and comb generation

Generation of coherent multi-carrier signals by gain switching of discrete mode lasers was reported in [39] in which a DFB discrete mode laser is modulated to generate comb lines and optical injection is also used to improve the linewidth of the comb lines. The linewidth measurement with self-homodyne coherent beating and the gain switching of discrete mode DFB can be used for generating comb lines. These set-ups are shown in figure 4.37.

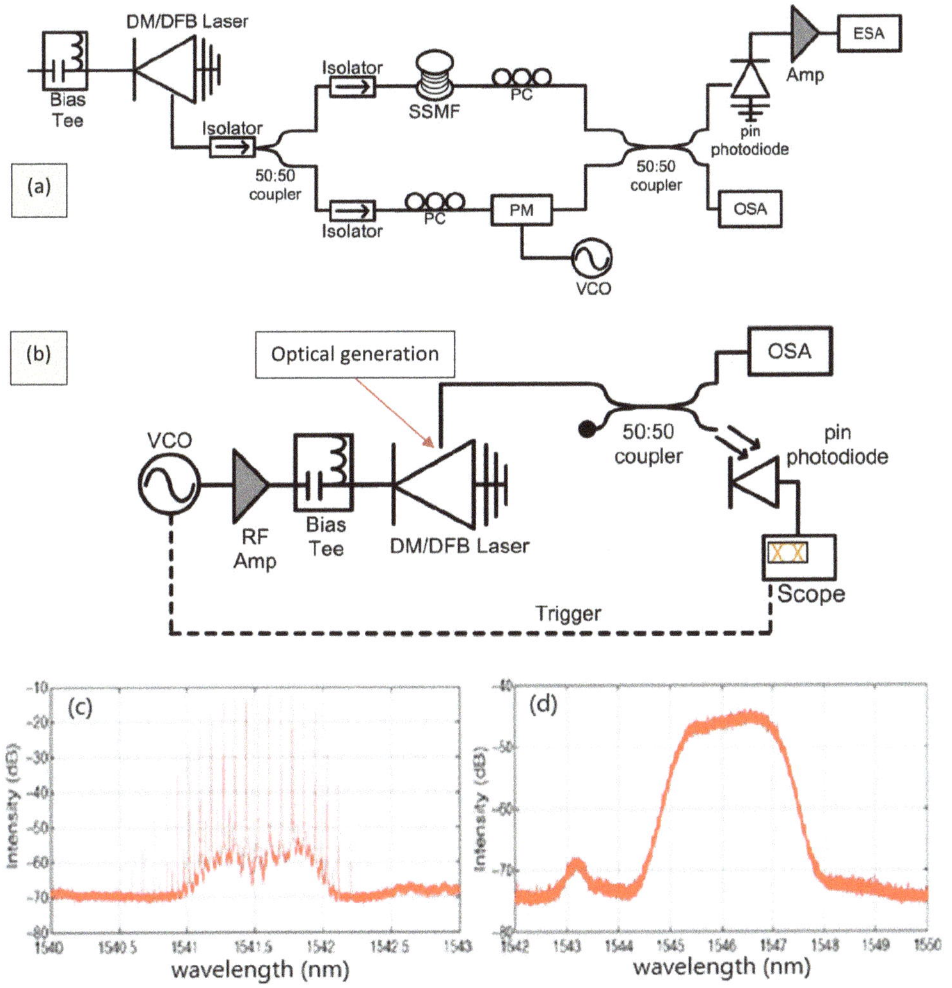

Figure 4.37. (a) Experimental set-up for linewidth characterization; (b) experimental set-up for gain-switched DM and DFB lasers; (c) optical gain-switched spectra for DM laser generating comb lines; (d) optical gain-switched spectra for DFB laser. DM-DFB = discrete mode distributed feedback laser; PM = phase modulator; RF = radio frequency VCO = voltage control oscillator; PC = polarization controller; ESA = electrical spectrum analyzer; OSA = optical spectrum analyzer.

The structure of the DM-DFB laser (DM = discrete mode) shown in figure 4.37(b) which can be generating the multi-comb lines shown in (c) and gain spectra in (d). This DM-DFB is then improved by an external injection of a narrow linewidth laser.

The gain switching is achieved by applying amplified RF waves of a 10.7 GHz sinusoidal feature whose power output of 24 dBm in combination with a dc bias current about four times the threshold current about 200 mA to the laser via a bias tee. The optical output of the laser source is split using a 3 dB FC to enable

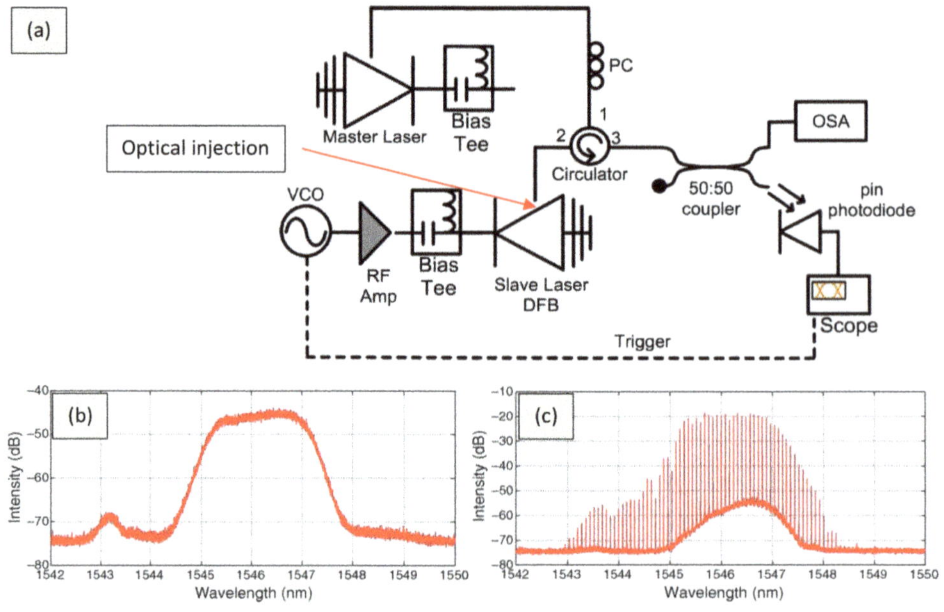

Figure 4.38. (a) Structure of externally injected gain-switched DFB laser with external optical injection arranged in master–slave coupled laser pair; (b) gain spectra of the DM-DFB; (c) comb line generation spectra of about 18 lines over 2.0 nm window of the C-band 1546 nm.

simultaneous temporal and spectral measurements. The characterization of the multi-carrier signal is carried out by using a high-resolution (20 MHz) OSA and a high-speed oscilloscope in conjunction with a 3 dB 50 GHz pin detector without any trans-impedance amplification (figure 4.38).

Compared with the comb laser whose comb lines are generated using dual optical modulators, as given in section 4.6.2, the comb line generation given in this section is much more compact and in complete integrated mode, thus less expensive and lower cost that can be applicable in low-cost transmission link such as those employed in DC interconnections and access networks.

4.7 Concluding remarks

This chapter has given a brief account of multi-carrier generation and their use in the ultra-high capacity transmission lines for long-haul, metropolitan networks and access networking as well as in DC interconnection and inter-DC connections.

References

[1] Buchali F 2010 Technologies towards terabit transmission systems *Proc. European Conf. Opt. Comm. ECOC 2010, paper We.6.C.1(September, 2010) (Torino, Italy)*

[2] Stojanović N, Xie C, Zhao Y, Mao B, Gonzalez N G, Qi J and Binh L N 2013 Modified Gardner phase detector for Nyquist coherent optical transmission systems *OFC/NFOEC Technical Digest, OFC 2013, Paper, JTh2A.50 (Los Angeles, CA)*

[3] Binh L N, Bangning M, Stojanović N, Xie C and Yang N 2013 Synchronous modulator incorporated re-circulating comb laser sources for Tbps superchannel transmission *Proc. OSA Congress, Conf. on Advanced Solid-State Lasers, Paper JTh2A.13 (Paris)* October 2013

[4] Lundberg L, Karlsson M, Lorences-Riesgo A and Mazur M 2018 Frequency comb-based WDM transmission systems enabling joint signal processing *Appl. Sci.* **8** 718

[5] Hu H *et al* 2018 Single-source chip-based frequency comb enabling extreme parallel data transmission *Nat. Photon.* **12** 469–73

[6] Smith G H, Novak D and Ahmed Z 1997 Technique for optical SSB generation to overcome dispersion penalties in fiber-radio systems *Electron. Lett.* **33** 74–5

[7] Wei R, Yan J J, Peng Y, Yao X, Bai M and Zheng Z 2013 Optical frequency comb generation based on electro-optical modulation with high-order harmonic of a sine RF signal *Opt. Commun.* **291** 269

[8] Mishra A K, Schmogrow R, Tomkos I, Hillerkuss D, Koos C, Freude W and Leuthold J 2013 Flexible RF-based comb generator *IEEE Phot. Tech. Lett.* **25** 701–4

[9] Xiao Y, Chen L, Li F and Hai-zhen H 2013 Novel full-duplex SSB WDM-RoF system with SLM technique for decreasing PAPR *Optoelectron. Lett.* **9** 309–12

[10] Xu Z, O'Sullivan M and Hui R 2010 OFDM system Implementation using compatible SSB modulation with a dual-electrode MZM *Opt. Lett.* **35** 1221–3

[11] Zhang Y, O'Sullivan M and Hui R 2010 Theoretical and experimental investigation of compatible SSB modulation for single channel long-distance optical OFDM transmission *Opt. Express* **18** 16751–64

[12] Nazarathy M, Marom D M and Shieh W 2009 Optical comb and filter bank (de)mux enabling 1 Tb/s orthogonal sub-band multiplexed CO-OFDM free of ADC/DAC limits, paper P3.12 *Proc. ECOC(September, 2009) (Vienna, Austria)* 20–4

[13] Zhang J, Yu J, Chi N, Dong Z, Shao Y, Tao L and Li X 2012 Theoretical and experimental study on improved frequency-locked multicarrier generation by using recirculating loop based on multifrequency shifting single-sideband modulation *IEEE Photon. J.* **4** 2249–63

[14] Li J, Li X, Zhang X, Tian F and Xi L 2010 Analysis of the stability and optimizing operation of the single-side-band modulator based on re-circulating frequency shifter used for the T-bit/s optical communication transmission *Opt. Exp.* **18** 17597–609

[15] Zhang J, Chi N, Yu J, Shao Y, Zhu J, Huang B and Tao L Generation of coherent and frequency-lock multi-carrier using cascaded phase modulators and recirculating frequency shifter for Tb/s optical communication *Opt. Exp.* **19** 12891–902

[16] Zhang J, Yu J, Chi N, Shao Y, Tao L, Wang Y and Li X 2012 Improved multi-carrier generation by using multi-frequency shifting recirculating loop *IEEE Photon. Technol. Lett.* **24** 1405–8

[17] Torres-Company V, Lancis J and Andrés P 2008 Lossless equalization of frequency combs *Opt. Lett.* **33** 1822–4

[18] Binh L N 2011 *Guided Wave Photonics: Fundamentals and Applications with MATLAB®* (Boca Raton, FL: CRC Press, Taylor and Francis Group)

[19] Wooten Ed L *et al* 2000 A review of lithium niobate modulators for fiber-optic communications systems *IEEE J. Sel. Top. Quant. Elect* **6** 69–82

[20] Campenhout J V, Green W M J, Assefa S and Yurii A 2009 Vlasov low-power, 2×2 silicon electro-optic switch with 110-nm bandwidth for broadband reconfigurable optical networks *Opt. Express* **17** 24020

[21] ALferness R C 1979 Polarization-independent optical directional coupler switch using weighted coupling *Appl. Phys. Lett.* **35** 748–51
[22] Dedic I 2010 56 Gs/s ADC enabling 100 GbE *Proc. OFC2010 Invited Paper, Digital Transmission Systems, Optical Fiber Conf.*
[23] Proakis J G 2000 *Digital Communications* 4th edn (Boston, MA: McGraw-Hill) Section 9.2 p 560
[24] Christodoulopoulos K, Tomkos I and Varvarigos E A 2001 Elastic bandwidth allocation in flexible OFDM-based optical network *IEEE J. Lightwave Technol.* **29** 1354–66
[25] Uematsu Y, Masuda A, Miyamura T and Hiramastu A 2012 Flexible virtualized optical transport networking technology *NTT Tech. Rev.* **10** 1–7
[26] Proakis J and Salehi M 2012 *Digital Communications* 5th edn (New York: McGraw-Hill)
[27] Binh L N 2008 *Digital Optical Communications* (Boca Raton, FL: CRC Press, Taylor & Francis)
[28] Proakis J G 2000 *Digital Communications* 4th edn (Boston, MA: McGraw-Hill) Section 9.2 pp 554–65
[29] Patricia Bower I D 2011 High speed converters and DSP for 100G and Beyond *Opt. Fiber Technol.* **17** 464–71
[30] Kikuchi K Performance analyses of polarization demultiplexing based on constant-modulus algorithm in digital coherent optical receivers 2011 *Opt. Express* **19** 9868–80
[31] Karinou F *et al* 2014 Directly PAM4 modulated 1530 nm VCSEL enabling 56G/λ data center interconnects *IEEE Photon. Technol. Lett.* **27** 1872–4
[32] Huhse D, Schell M, Bimberg D and Tarasov I S 2014 Generation of electrically wavelength tunable (delta lambda =40 nm) singlemode laser pulses from a 1.3 um Fabry–Perot laser by self-seeding in a fibre-optic configuration *Electron. Lett.* **30** 157–8
[33] Schmeckebier H, Fiol G, Meuer C, Arsenijević D and Bimberg D 2010 Complete pulse characterization of quantum-dot mode-locked lasers suitable for optical communication up to 160 Gbit/s *Opt. Express* **18** 3415–25
[34] Yole Inc 2014 A view on the silicon photonics market, *Report*
[35] Smith B T, Feng D, Lei H, Zheng D, Fong J and Asghari M Fundamentals of silicon photonic devices http://mellanox.com/related-docs/whitepapers/KOTURA_Fundamentals_of_Silicon_Photonic_Devices.pdf
[36] Gardes F Y, Thomson D J, Emerson N G and Reed G T 2011 40 Gb/s silicon photonics modulator for TE and TM polarizations *Opt. Exp.* **19** 11804–14
[37] Vlasov Y A 2012 Silicon integrated nanophotonics: road from scientific explorations to practical applications, CLEO'2012 *Plenary Paper*
[38] Ujjwal Y and Thangaraj J 2019 Generation of an ultra-flat and power efficient optical frequency comb by cascading of electro-absorption and single drive Mach–Zehnder modulator *Opt. Eng.* **57** 126106
[39] Anandarajah P M, Maher R, Xu Y Q, Latkowski S, O'Carroll J, Murdoch S G, Phelan R, O'Gorman J and Barry L P 2011 Generation of coherent multicarrier signals by gain switching of discrete mode lasers *IEEE Photonics J.* **3** 112–22

Chapter 5

Photonic signal processors

The routing and processing of channel signals require several high-speed transmission channels and their routing to services and storage. There are several functions in such routing and processing of switching and multiplication as well as addition or subtraction of ultra-high-speed signals. Photonic signal processors can implant several manipulation functions in integrated optics and photonics in the optical domain, which will enhance the speed and provide more accurate processing.

This chapter gives a brief overview of processing signals in optical and photonic domains. In addition, the devices and processors for the fifth-generation wireless networks (5G networks) which require substantial high capacity optical transport techniques. From this view we give again a brief description of optical transmission in the access layer.

5.1 Optical transformed channels and transmission: spectral domain processing

Photonic processing and high-speed data centres (DCs) will revolutionize data processing in modern massive data transport and usage in society. In today's digital age, data is constantly being generated and processed at an unprecedented rate. With the rise of cloud computing, artificial intelligence, and the Internet of Every Thing (IoET), the need for faster and more efficient data processing has become paramount. Traditional electronic processing has limitations in terms of speed, power consumption, and scalability. This is where photonic processing comes in as a game-changer. So, what is photonic processing?

Photonic processing is a new approach to data processing that uses photons instead of electrons to transmit and process information. It uses optical fibres, lasers, and other integrated photonic components to transmit, manipulate, and detect light signals. Photonic signal processing is based on the principles of photonics, which is

doi:10.1088/978-0-7503-5849-1ch5

the study of the behavior and manipulating techniques of photons and properties of lightwaves (not light) as the photonic waves are purely electromagnetic waves in coherent senses.

Photonic signal processing offers several advantages over traditional electronic processing in high-speed DCs, including the following. (i) Speed: photonic processing is much faster than electronic processing. Lightwaves travel at a speed of approximately 300 000 km s^{-1}, while electrons travel at a speed of only a few hundred meters per second. This means that data can be transmitted and processed much faster using photonic processing. Furthermore, the input channel signals are in the optical domain, so processing is in the photonic domain in which it is necessary to transform the optical signals. (ii) Energy efficiency: photonic processing is much more energy-efficient than electronic processing. Since photons do not have mass, they require much less energy to transmit and process than electrons. This means that DCs using photonic processing can save a significant amount of energy and reduce their carbon footprint. Integrated photonics enhances this aspect so that only a minute amount of energy is consumed. (iii) Scalability: photonic processing is highly scalable, which means that it can be easily integrated into existing DC infrastructure. It also allows for the creation of much larger DCs that can handle much more data than traditional electronic processing; (iv) Bandwidth: photonic processing offers much higher bandwidth than electronic processing. This means that more data can be transmitted and processed simultaneously using photonic processing, especially with multiplexing of wavelength carriers and their modulated data sequences.

Applications of photonic processing in high-speed DCs: photonic processing has several applications in high-speed DCs, including the following. (a) Data transmission: photonic processing can be used to transmit data over long distances at high speeds. This makes it ideal for applications such as high-speed internet and cloud computing. (b) Data processing: photonic processing can also be used for data processing in high-speed DCs. It can be used for applications such as machine learning, artificial intelligence, and big data analytics. (c) Quantum computing: photonic processing is also being explored for use in quantum computing, which is a new approach to computing that uses the principles of quantum mechanics. Photonic processing can be used to create and manipulate quantum states, which are the building blocks of quantum computing and data security transport, especially the quantum keys.

Therefore, photonic processing is a revolutionary technology that has the potential to transform high-speed data processing in DCs. Its speed, energy efficiency, scalability, and high bandwidth make it ideal for applications such as cloud computing, artificial intelligence, and quantum computing. As data continues to grow at an exponential rate, photonic processing will become increasingly important in enabling faster and more efficient data processing.

In processing there are two commonly known domains in electrical or photonic systems, temporal or spectra or time or frequency domains. This chapters deals with photonic processing in the spectral domain and temporal domain including all processing using neural optical networking.

5.1.1 Optical Fourier transform (OFT) based structure

Super-channel (SC) transmission system can also be structured using optical fast Fourier transform (OFFT) as demonstrate in [1, 2] and shown in figure 5.1 in which Mach–Zehnder delay interferometer (MZDI) components (see figure 5.1(a)) act as spectral filter and splitter (see figure 5.1(b)), the optical FFT. The outputs of these MZDIs are then fed into coherent receivers and processed digitally, as in figure 5.1(c), with the electro-absorption modulator (EAM) performing the switching function so as to time de-multiplexing the ultra-fast signal speed to lower speed sequence so that the detection system can decode and convert to the digital domain for further processing.

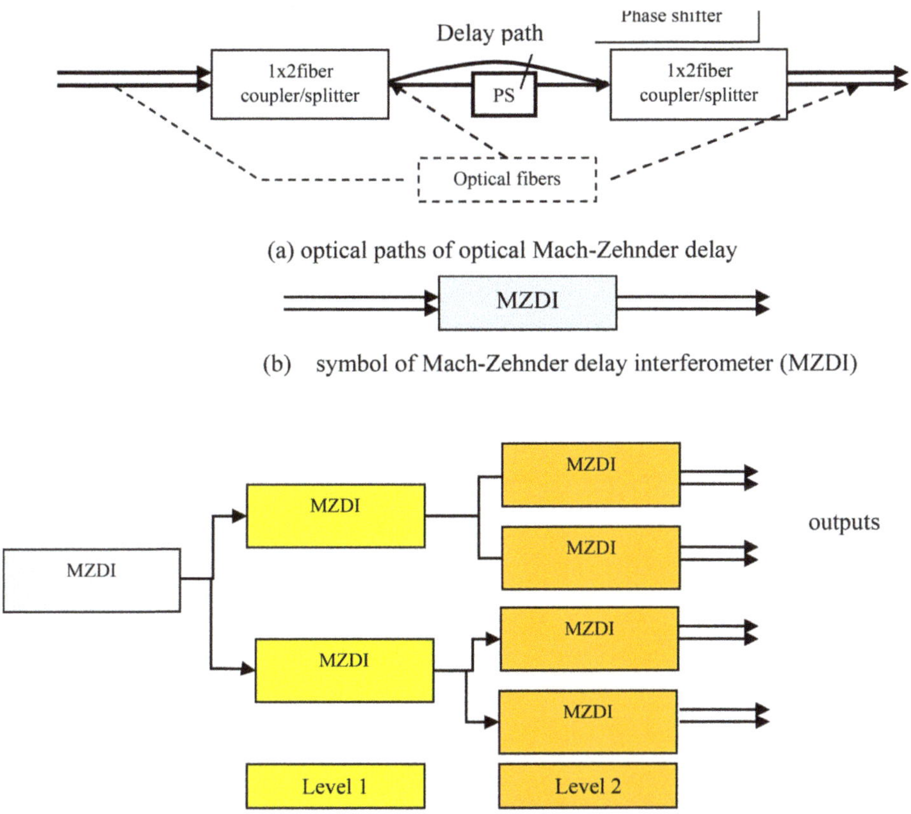

(a) optical paths of optical Mach-Zehnder delay

(b) symbol of Mach-Zehnder delay interferometer (MZDI)

(c) 1x8 fiber optic FFT or 8 points optical FFT

Figure 5.1. Operations by guided wave components using fibre optics. (a) Guided wave optical path of an MZDI or asymmetric interferometer with phase delay tunable by thermal or electro-optic effects; (b) block diagram representation, and (c) implementation of OFFT using cascade stages of fiber optical MZDI structure. EAM/S = electro-absorption modulator/sampler used for de-multiplexing in time domain. Note also phase shifters employed in MZDIs between stages to tune exact delay between two branches. Insets are spectra of optical signals at different stages as indicated for the OFFT (serial type). EAM = electro-absorption modulator; PC = polarization controller, TMZDI (τ) = tunable Mach–Zehnder Delay Interferometer (dual-output) with τ = 20 ps delay difference between paths. EAS = electro-absorption sampler.

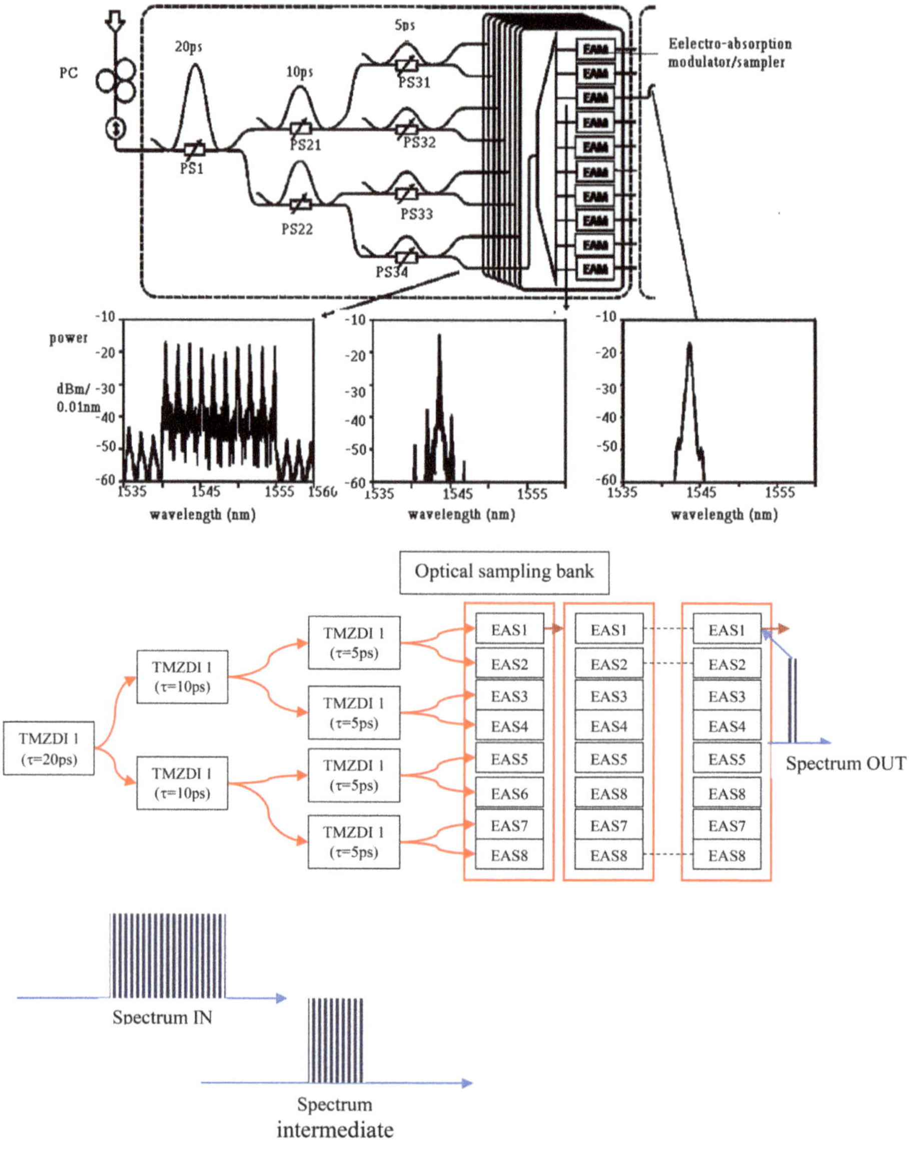

(d)

Figure 5.1. (Continued.)

The spectra of SCs at different positions in the transmission system can be seen in figures 5.2(a)–(c). It is noted that the pulse shape is Nyquist and the sub-channels are placed close and satisfying the orthogonal condition, thus the name optical OFDM (orthogonal frequency division multiplexing) is used to indicate this SC arrangement.

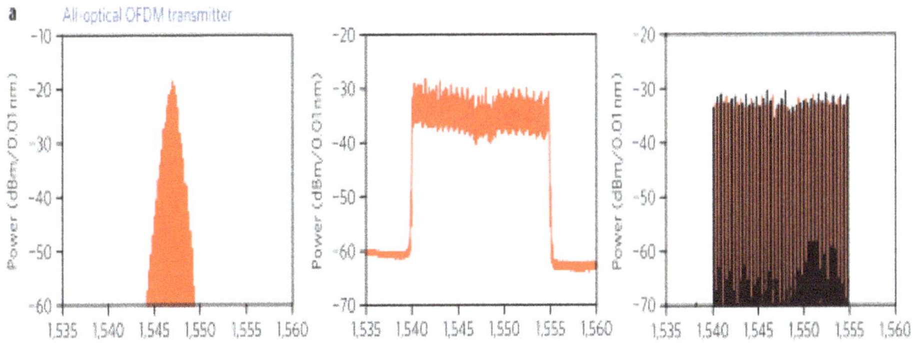

Figure 5.2. Spectrum of SCs and demultiplexed channels. All sub-channels are orthogonal and thus the name OFDM (orthogonal frequency division multiplexing). (a) At output of transmitter, (b) after fiber propagation and (c) after polarization de-multiplexing. Reprinted from [1], copyright (2011) with permission from Nature.

5.1.2 Optical Fourier processor

The processing of SCs can be considered as similar to the digital processing of individual sub-channels except when there may be crosstalk between sub-channels due to overlapping of certain spectral regions between the considered channel and its adjacent channels.

Thus, for the Nyquist quadrature phase-shift keying (QPSK) sub-channel the digital signal processing (DSP) would be much the same as for QPSK DWDM (dense wavelength division multiplexing) for 112G described above with care taken for the overlapping either at the transmitter or at the receiver. The error vector magnitude is a parameter that indicates the scattering of the vector formed by I- and Q-components departing from the center of the constellation point. The variance of this error vector magnitude (EVM) in the constellation plane is used to evaluate the noises of the detected states, thence the *Q*-factor can be evaluated with ease and thus the BER by using the probability density function and the magnitude of the vector of a state on the constellation plane. The BER of the sub-channels of the OFDM SC is shown in figures 5.3(b) and (c) for different percentage overloading due to FEC. The loading factor is important as this will increase the speed or symbol rate of the sub-channel one has to offer. The higher this percentage, the higher the increase in the symbol rate is, thus requiring high-speed devices and components.

In the receiver of the optical OFDM SC system of reference [1], to judge the effectiveness of the optical FFT receiver, three alternative receiver concepts were tested for a QPSK signal. A QPSK signal is chosen because it was not possible to receive a 16-QAM signal with the alternative receivers owing to their inferior performance [1]. First, a sub-carrier with a narrow bandpass filter is used to extract a sub-channel. The filter passband is adjusted for the best performance of the received signal (figures 5.3 and 5.2(a)). The selected filter bandwidth is 25 GHz. The constellation diagram shows severe distortion. When using narrow optical filtering, one has to accept a compromise between crosstalk from neighboring

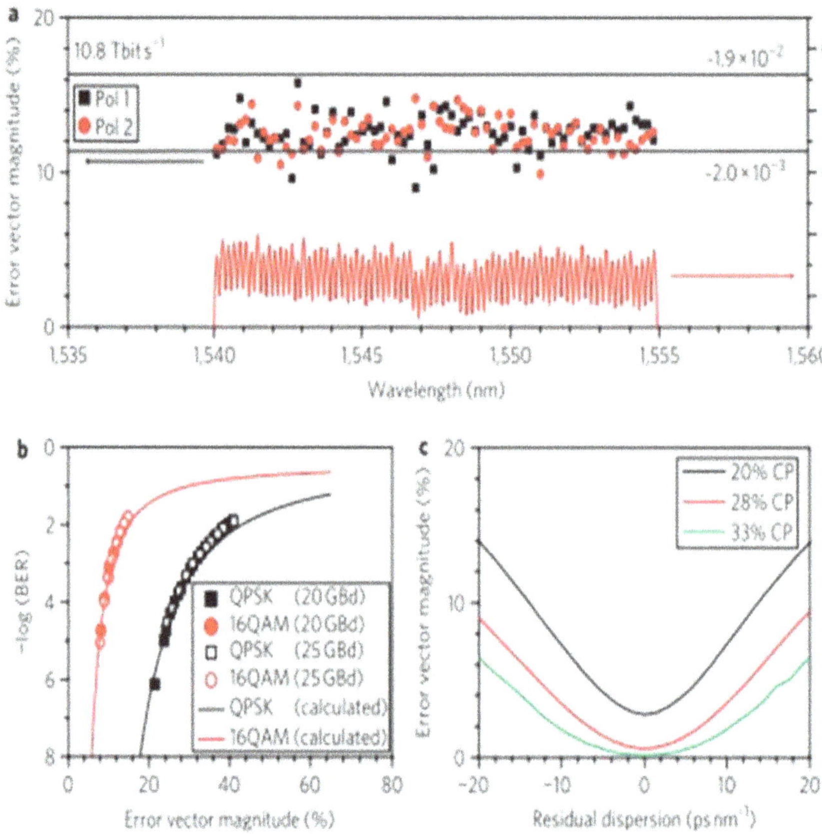

Figure 5.3. All-optical 10.8 Tbps OFDM results. (a) Measured EVM for both polarizations (symbols) and for all sub-carriers of the OFDM signal decoded with the all-optical FFT. The estimated BER for all sub-carriers is below the third-generation FEC limit of 1.9e–2. The optical spectrum far left race is drawn beneath. (b) Relationship between BER and EVM. Measured points (symbols) and calculated BER as a function of EVM for QPSK and 16-QAM. (c); tolerance towards residual chromatic dispersion of the implemented system decoded with the eight-point FFT for a cyclic prefix of 20%, 28% and 33%. Reprinted from [1], copyright (2011) with permission from Nature.

channels (as modulated OFDM sub-carriers necessarily overlap) and inter-symbol interference (ISI) owing to the increasing length of the impulse response when narrow filters are used. Narrow filters can be used, however, if the ringing from ISI is mitigated by additional time gating. The reception of a sub-carrier using a coherent receiver is then performed. In the coherent receiver, the signal is down-converted in a hybrid coupler, and detected using balanced detectors and sampled in a real-time oscilloscope. Using a combination of error low-pass filtering due to the limited electrical bandwidth of the oscilloscope and DSP, the sub-carrier is extracted from the received signal. This receiver performs better than the filtering approach, but a larger electrical bandwidth and sampling rate of the ADC and additional DSP would be needed to eliminate the crosstalk from other sub-carriers and then to achieve a performance similar to that of OFFT.

Thus optical OFDM may offer significant advantages for SC but additional processing time would be required, while for Nyquist QPSK SCs allow better performance and less complexity in the receiver DSP sub-system structure.

5.2 Photonic signal processors

Integrated optic technologies have progressed tremendously over the years, especially since the beginning of the millennium and after the dotcom problem. The baud rate has now reached 100 GBaud and even higher 200 GBaud via the time multiplexing. Thus, under coherent transmission and reception techniques nearly and higher than 1.00 Gbit s^{-1} are offered for metro and short-haul or interconnection in DCs.

At these baud rates the operation of the transmitter and receiver can be better by operating all in the photonic domain, including the cross-connect integrated circuits as the electronic switching speed has reached the limit and the power consumption will increase drastically when such baud rate and bit rates are processed in the electronic domain.

This section gives an overview of photonic signal processors (PSPs). By photonic processing we mean the manipulations of photons in both active and passive domain in coherence while all-optical indicates highly passive. A schematic diagram of the photonic processor is given in figure 5.4.

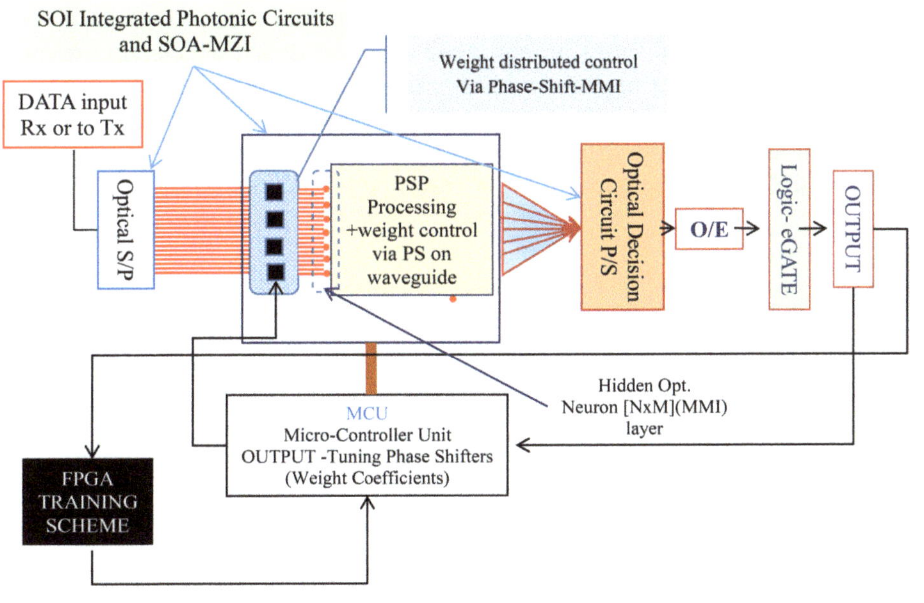

Figure 5.4. Schematic of PSP principles. Labels: PS = phase shifter; MCU (Micro-controller unit); eGate = logic gate in the opto-electronic domain; S/P = serial-to-parallel conversion; Rx/Tx = receiver or transmitter; SOI = silicon-on-insulator (oxide); MMI = multi-mode interferometer. FPGA = field programmable gate array; O/E = optical-to-electrrical converter; SOS = semiconductor optical amplifier; MZI = Mach–Zehnder interferometer. PSP = photonic signal processoor; MCU = micro-control unit (low sampling speed); NxM = optical 2D matix of order N and M.

5.2.1 Generic deep learning neural network photonic signal processor (DLNNPSP)

In practice, all-optical signal processing is workable provided that one can determine exactly the optical guided path. The OFT processor is a linear processing system with minimum use of optical decision circuitry. This section briefly presents the photonic signal processor using optical neural networks with deep learning with optical decision maker circuitry for decision feedback and tuning the optical neural network so that convergence can be achieved. This section gives a proposed photonic signal processor whose structure consists of a central optical neural network (ONN) in association with optical sampler and optical decision circuitry so that feedback can be used to tune the weighting coefficients of the ONN. The summation and subtraction functions that can be implemented are given in the optical domain by an MMI (multi-mode interferometer) and an integrated phase shifter so as to change the coupling coefficients, hence the variation of the total waves at the output port of the MMI. See table 5.1 below—as a table of abbreviated terms.

The motivations for processing in the photonic domain can be stated as follows. (i) Firstly, electronic DSP at ultra-high sampling rate, e.g. 200 GSymbols/sec (GSym s^{-1}) for 100 GBaud would require very high power consumption and imitation on operating speed at which electronic DSPs are facing considerable difficulties, especially at 200 GSymbols s^{-1} and the ultra-high power consumption (note: power consumption in electronic sampling increases by a cubic factor of the sampling increase factor). (ii) Secondly, processing speed in the optical domain can be much faster than that by the electronic DSP on conditions that accurate delay in optical domain can be fabricated for sampling speed of 1–2 ps delay, which can presently be implementable in Si photonics technology. (iii) Thirdly, the advances of integrated photonic devices, especially in Si photonics based on silicon material, waveguide and active as well as passive device structures for modulation, routing, splitting etc, that make it possible for implementations of several optical functions.

Table 5.1. Abbreviated terms used in this section.

FF = feed forward	SiIP = silicon integrated photonics
FB = feedback	SOI = silicon-on-insulator
DFB = decision feedback	SOA = semiconductor optical amplifier
ODFB = optical decision feedback	OLG = optical logic gate
O/E = optical-to-electronic conversion	MZI = Mach–Zehnder interferometer
eDSP = digital signal processing (electronic)	NOT Gate = optical inverter
MCU = micro-control unit	AND/NAND = logic gate AND and not AND
ONN = optical neural network/networking	XOR = exclusive OR gate
OSP = optical signal processing	OC = optical coupler
PSP = photonic signal processing/processors	
S/P = serial-to-parallel conversion/converter	
P/S = Parallel-to-serial conversion/converter	

This section proposes a generic PSP structure using ONN and FF decision optical logic circuit. This idea is to extend the basic structure to including the optical feedback from the output of the OLG. Such optical feedback will minimize the complexity of the training scheme and accelerate the convergence of the ONN to desired output so that the determination of the signal level can be obtained. We can refer to figures 5.4 and 5.5 and a more detailed description of the operational principles in the next sub-section.

The optical data are derived from either transmitter (Tx) or receiver (Rx in optical domain) or even the outputs of all-optical interconnectors or switches. The optical data sequence is sampled and converted to parallel form via the use of optical delay lines of different delay time in optical guided wave lines. This block is indicated as 'optical S/P'. Once the data channels in are parallel form. These optical paths are fed into a PSP unit which is a sub-system of the PSP. The photonic processing is implemented using an optical neural computing network. All manipulations are implemented on SOI, which is the silicon photonic devices fabricated in a Si thin film guided medium which can be activated via the electro-optic effects in the C-band silica fibre. These SOI integrated optic devices are now quite popular in the field of integrated photonics. The operation of this ONN is described later in this section. In principle optical neural computing is composed of two main functions, one is to distribute one path to all other paths with the weightings or magnitudes of optical fields differently controlled. These weightings are controlled and adjusted continuously till the convergence of a target level which is decided by an optical decision

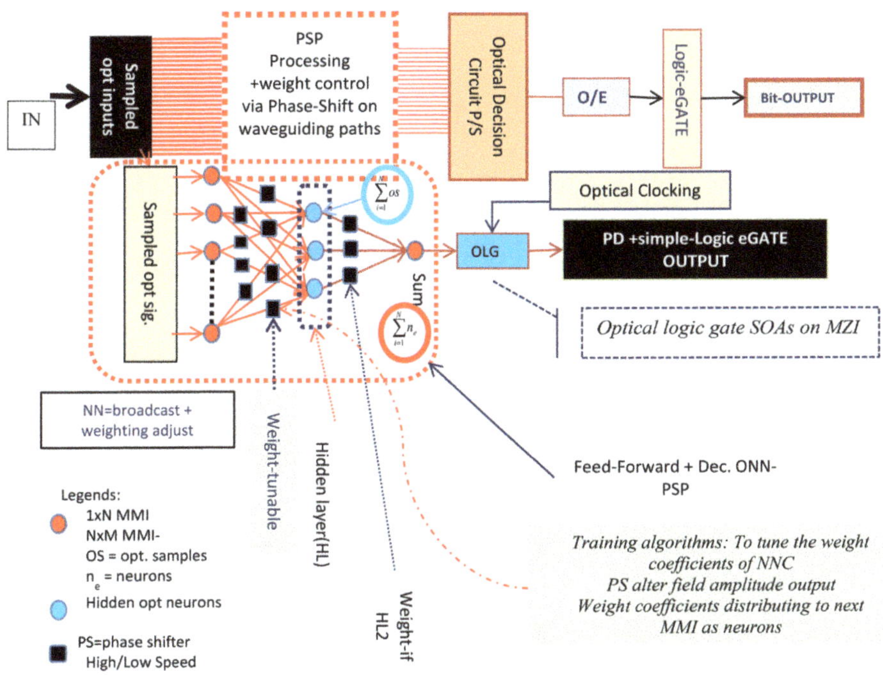

Figure 5.5. Schematic of a PSP system by ONNC.

circuit which derives from the summation of all the optical paths to a nonlinear optical opto-electronic sub-system, the O/E converter and the eGate or electronic gate which is set at a level for decision of 'high' or 'low'.

The ONN can be trained via an FPGA (field programmable gate array) sub-system which outputs electronic signals to an MCU (micro-control unit) whose output signals are in voltage or current levels feeding to the electrodes of the PSs (phase-shifters) to tune the optical phase via the ferro-electric effects or electro-optic effects if higher speed is required. These phase shifting actions, in turn tune the coupling coefficients of the outputs of the MMI. This is the physical insights of the MMI and the central operating principles of the ONN. The control can be very fast with specific processors whose sampling rate can be in the pico-seconds and the electro-optic phase modulators using integrated Si photonics.

Again, we can summarize the operations of the PSP and ONN in the following.

5.2.2 PSP operating principles

Schematic of generic PSP system principles:

- Equivalent DAC in optical domain: referring to figure 5.6, the optical sampling concept and optical S/P conversion without memory and a FIFO (first-in first-out) occur for sampled optical signals—the optical discrete samples.
- PSP of parallel sampled optical inputs: ONNC algorithms.
- Neural net computing (NNC) optical algorithms: feed-forward NN (FF-NN) w/o decision, FF-NN-decision, FF-NN-decision FB (deedback), optical reservoir NN computing system/chip. Only FF-ONN is given in this patent idea application.
- Optical decision circuits—SOA-MZI.

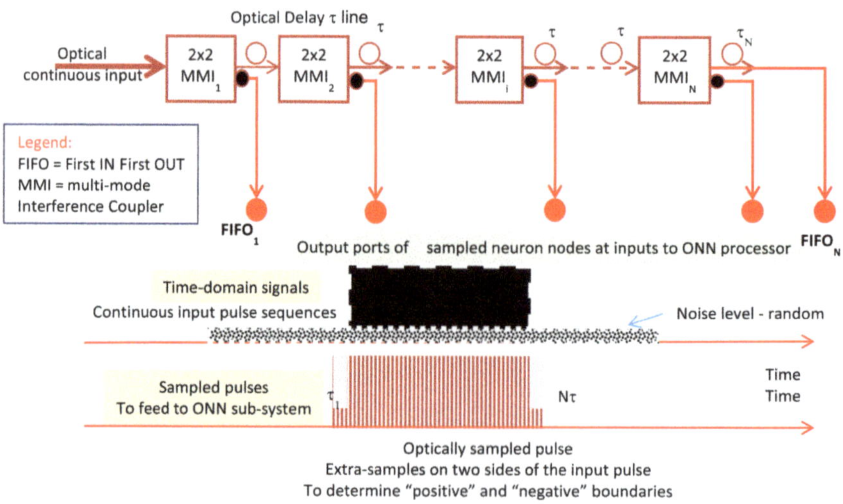

Figure 5.6. Optical analogue sampling by MMI and delay lines—equivalent to sampler in DAC/ADC. Label: FIFO = first-in first-out.

- Algorithms of deep NN learning (that is with back propagation) and possible applications to PSP—to be applied for patents after the approval of the Patent Committee on this idea. Such as those listed in the fourth point, namely FF-decision FB equalization PSP, adaptive FF equalization PSP, etc.

5.2.3 ONN operating principles

- Input optical signals (including embedded noises) are converted to parallel by optical sampling (the optical sampler—OS and S/P conversion) in the optical domain by using optical delay and split and coupling out by MMI coupler—similar to transversal filtering in the electrical domain. These principles are shown in the next two slides. So these samples are equivalent digital optical signals but in optical analogue features.
- The optical sampled signal waves are fed into a PSP processor consisting of hidden optical neuron layer(s) or an MMI with controlled phase via the MCU so that the weighting coefficients can be varied/tuned. (See generic structure of ONN for PSP—given in figure 5.5.)
- After ONNC by various processing algorithms the outputs of optical neurons are fed into a summation neural and fed into an optical logic gate (OLG) so as to determine that the processed inputs can be a '1' or '0' for NRZ or '00', '01', '10' or '11' of PAM4 signals for one-directional signal.
- OLG can consist of two SOAs embedded in an MZI with three coupled inputs which are allocated for optical signals, control signals and optical clocking signals.
- In the case of complexed signals then there would be another $\pi/2$ optical phase shift on the split optical path ($\pi/4$ PS) to another ONN and optical logical gate.
- Training schemes of different ONN structures are to be conducted so that fast convergence of the optical processing can be achieved. A look-up table could be used if compensation of nonlinear impairment effects is required for fast convergence to the targeted output.
- Our proposed idea of a PSP system is highly stable as there is no self-optical feedback, which is positive feedback, and oscillation can happen. But without self-feedback our PSP system is memoryless and highly stable.

The operations of optical logic gates can be described as follows (figure 5.7).

- OLG consists of two SOAs embedded in the optical paths of an MZI. Thus the operation relies on the phase changes in the SOA under signal and control signal inputs as well as the third input which is considered as a clocking signals as shown in the previous slide.
- Data Signal 1 and Data Signal 2 are coupled into the MZI paths via an optical coupling region, the planar multi-mode interferometric waveguide. Two control signals can be used to feed into the MZI. First the control signal can act as an ENABLE to allow the output of the MZI-SOA to operate or a clocking optical signal can be used to clock the data sequence out.

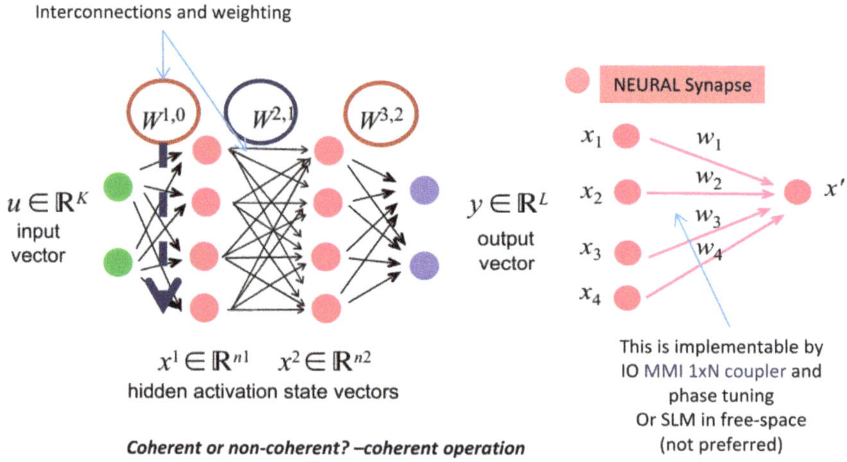

Figure 5.7. Generic model ONNs: functions and technology. Label: SLM = spatial light modulation.

- The input signal which comes from the total sum of all the optical samples which are processed by the ONN. IF the signal amplitude is large enough to compare logically with reference optical signal (Signal 2) then an XOR or AND operation can be performed. This is the nonlinear optical logic operation as required for a decision making in the optical domain.
- The novelty of the use of this OLG is that it simplifies the optical decision circuitry or using only one optical device unlike the other two recently reported schemes [3, 4] in which PD arrays must be used.
- Several logic gates can be formed by this structure. The reference optical signal 2 can be set after the training scheme as indicated in previous slide. These OLG will be used in the coming patent idea submissions.
- Speed of logic operations can now reach 20–30 GHz and expect to be higher pending on the speed of the phase changes of SOA.
- Possible Logic Operations using MZI-SOA: XOR, AND, NAND, NOT.

5.2.4 Remarks

In this chapter we discussed and demonstrated a 5G transport scheme that can be implemented as an evolutionary step to support future data centres (DC). The successful realization of this transmission link demonstrates the potential of the photonic integrated circuits to be implemented in 5G infrastructures. Further, the security transport techniques for data and secret keys are given and will, in more details, be presented in a later chapter.

5.3 5G optical transport networking: from photonic devices to processors

In this section we present the urgency of optical transport networking evolution for 5G delivery and ultra-broadband services to users and communities of users, in the following: (a) cloud core, metro-cloud core and edge cloud networking structure with optical-software defined networking and network function virtualization (SDN) and SDN/network function virtualization (NFV); (b) photonic enabled technologies including principal devices and photonic processors; (c) security aspects and transmission technology for secret keys in co-transmission of massive data transport.

5.3.1 Introduction

5.3.1.1 5G infrastructure evolution

In recent years, an evolution of the traditional radio access network (RAN) with a base station (BS) at each cell tower site (1G/2G), where all the baseband and radio processing is integrated, to a separated baseband and radio part of the BS (3G/4G) into a baseband Unit (BBU) and a remote radio head (RRH) (or RRU, unit) has occurred. Cloud-RAN (C-RAN), where the RRHs are connected to a BBU pool, is expected to play an important role in 5G networks reducing CAPEX (capital expenditure) as well as improving performance and energy efficiency via coordinated multipoint (CoMP), capacity sharing and optimization. Mobile front-haul (FH) is a key network element in the C-RAN architecture, to connect centralized BBUs with RRHs. Mobile FH will need to meet the strict requirements of the key technologies that will be the 5G enablers, such as massive MIMO, mm wave and ultra-dense networks. Only the optical fibre-based mobile FH can meet the high-capacity demands of 5G and thus 5G not only brings big opportunities but also great challenges for the optical communication industry. The potential revenue will be greatly affected by the architectures and interfaces that will be used and a careful investigation and roadmap strategy is necessary at this early stage of 5G deployment. DCs and 5G will drive growth in the optical communication market and the optical equipment market will further grow ~4.5% toward 2022 accordingly (according to the latest forecasts). The driving forces for the 5G growth will be the preparation for 5G/enhanced mobile broadband (eMBB) (phase I) and the needs for machine-to-machine (M2M) and Internet of Things (IoT) applications (phase II) in the optical FH and shift from data video to augmented reality/virtual reality (AR/VR) and applications like 3D-holographic imaging for consumers or vehicle-to-the-X (V2X) and smart factory/city. Optical communication technology will be the enabling technology not only for FH but also for back-haul (BH) and mid-haul (MH) (or X-haul), note also figure 5.8.

5.3.1.2 Evolutionary distributed data centers

It is expected that the optical transport networks will evolve to 5G cloud-based network topology and can be visualized as shown in figure 5.9. The evolution can be expressed as a multi-step process as follows: (i) massive evolution of transport network from IP-based architecture of 4G to cloud-based architecture for 5G, driven by changes in ultra-high capacity and dynamic demand, wireless and core network architecture of evolutionary traditional telecom and cloud DC networking; (ii) aggregation of DC interconnect to metro core networks. The low latency and

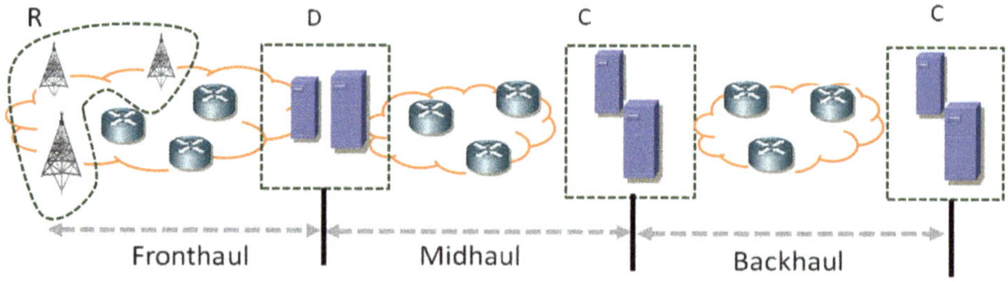

Figure 5.8. Optical communications as the enabler for 5G X-haul. R = regional, D = data storage data centers; C = local data storage, DCs incorporating photonic routers and add/drop multiplexers and demultiplexers.

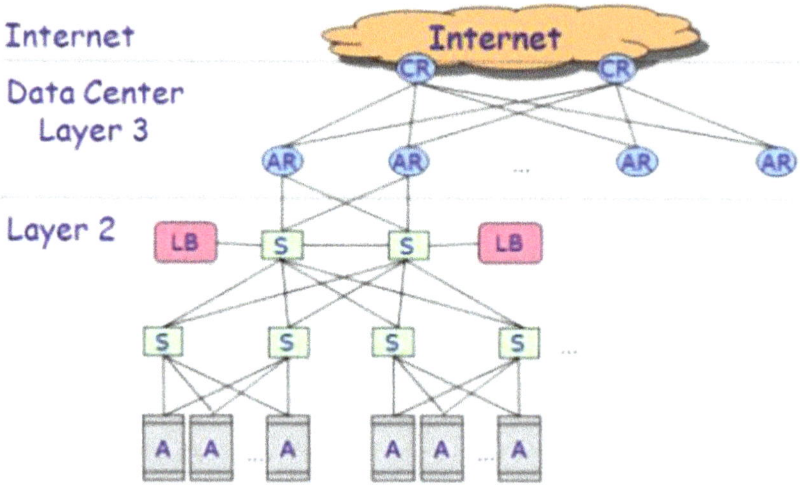

Figure 5.9. Schematic of the generic three-tier layers DC in interconnection with 5G global internet structure whose details are shown in figure 5.10. Labels: CR = Level-3 core router; AR = Level-3 access router; S = Level-2 switch; LB = load balancer; A = rack of 20 servers with top rack switch. Routers are photonic ADD/DROP muxers or demuxers.

wideband services of 5G will drive the down shifting of the backbone node to metro node. Alternatively, the whole terrestrial or intercontinental networks can be viewed as a metro transport network in which low cost, low latency and ultra-high capacity must be met. These are the new challenges to be met by the DC networking infrastructure or telecom carriers or effectively all of 5G service providers (5GSP); and (iii) as far as the RANs, shown in figure 5.10, are to be densely distributed, metro D-RAN (distribute RAN) layers are formed and flattened on which the cloud data units and centralized units can be aggregated.

Cloud-RAN can enhance the gain of the base station in a 4G network by 10 dB based on test results in Japan and more obviously in the 5G environment. This thus requires a unification of the CPRI and eCPRI. The OLTs (optical line terminals) are evolved into cloud-based mobile engines (CMEs) and receiving or transmitting to

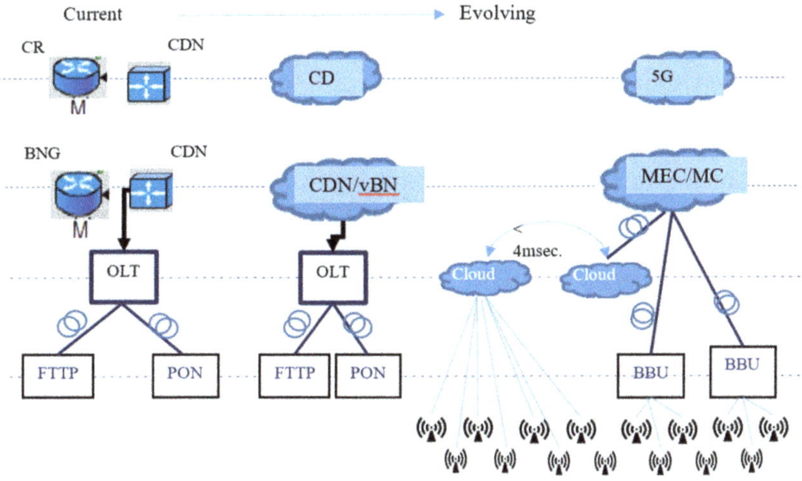

Figure 5.10. Evolution of optical transport networks to cloudified networks for 5G. Labels: CDN = content delivery networks; FBB = fixed broadband; CDN = current delivery network; vBNG = variable business network group; OLT = optical line terminal; MCE = mobile cloud engine; BBU = baseband unit; FTTP = fiber to the premises; PON = passive optical networks.

new 5G core nodes, the cloud-based topological network with several OTTA and mmW MIMO antenna sites. 5G network architecture is trending towards common and dynamic broadband and metro private line networking as well as multi-service transport.

5.3.2 Optical transport networking evolution for 5G delivery

5.3.2.1 State-of-the-art, current challenges and the role of novel technologies

Currently, the radio-optical medium CPRI is the standard interface carrying the data between the RRH and the BBU pool. In the CPRI interface a wireless signal is digitized using a 15-bit resolution for each I/Q component of the complex waveform. For a two sector, 4 × 4 MIMO channel of 20 MHz (4G), based on the CPRI-7-option and a sampling rate of 30.72 MS s^{-1}, a bit rate as high as 9.8304 Gb s^{-1} is required. For a potential 5G scenario, as an example 200 MHz bandwidth, 64 × 64 M-MIMO, and three sectors, there will be a need for 240 (!) CPRI-7 like requirement (equivalent to 2.4 Tb s^{-1}). As a result, bandwidths well above 100 Gb s^{-1} might be required from the optical network. Disaggregated RAN is an evolution of C-RAN to relax the bandwidth requirements at the front-haul using functional splits. eCPRI is an enhanced version of CPRI to support more efficient interfaces as an onset for 5G development. eCPRI can support 5G 64 × 64 MIMO scenarios (25 Gb s^{-1}). The eCPRI Interface specification can be supported by ethernet-switched or IP-routed front-haul networks, or a similar type of transport networks. However different technical challenges emerge depending on where the split is, leading to a trade-off between front-haul bandwidth versus latency requirements, complexity and cost of RRH. In particular, front-haul latency needs to meet the requirements for the end-to-end latency on a 5G network which can be in the order of 10 ms (cloud-assisted driving) down to 0.25 ms (high-frequency trading). A higher layer split (HLS) has already been agreed, but no consensus has been reached (at this time) for a lower-layer split (LLS).

Different optical network solutions/interfaces can be applied to meet the 5G challenges at the front-haul, like high capacity and flexibility and also latency and low power consumption. In all cases, current network evolution and optimization is needed. Optical transport network solution (OTN/WDM) is a robust solution which can provide reliable and broadband front-haul delivery to eMBB, M2M/IoT and cloud using high capacity DWDM. OTN is simple, reliable and transparent for CPRI/eCPRI, however, an evolution to the access ring is needed, or relevant approaches (figure 5.11) need to be investigated. Ethernet (Eth-RoE and IP/Eth-eCPRI) can be a promising option to support functional splits. Packet switched networks based on ethernet look like a promising approach for 5G as they offer resource sharing, and are also reconfigurable. Ethernet is a standard technology, widely adopted, and can be easily upgraded to include software defined networking and network function virtualization (SDN/NFV) aspects. However, low delay requirements must be met in all cases, e.g. packet delay and variation is dependent on traffic load, and synchronous ethernet will be needed. Last but definitely not least, passive optical network (PON) can reduce the cost of fiber deployment and reuse the current optical distribution network (like OTN/WDM), using SDN/NFV to optimize the network according to applications. However, PON capacity needs to be upgraded (25G to 50G or 40G/NGPON2); latency issues need to be met using a low latency dynamic bandwidth allocation (DBA) algorithm and protection-multi-wavelength allocation with, e.g. NGPON2, might be required. High risk technologies such as radio over fiber (RoF) can be considered in the long term (5G+) as

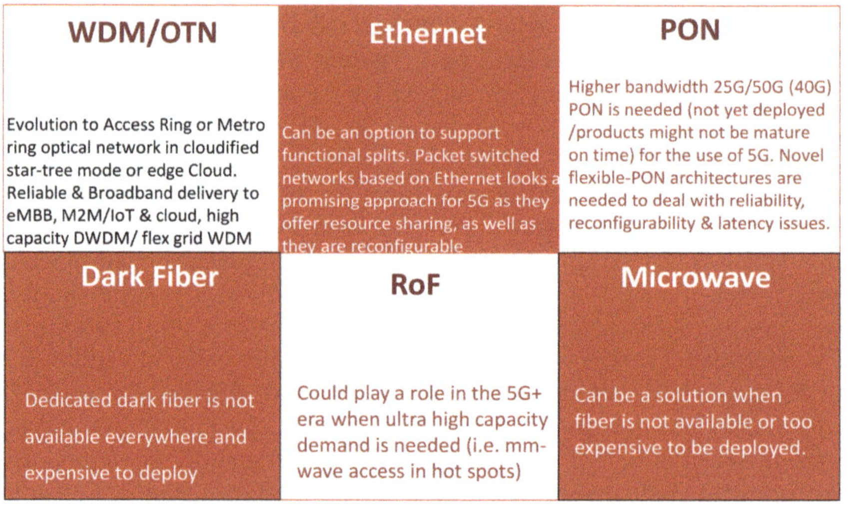

Figure 5.11. 5G optical front-haul, highlight characteristics of core and optional solutions. Dark fibers are reserved fibers in a fiber-bundle cable. RoF = radio over fibers; PON = passive optical networks, OTN optical transport networks with WDM (wavelength division multiplexing).

digital RoF (CPRI-like) cannot cope with mm wave technology (ADC etc) and RoF needs only the bandwidth required by the wireless channel. In all cases, the introduction of WDM technology will need resource sharing. Putting as many technologies with similar requirements together will increase the production volume and thus reduce the cost.

5.3.2.2 Photonic-enabled technologies: principal devices and photonic processors. In order to cope with the 5G revolution demands from an optical communication viewpoint, breakthrough technologies in optical interfaces and components will be needed. In particular, advanced technologies like silicon photonics (SiPh) can be the cost efficiency enabler for C-RAN based on optical front-haul and WDM. SiPh offers compatibility with the CMOS technology and will enable the low cost use of the optical transport in 5G access. SiPh devices like C-band reconfigurable optical add-drop MUX (ROADMs), or 25G externally modulated lasers (EMLs) are able to support the 25G rates, or in the future 50G and 100G must be available. ROADMs based on SiPh will lead to cost reduction compared to the conventional ones. DCI, OTN, 5G and PON, all share 25G requirements (figure 5.12) so SiPh devices should target the speeds of 25G and beyond and will enable a low cost solution for C-RAN based on optical front-haul. Moreover, graphene can transform SiPh to a low cost, CMOS compatible, outperforming the III–V equivalent platform, but big investment is needed to transform prototypes to industrial products.

The frequency bands for 5G, especially 5G+ besides the 2–6 GHz, are located in the microwave and millimeter wave (mmW) centralized at 18.6 GHz; 28.6 GHz or 56.8 GHz and 90 GHz. The free-license band of 7.0 GHz of the 58.6 GHz data channels can be delivered to mmW antennae which is MIMO integrated horn, via optical guided medium and converted to mmW electronic frequency region and amplified to drive mmW MIMO antenna elements.

Connecting devices range from massive capacity premises such as education campus DCs and direct access by academic departments, machine-to-machine communications, FTTP, fiber or OTTA, as shown in figure 5.13, in which the

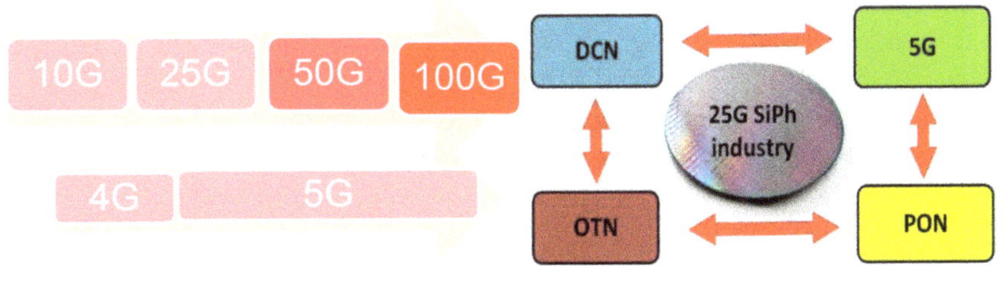

Figure 5.12. Components requirements and resource sharing drive optical technology. Labels: SiP = silicon-based photonics (integrated); 5G = 5th generation wireless network; DCN = DC networks; PON = passive optical networks, OTN = optical transport networks. 100G = 100GSym./sec transmission system, 50G = 50 Gsym./sec transmission systems, and so on to 10 GSym.sec.

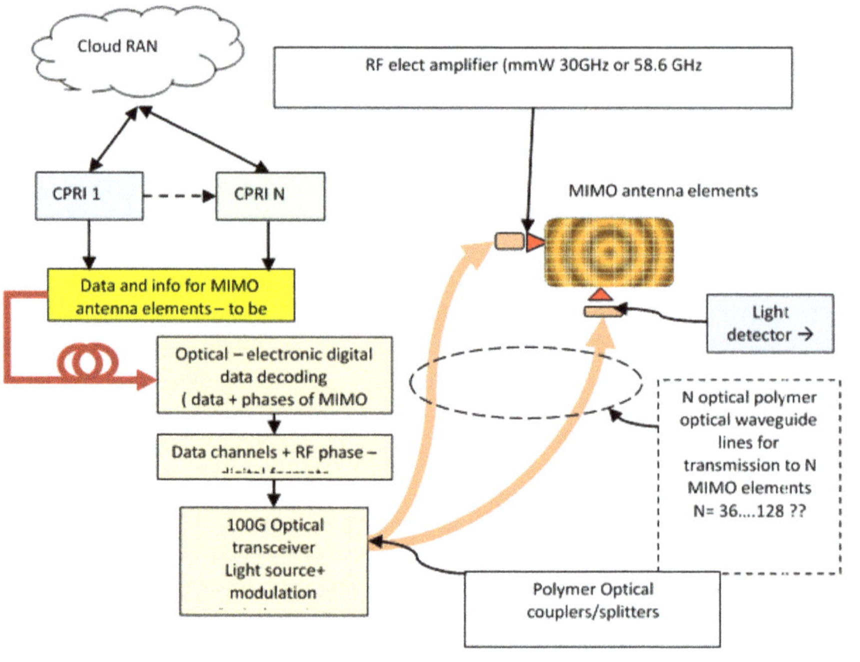

Figure 5.13. Integrated photonic processor in an OTTA optical to the mmW MIMO antenna.

data communication link delivers data to the mmW antenna elements where the optical signals are received and mixed with an LO laser in the photodetector. The frequency of the LO is tuned to a frequency difference with respect to that of the data carrier of mmW center frequency, e.g. 58.6 GHz. This results in opto-electronic conversions of optical signals to the mmW band. The mmW power amplifier delivers the signal to the antenna element, an mm size horn antenna embedded in a printed circuit board. This is the optical to MIMO antenna device for mmW band wireless carrier. This scalable design can have a bank of receivers and mixers and mmW RF electronic drivers. A bank of photonic phase shifters is used to generate a pattern of phase delay so that beam steering can be implemented. These phase shifters must be modulators in integrated optic structure so that fast and high-speed beam steering can be done. Furthermore, these OTTAs are passive optical networks for distribution and hence enable sharing of capacity and cost. The integrated photonic processor can be implemented in SiPh to keep the costs sufficiently low. Beam steering phase modulators can be embedded using p–n junction by ion implantation, operating in junction reverse mode whose speed can be in tens of GHz.

With the 7 GHz bandwidth of the free-license band and 500 MHz per sub-band for data allocation. High level complex modulation can be used as in common RF wireless. So that more than 10 Gbps can be delivered to MIMO antenna elements. With an antenna gain of about 22–0 dB for a 56 × 128 element array of the MIMO antenna we can deliver more than 1.0 Tbps capacity covering an area of radius of 500 m by beam steering to, for example, clients of a crowded stadium.

5.3.3 Security aspects and transmission technology for secret keys in CO-transmission of massive data transport

5G RoF networking needs to be flexible, dynamic and with the ability to handle high capacity. A converged SDN/NFV-based optical transport solution, will support a large number of services in parallel on a dynamic network operation to significantly reduce costs. SDN will allow the separation of control and data planes to reflect the integration of core, metro, access and 5G on a single SDN-based network platform. However, it is likely that a full-scale deployment of an SDN-based architecture will take more time than expected limited by capacity, compatibility, security and transformation cost issues, as well as orchestration and management, which needs to be optimized. In particular, secure communication will get more attention in optical-SDNs, however classical key distribution is subject to attack algorithms, computational power consumption and in the long term will be more relevant with the rise of quantum computers. Quantum key distribution (QKD)[1] can be considered as an additional layer of security and can be used in the 5G back-haul or the front-haul optical network. Continuous variable (CV)-QKD can integrate QKD using a legacy network (like OTN) without the need for dark fiber infrastructure, with the additional advantage of using standard devices with minor modifications. Optical-SDN in 5G back-haul/front-haul, i.e. decoupling of control and data layer provides a unique case for QKD. Quantum keys can be used to secure the control layer, and integration of QKD into an SDN network can greatly reduce the CapEx as the transmitter can be shared among different receivers, thus fewer QKD devices will be needed.

The CV-QKD is a case of power limited in transmission capacity as considered by Shanon in 1947. On the other hand, the transmission of possible capacity of maximizing the available system bandwidth is bandwidth limited cases. We present these two scenarios in chapter 7.

References

[1] Hillerkuss D *et al* 2011 26 Tbit/s line-rate super-channel transmission utilizing all-optical fast Fourier transform processing *Nat. Photon.* **5** 364

[2] Binh L N 2013 *Guided-Wave Fourier Optics, Chapter in Guided Wave Optics and Photonic Devices* ed S Bhadra and A Ghatak (Boca Raton, FL: CRC Press)

[3] Photonics.com 2018 Silicon chip distributes optical signals for potential use in neural networks https://photonics.com/Issues/Photonicscom_August_2018/i1040

[4] Chiles J, Buckley S M, Nam S W, Mirin R P and Shainline J M 2018 Design, fabrication, and metrology of 10 100 multi-planar integrated photonic routing manifolds for neural networks *APL Photonics* **3** 106101

[1] See also chapter 7 for more detail.

IOP Publishing

Transmission, Processing, and All-Optical Routing for Ultra-High Capacity Data Center Networking (Second Edition)

Le Nguyen Binh

Chapter 6

All-optical routing and switching for DC networking

Data center networking (DCN) has reached the stage in which the transport speed of channels is required to be at the highest level possible. This can be achieved, under the bandwidth-limited aspect of Shannon's channel capacity [1], by increasing the symbol rate and modulation of higher level and multiplexing of carrier frequency and polarization diversity, as shown in figure 6.1. The switching and routing of these ultra-high-speed signals for distributing and accessing by server networks are important, and novel technologies must be available to perform these functionalities. The transmission and switching under a noisy practical system can be written as

$$\text{Capacity} = \text{bandwidth} * \log 2(1 + \text{SNR})\text{bits s}^{-1} \tag{6.1}$$

where SNR is the signal-to-noise ratio in the electrical domain, thus the optical SNR (OSNR) is a square factor of SNR if the detection is photo-direct detection, and directly proportional with a multiplication factor of the amplitude of the local oscillator (LO), if under coherent detection. The noise of the switching devices is critical in such switching and routing systems.

This chapter describes the optical switching and routing technologies for inter- and internal DCNs as well as inter-networks for global terrestrial, deep space and/or submarine networking. The main photonic components are presented, especially the micro-electro-mechanical switches (MEMS) and their integration with other photonic components for a reconfigurable add/drop multiplexer (ROADM) with colorless, directionless, contention-less (CDC) capability. These CDC-ROADMs can assist in configuring many possible routings, transferring and switching of wavelength channels to desired connections.

doi:10.1088/978-0-7503-5849-1ch6

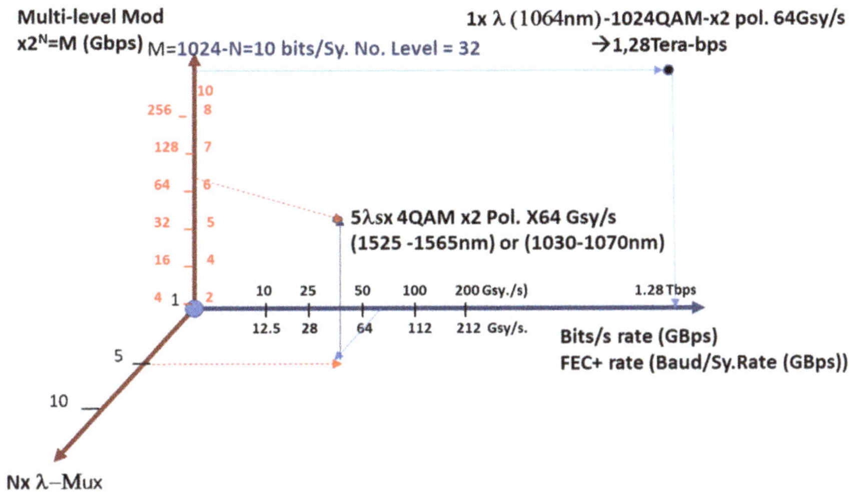

Figure 6.1. Optical multiplexing and higher order modulation formats to reach aggregate capacity to multiple tera-bits s^{-1} (Tbps).

6.1 Introduction

As our world becomes increasingly digital, DCs have become critical infrastructure for many businesses, personal social interactions and organizations. DCs are where vast amounts of data are stored, processed, and transmitted, making them a vital component of modern technology, especially in massive optical transmission and networking technologies.

One essential aspect of DC operations is networking. DCN involves connecting various components of the DC, such as servers, storage systems, and network devices, to enable efficient data transfer and communication. In this section, we'll explore some of the key considerations for designing and operating DC networks.

First, let's start with some of the basic components of a DC network. At its core, a DC network typically consists of switches, routers, firewalls, and load balancers. These devices work together to move data between servers and storage systems, manage network traffic, and provide security.

One of the primary challenges of DCN is dealing with the sheer scale of the infrastructure, the optical communications, routing and switching amongst sever racks. Large DCs can contain thousands of servers, each of which requires ultra-fast and ultra-high capacity connectivity to the network, the photonic technologies. This requires a robust network architecture that can handle high traffic volumes and provide sufficient bandwidth.

To address these challenges, many DCs use a hierarchical network design that separates the network into multiple layers. At the top of the hierarchy are core switches that provide high-speed connectivity between different parts of the DC, tera-bits s^{-1} speed connectivity. Beneath the core switches are distribution switches that handle traffic between different areas of the DC. At the bottom of the hierarchy are access switches that connect individual servers and storage systems to the network.

Another critical consideration in DCN is security. DCs often contain sensitive data, and it is essential to protect it from unauthorized access. This requires using firewalls to control traffic and prevent unauthorized access, as well as implementing strong authentication and access controls. The security aspects are not treated in this chapter but the optical technology that creates the secure transmission, is presented.

In recent years, software-defined networking (SDN) has emerged as a popular approach to DCN, and also optical SDN (OSDN). OSDN allows for greater flexibility and agility in network design and management, making it easier to adapt to changing networking needs. With SDN and OSDN, network operators can programmatically manage the network, rather than relying on manual configuration.

Currently the base rate of DCN is reaching toward 224 Gbps and much higher towards 1.6 Tbps (Tera-bits s^{-1}) per wavelength carrier, with DP-16-QAM (quadrature amplitude modulation) or DP-256QAM, whose signal eye diagrams of one polarized channel under coherent reception technique are shown in figure 6.2. Thus, the ultra-high speed and ultra-high capacity is under extensive upgrading to very high capacity massive data transport within and between DCs. Similarly, the aggregate bit rates can be increased when the dimension in one direction is increased, for example from QSK to 16-QAM, as shown in figure 6.3 with different symbol rate of 50 GSym s^{-1} and 100 GSym s^{-1}. However, the OSNR must be increased accordingly because the bandwidth of the signal increasing the corresponding noises of the system is doubling. We can see in figure 6.3 that when doubling from 50 GSym s^{-1} to 100 GSym s^{-1} the OSNR is increased to 25 dB from 12 dB for the same pre-FEC (forward error coding) BER (bit error rate) of 2e−2.

In conclusion, DCN is a critical aspect of modern photonic technology infrastructure. It involves designing and managing complex networks that can handle high volumes of traffic while providing robust security. By using a hierarchical network design, implementing strong security measures, and leveraging emerging

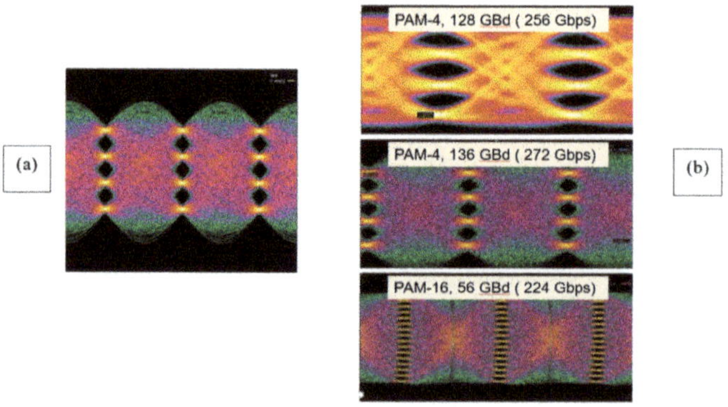

Figure 6.2. Typical eye diagrams of channels modulated by (a) dual polarization QAM (PAM-4 in each dimension) modulated signals, thence DP-16-QAM of coherent reception transmission systems; and (b) different bit rates of networking techniques plus forward error coding overhead, including PAM-16 (PAM-4 in one dimension) or 256QAM (16PAM in one dimension) modulation format.

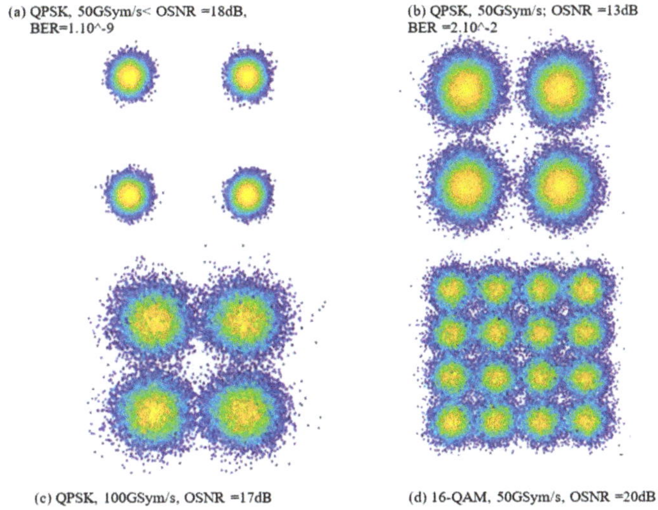

Figure 6.3. Typical constellation diagrams of QPSK and 16-QAM (a and b) 50 GSym s^{-1} and OSNR = 18 dB BER = 1.10^{-9}, and 12 dB BER = 2.10^{-2}; (c) QPSK with 100 GSym s^{-1} with OSNR = 17 dB → BER = 2.10^{-2}, and (d) 50 GSym s^{-1} 16-QAM with OSNR of 20 dB.

technologies like OSDN/SDN, organizations can build and operate DC networks that meet the demands of the digitalization age.

DCs are the backbone of the modern digital world. They are the foundation of the cloud computing infrastructure that powers most of today's digital services and applications. As such, DCs must be designed and built in a way that maximizes performance, reliability, and efficiency. This chapter, firstly provides an overview of the structure of DCs and how they are designed to ensure optimal performance.

6.1.1 DC networking

At the core of any DC is its server racks. These racks are the physical structures that house the servers, storage, and networking equipment. The racks are typically organized in aisles and arranged in a configuration that allows for easy access. They store, process, and transmit vast amounts of data, powering everything from websites to online banking. But what exactly is the structure of these DCs? At the most basic level, a DC is structured with computer systems and associated digital components, e.g., telecommunications and storage systems. In addition, it includes backup power supplies, redundant data communications connections, cooling and protection controls and various security devices. DC can be structured in different sizes and configurations depending on demand capacity and locations, e.g., at the edge or center of networks which are very very large, that is, intercontinental and inter-cities. The smallest DCs are often referred to as server closets, and are typically located within an office building. Larger DCs are usually located in dedicated buildings, and can range from a few hundred square meter to hundreds of thousands of square km. The DC internal structure depends on its size and purpose, the transfer and communication speed can reach multi-Tbps per wavelength carrier and guided via single-mode fibers (SMFs). Several lightwave carriers can be multiplexed and de-multiplexed in a single strand of fiber [2], thus the capacity of information can reach 50 Tbps and even higher per fiber if the transmission carriers are located in the S- and L-(short and long) spectral bands. Smaller DCs may consist of just a few racks of servers, while larger ones may include multiple rooms, each with its own cooling and power systems. DCs may also include a variety of networking equipment, such as routers, switches, and firewalls.

DCs are typically organized into a hierarchical structure. At the top-level is the core layer, which consists of the most important servers, such as web servers and database servers. This layer is often considered as the DC central sub-system.

The next layer is the distribution layer, which consists of networking equipment that connects the core layer to the edge layer. The edge layer consists of servers and other devices that provide services to end users. Finally, the access layer consists of the devices that end users use to access the services provided by the DC.

DCs are designed to provide a secure, reliable, and efficient environment for storing and processing data. By understanding the structure of a DC, organizations can ensure that their data is protected and that their systems are operating efficiently.

Driven by DCN and cloud computing, Internet of Everything (IoE), intelligent wireless access devices, and massive data processing, high-quality-demand requirements in terms of ultra-wide bandwidth, low delay-time, and massive inter-connectivity are imposed on the inter-connections and communications within the DC inter-networking (DCI). Traditional intra-DCN based on electronic switches is organized in a multi-level topology, and therefore severely affected by the communication and switching speed and structures, hence it has low servicing efficiency. Thus novel architectural and technological developments are required to provide scalable DC global networking.

Agile and transparent optical switching with wavelength selection capability are considered as an attractive technology, and realization with micro-optics and lightwave switching devices can provide data-rate and data-format flexible operations. Furthermore, this can minimize the power consumption in digital signal processing (DSP)-based transceivers as well as modulation format-dependency. This can be termed as all-optical switching and routing that offers efficient and flexible bandwidth utilization as well as spectral channel granularity, especially from 6.25 to 75 GHz and 225 GHz, which can be tunable. All-optical selective switching and reconfigurable optical add/drop multiplexing/de-multiplexing (ROADM) technologies are described in this chapter for DCI networks. DCI networking can be highly scalable to meet the expansion of DC operations.

Firstly, technology of MEMS is presented with a minor fabrication process. Secondly, we discuss the use of MEMs in the design of all-optical routing systems and their integration in all-optical networking.

6.1.2 Power consumption

A medium DC can be considered as a facility that has between 5 and 50 racks of servers and other IT equipment, and consumes a wall-energy power consumption between 10 and 100 kW. The power consumption of a medium DC can vary greatly depending on the type of equipment and the amount of load it is under. Generally speaking, the power consumption of a medium DC can range from 10 kW to 100 kW, depending on the number of racks and the type of equipment.

The power consumption of the servers and other IT equipment in a medium DC is typically the largest contributor to the overall power consumption. Servers typically consume between 0.5 kW and 4 kW of power, depending on the type and number of servers. Additionally, storage devices, such as SANs and NASs, can consume between 0.5 kW and 2 kW of power.

The cooling system in a medium DC consumes a major part of the energy. We note that an average digital signal processor consumes at least 7 W of power, thus 70 W wall-plug power, and here there are several thousands of processors in a medium DC. Cooling systems can consume power of up to and higher 10 kW, depending on the type and number of cooling systems.

The power consumption of the network and other infrastructure in a medium DC can also vary greatly. Network switches, routers, and other networking equipment can consume between 0.5 kW and 3 kW of power, depending on the number and type of processing devices.

Finally, the lighting power consumption in a medium-size DC can also vary greatly and consumes between 0.5 kW and 2 kW energy.

In conclusion, the power consumption of a medium DC can vary significantly depending on the types and number of equipment, as well as the servicing load. In general, a medium DC can reach a wall-plug power of 10 kW to 100 kW.

6.1.3 MEMS, WSS and ROADM

In order to reduce the power consumption and ultra-fast and ultra-high capacity communications within the DCs, a number of optical sub-systems must be used such as MEMS, WSS (wavelength selective switches) and ROADM. MEMS can provide switching on a large scale. However, no specific wavelength can be assigned by MEMS. WSS is a selectable switch that can be assigned with wavelength carrier and channel signal bandwidth to desired input and output ports. A combination of MEMS and ROADMs switches can be formed for structures with a photonic system that can allow adding (injection) and dropping of selected wavelength channels from input streams of data channels to their destinations. This would allow flexible, agile and scalable DC structures.

This chapter describes the structures and functionality of DCs that employ these photonic sub-systems, i.e. optical switching and manipulation in the photonic domain, of channel paths to and from ultra-high speed servers and associated speed service layers to provide agile, flexible and scalable DCs.

6.2 Optical MEMS, WSS and routing in optical domain

6.2.1 Remarks

More than 50 years ago, the spatial light modulator array [3] and silicon torsion mirror [4] were invented, and these created an opening of new applications in optical networking and processing. This is presently known as the optical MEMS, which is a key device and technology for the 'dynamic' optical signal processing for optical transmission systems and networking, especially as the wavelength selective add and drop sub-systems which can direct and route optical channels of different wavelength to their desired destinations, that is, optical switching and add or drop to and from desired optical channels. The telecom network accelerated in the late 1990s and early 2000s due to this technology maturation. A wide range of optical MEMS systems and ROADM have been extensively deployed to meet the high for demand optical networking, especially in routing and interconnection to form flexible and agile DC global networks. MEMS are employed extensively in many applications, such as optical switches, optical cross-connect (OXC), optical filters, dispersion compensators, spectral equalizers, spectrometers, tunable lasers, and other dense-wavelength-division-multiplexing (DWDM) networking devices such as wavelength add/drop multiplexers, WSSs, and wavelength selective optical cross-connect.

A WSS is a switching device used in WDM optical communication systems. It is an all-optical device which can switch optical signals between multiple paths based on their wavelengths. The WSS works by using an array of optical filters to selectively route optical signals with a specific wavelength to a specific output port. The WSS allows multiple signals with different wavelengths to be sent over the same fiber. It is also used in ROADM systems, which allow optical signals to be added or dropped from a fiber network. WSSs can also be used in optical amplifiers and optical cross-connects. They are used to improve the efficiency and flexibility of optical networks by allowing more signals to be routed to different paths.

WSSs can selectively route different wavelengths of light through an optical network. They can dynamically allocate data channel passband to different paths within the network.

In DCN, WSSs can create flexible and scalable optical interconnects between switches and servers. They can support high-speed data transmission over long distances, allowing DC operators to manage traffic efficiency and reduce channel delay times. WSS can improve network reliability and reduce network failures by providing redundancy and fault capabilities. In general, WSS can be the critical component of massive data DCN that would enable agile, efficient and flexible optical networking.

The next section briefly describes the MEMS technology, followed by the system applications of MEMS in optical routing and switching. A typical fabrication of MEMS can be referred to in section 6.7.

6.2.2 MEMS and nano-technology

MEMS employs the electromechanical dynamics in micro-structures that combines mechanical and electrical components on a single microchip which can only be produced by micro-electronic technology.

MEMS are increasingly being used in WSSs due to their small size and low power consumption. WSSs are used to route optical signals based on their carrier wavelength, allowing for efficient data transmission. MEMS-based WSSs can switch optical signals with a high degree of accuracy and speed, thus, making them ideal for high-speed optical communication networking.

MEMS-based WSSs are also being used in optical amplifiers, commonly used to amplify optical signals in the photonic domain. These amplifiers are used in optical networks as power boosters to enhance the strength of light signals, allowing for more efficient transmission of data.

MEMS-based WSSs are also being used in optical communications systems. These systems are used to transmit data over long distances, allowing for faster data transfer rates. This technology is also being used in optical computing networks, which are used to interconnect computers and other devices over long distances. The main technology that allows the possibility for fabrication and realization of MEMS is nano-deep ultra-violet (nano-DUV) technology.

Nano-DUV technology deals with the creation of materials, devices and systems in the nanometer scale (1–100 nm) through manipulating matter using deep UV lithography. At that scale and exploiting novel properties arising because of the nano scale, nanotechology has an impact on lightwave beam path switching, routing and steering. It enables the fabrication of numerous optical switches and routers, and devices for terrestrial and space applications in optical communications and networking.

The fabrication of MEMS using such nano-lithography technology, as outlined in section 6.7 (appendix), involves several steps. In short, firstly, the desired material is chosen and a pattern is created on a substrate. This pattern determines the shape and size of the MEMS device. Next, a thin film of material is deposited on the pattern to form the device. This step is implemented by chemical vapor deposition. After forming the device structure, plasma dry etching is used to form the final desired features. Then the device is tested and packaged.

DUV laser nano-litho-technology enables the fabrication of MEMS with high precision, allowing for the creation of complex devices with nano-scaled features. This technology is used to create components with high levels of electrical conductivity, which is important for the operation of MEMS devices.

Overall, nanotechnology is an important tool for the fabrication of MEMS devices. It allows for precise control over the structure and composition of materials, enabling the creation of complex switching and routing devices with nano-resolution features. This technology is also used to create components with high levels of electrical conductivity and durability, making it an ideal choice for MEMS applications.

The fabrication can be referred to in section 6.7.

DUV-nanotechnology integration can certainly improve MEMS devices. In high speed and massive data transfer and communication for optical networking, MEMS can act as signal beams steering and switching with low consumption of power and low insertion loss due to free-space reflection of lightwaves. The free-space mechanism is advantageous as compared to their integrated optics counterparts when the number of beams is high.

6.2.3 Optical MEMS and applications

MEMS integrated with surface micromachining, have significantly influenced the development of miniaturized optical–mechanical micro-devices. They create a substantial impact in the field of integrated and free-space optics and related devices, which contributes to many technological possibilities for optical networking. The precision mechanics of MEMS, microfabrication techniques, and their optical functionality make possible a wide variety of micro-mechanical movable and tunable mirrors, lenses, filters, and other optical structures in several applications such as in photonics, networking, electrostatic, magnetic, thermal, and pneumatic actuators as well as providing precise mechanical dynamics and control. A large number of electromagnetic field modes of associated structures can be integrated with micro-beam-steering and diffractive optical MEMS to provide precision micro-mechanical dynamics. These functions can be utilized in integrated optical switches, filters, and optical in-line dispersion compensators which are essential elements in optical long-haul and short-haul with ultra-high capacity networks.

MEMS potential in integrated silicon photonics, associated with electronics and mechanics offers great advantage in several fields of bio-photonics and optical networking instrumentation, where the integration of miniaturized optical detection systems enables agile and inexpensive systems. The precise dimensions and alignment of MEMS devices, combined with the mechanical stability due to their miniaturization, make optical MEMS switches well suited to a variety of challenging measurements, especially in DC networks and several points in optical local or global networks. Micro-optical free-space systems, such as satellite optical technology, also benefit from these nanostructures.

6.2.4 Bandwidth tunable optical filtering

A fabrication process which integrates silicon MEMS actuators with silica optical structures can be used to form integrated optic tunable filtering in the photonic domain. The suspended silica optical waveguides can be controlled and activated by a Si electrostatic comb drive actuator, that is, for example with a maximum displacement of 8 μm at 50 V bias voltage.

Optical tunable-bandwidth filters can be highly useful in optical communication systems and networks, especially in optical signal processing. The tuning of these optical micro-devices can be further accomplished with an additional bulk micro-positioning stage. The appendix of section 6.7 gives an outline of the fabrication steps of optical MEMS.

6.3 Optical routing for DCI and DCN

6.3.1 Introductory remarks

Coherent detection [5] has demonstrated significant improvement in reception and much improved and higher OSNR[1] (defined with a noise bandwidth of 0.1 nm) to optical networks. Thus, the coherent communications systems can extend the transmission reach and enabling reconfiguration of network architecture, from long-haul to fiber to the home or to the building throughout global networks.

[1] Readers can refer to the measurement techniques of OSNR of polarized multiplexed channelks from JDSU Viavi, 'Measuring In-Band OSNR on Polarization-Multiplexed Signals', https://www.viavisolutions.com/en-us/support/files/I-OSNR-with-PolMux.pdf.

This section gives further fundamentals of coherent optical reception where the receivers can be tunable to a desired carrier wavelength, thus selecting the desired channel and rejecting all other channels. This wavelength selectivity can be used to provide add/drop functionalities of selective optical carrier channels to form a ROADM. The optical nodes can be considered as the intersection points at which a number of channels of different wavelength carriers can be injected (or added) in and at the same time with another set of wavelength carriers that can be dropped out of the nodes, to their destinations. At the same time a number of other different wavelength channels can be transmitted through or by-pass to the next node points.

Readers can refer to the appendices of this book on optical fibers including propagation properties (appendix A), transfer function of optical fibers and thence their impacts on optical networking. Furthermore, typical estimation of power budget and estimation techniques for signal parameters such as the root mean square (RMS) etc are presented in appendix B. These appendices have been summarized as references from appendices of a number of books published by the author.

Such a ROADM can provide flexibility on the design and implementation of network architecture on demand (AoD), with necessary north, east, west and south (NEWS) optical interfaces, so that the ROADMs can be deployed as part of a disaggregated interconnecting and transponder in the optical domain in which various network elements can be connected by different vendors' products.

In an open line system (OLS) of DC networks, signals can be injected into the network through the add ports of a ROADM. After traveling through several line degrees and long-haul fiber core transmission lines incorporating in-line amplifiers, the channel signals eventually reach their destination. They are transmitted to client terminal equipment via the ROADM drop interface in the optical domain. This offers significant advantages as no conversion back to the electrical domain. Thus, these add/drop modules can perform in a layer of wavelength filtering and manipulation via the fixed multiplexers and de-multiplexers.

The advent of WSS technology has opened new technological approach in optical switching and networking techniques. WSS is a key component of ROADMs, which are used to manage and optimize optical networks using multi-wavelength multiplexing.

WSS technology enables ROADMs to switch wavelengths of light on an optical fiber, allowing for more efficient network management. As a result, ROADMs are becoming increasingly popular for managing large-scale optical networks. WSS technology works by using an array of optical filters to select one or multiple specific wavelength lightwaves carriers, then switch and route them to different output fiber ports. These wavelength channels can thus be switched to other fiber ports, allowing for efficient routing of data channel in massive capacity. WSS technology also allows for the dynamic addition and removal of wavelengths channels, allowing increased flexibility in network management. The optical filtering capability of WSS with resonant cavities can reduce optical noise in an optically amplified ROADM.

The benefits of WSS technology in ROADMs are numerous. WSS technology allows for more efficient network management, as it allows for the dynamic addition and removal of wavelengths. Additionally, WSS technology enables ROADMs to switch wavelengths of light on an optical fiber, allowing for more efficient routing of data. Finally, WSS technology enables ROADMs to support a wide range of services, from voice and video to data and storage.

Overall, WSS technology is a key component of ROADMs, and is essential for efficient network management. WSS technology enables ROADMs to switch wavelengths of light on an optical fiber, allowing for more efficient routing of data. Because WSS technology allows for the dynamic addition and removal of wavelengths, allowing increased flexibility in network management, ROADMs are becoming increasingly popular for managing large-scale optical networks.

A recent ROADM design typically used a WSS for the add/drop in order to provide the desirable set of flex-grid, colorless, directionless, contention-less (CDC) properties for ROADM to be realized, so CDC-F by integrating MEMS to the system, as shown in figure 6.4.

The fiber arrays are employed as signals inputs as well as collecting ports at the output. The optical signals are manipulated by the MEMS and associated optical components in the middle. A lens for alignment of the fiber beam arrays to the MEMS mirrors through which the signals are manipulated and steered to assigned positions on a grating and spherical mirror and thence the LCOS (liquid crystal on silicon) which acts as an optical switch and phase manipulator for resonance, hence optical filtering depending on the cavity length. The grating ensures the de-multiplexing of channels of different wavelength or optical frequencies. Therefore, manipulating of wavelength multiplexed channels can be conducted so that different wavelength channels can be implemented, and later the ROADM which is a highly important component for flexi- and reconfigurable agile optical networking.

The structure of the multi-input/output ports ROADM are simplified as shown in figure 6.4(a).

The testing of a WSS under different passband characteristics using broadband signal generator and QPSK coherent modulator are given in figure 6.5(a), as well as two single lightwave carriers tunable to the edge of the optical passband of the WSS filters (figures 6.6 and 6.7).

Under certain power budget limits on the number of allowable channels and their optimum power, the sensitivity of coherent receivers can be adjusted in a ROADM system. Using power splitters/couplers in the add/drop ROADM modules can further improve the ROADM features such as filter-less or semi-filter-less networks.

Wavelength carriers are classified into ITU-T-standard[2] grid size of 12.5, 50, 75, 100, 150 and/or 300 GHz spacing whose bit rates can be extended accordingly so that the aggregate DWDM capacity can be increased, respectively, as shown in figure 6.9. It is noted that super-C-band can be used increasing the whole C-band capacity to 110 channels of 100 GHz spacing. The corresponding baud rate (bits per secon, bps) can be 250 Gbps to 1.6 Tbps for such grid using multilevel modulation techniques, as presented in chapter 2. The transmission distance can vary

[2] ITU-T = International Telecom Union, Telecommunications.

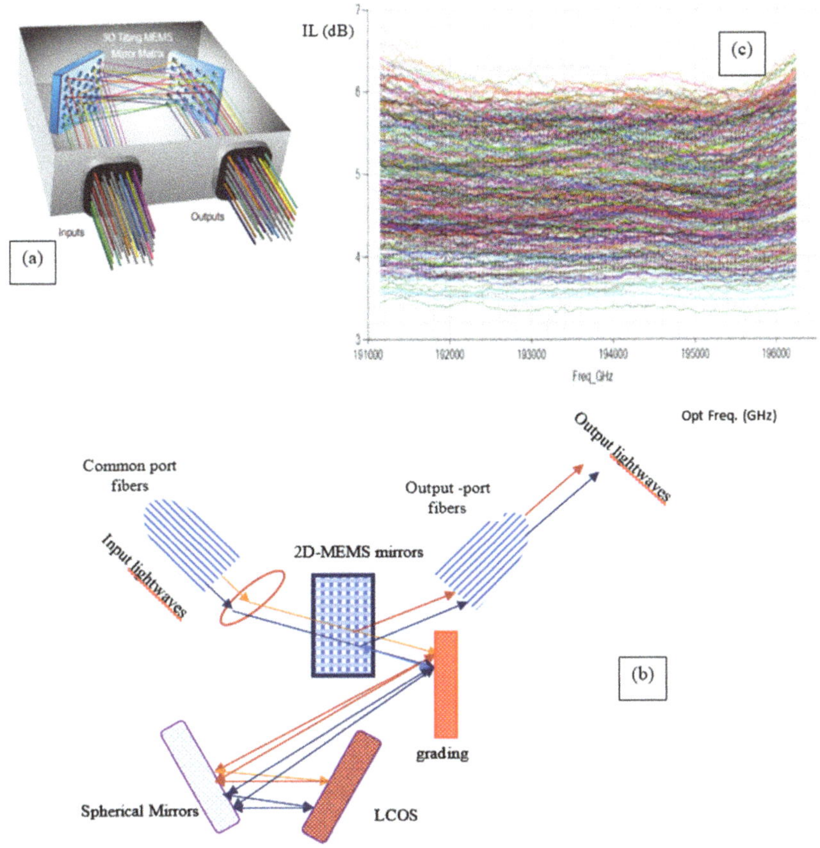

Figure 6.4. (a) Optical cross-connect using two-dimensional MEMS interconnection. (b) Optics of 8 × 28 WSS by Lumentum (top) and (bottom) an insertion loss measurement (in dB) of a contention-less directionless, colorless and flexible grid (CDC-F) Twin 8 × 24 WSS, showing all frequencies and all combinations of the eight common ports switched and routing to 24 add/drop ports. Copyright 2018 IEEE. Reprinted, with permission, from [6]. (c) Insertion loss of ROADM port to port (reference: Lumentum Inc. commercial product). LCOS = liquid crystal optical switch.

accordingly. In a DC the distance can be from a few hundred meters to a few kilometers only and the rates and optical connect distance as well as traffic can be ultra-dense and ultra-high speed under SMF transmission media.

When talking about information data streaming, two key terms that come up are channel bandwidth and grid size. These two metrics are important for understanding the quality of streaming of data channels and how it will be experienced by network users, and then in turn by data users and streamers of video, data etc.

Channel bandwidth is the amount of data that can be transferred over a given channel. It is usually measured in bps and then giga-bps and tera-bps. The higher the channel bandwidth, the more data can be transferred over the channel in a given amount of time. This means that higher channel bandwidths can provide higher quality data channel streaming.

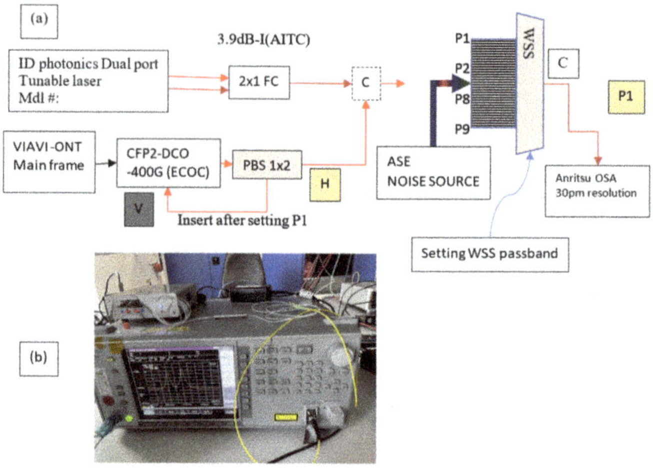

Figure 6.5. Test using wideband signal generator and optical QPSK modulator, with 400G channel (optical) via an optical PBS unit where the V_component is feedback to the optical receiver and the H-component is fed into the common port to drop to the WSS port 1 and then the OSA to observe the spectrum. The reverse is also true. (a) Set-up block schematic; (b) photo of set-up for monitoring dual laser generated spectrum prior to launching into WSS. C = common port of WSS, where C is the output to direct any input port signal to C with a passband to be specified. If C is used as input then one can direct to any specific port as assigned with the passband bandwidth and central wavelength.

Grid size is the number of Hz and bits in a given channel frame. It is usually measured in GHz spacing between transmitted channels in the optical spectrum of silica fibers, the traditional C-band, the extended C-band and the super-C-band as well as the L-band and S-band, the O-band (around the 1310 nm region).

The wider the grid size, the more data that can be transferred over the channel. This means that higher grid sizes can provide higher transmission rate. The aggregate bit rate per channel is the total amount of data required transferring over the channel in order to render the data channels. It is usually measured in GHz for ultra-high capacity transmission. The higher the aggregate bit rate per channel, the more data that can be transferred over the channels.

Understanding the relationship between channel bandwidth, grid size, and aggregate bit rate per channel is important for minimizing the crosstalk between DWDM channels. By choosing the right combination of these three metrics, DWDM channels transmission operators can ensure that users have the best possible effectiveness in spectrum efficiency. Figure 6.8 shows a typical band of all three C-bands, the traditional C-band to super-C-band and the bandwidth of the channels their grid can support without crosstalk.

Figure 6.9 illustrates the mixing of different channels of different bandwidths on a flexi-grid spectrum.

Flexi-grid is a technology used in optical fiber communication networks that allows for greater flexibility in the use of available bandwidth. It works by dividing the available spectrum into smaller channels, each of which can be used for different applications. This allows for more efficient use of the available spectrum and

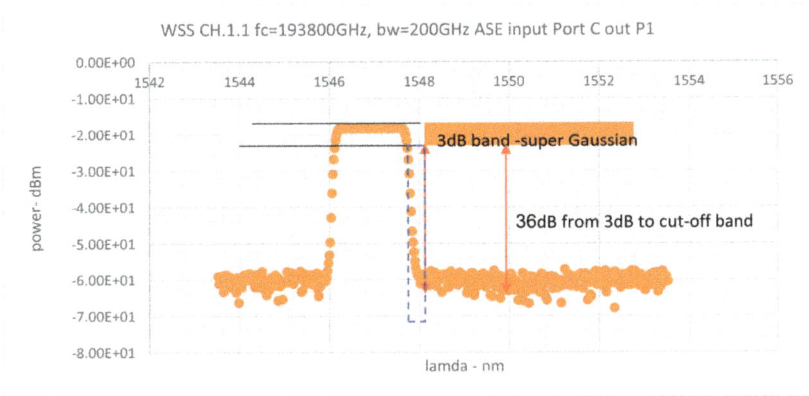

(a) WSS passband characteristics – fc =193800GHz input source ASE broadband source. WSS P1 setting fc =193800GHZ BW =200GHz. 1.5 unit x 0.4nm = 0.6 nm or 75GHz.

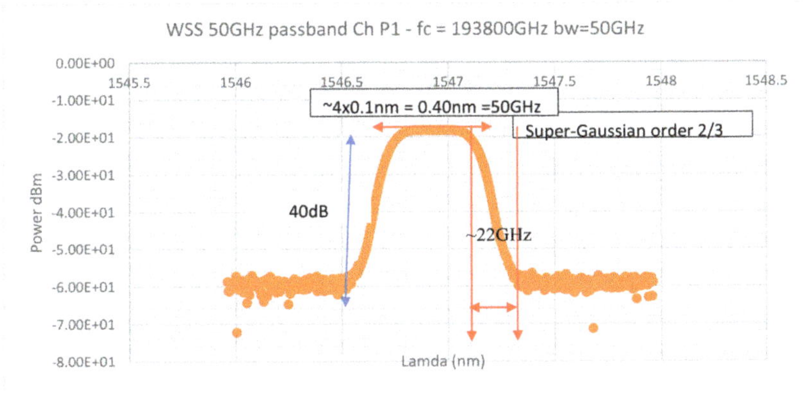

(b)

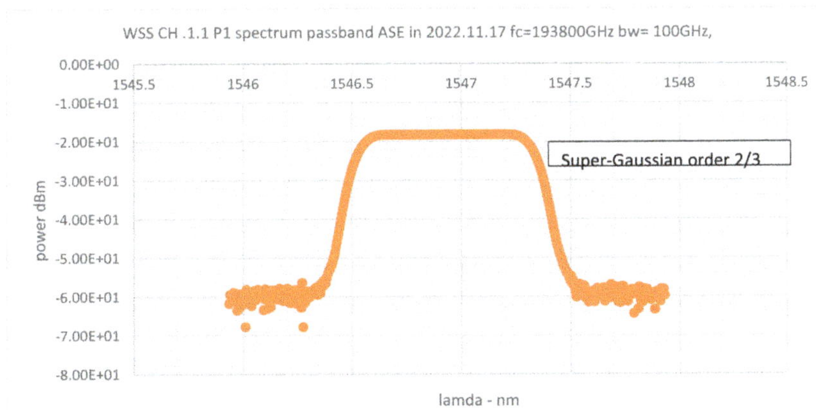

(c)

Figure 6.6. Spectrum of 1 × 9 WSS under setting of common port as multiplexed input channels and output ports as filtered and switched channel under different bandwidth (a)–(e) optical channel at output of WSS channel port 1 with input common source is a broadband noise source, bandwidth from 200 GHz to 12.5 GHz, respectively. Labels: BW = bw = bandwidth of passband. *fc* = center frequency of the WSS passband.

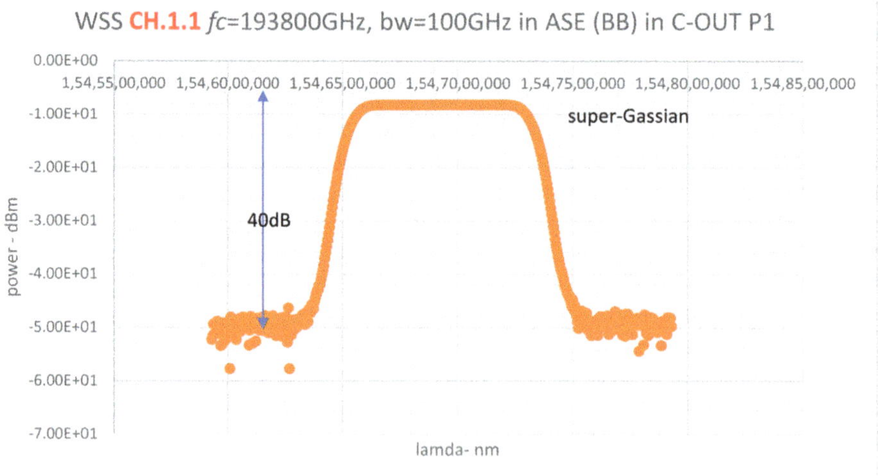

(d) WSS passband of bandwidth of 100GHz with input ASE broadband source. WSS P1 setting fc=193800GHz BW=100GHz, attenuation=0. CH1.1 indicates port number 1 of the WSS module number 1 of a twin selective switch.

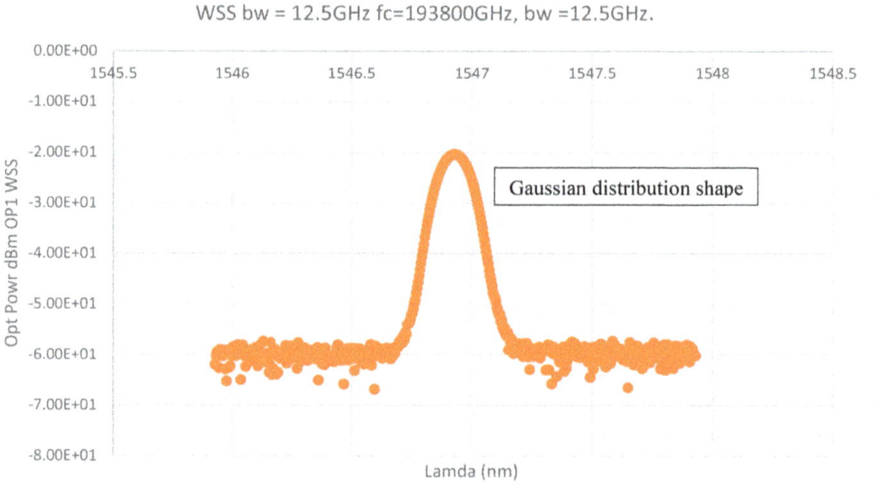

(e) WSS passband 12.5GHz with ASE broadband input source – channel P1 bw = 12.5GHz; input WSE wideband source at -20dbM power flat level. Span = 3nm or 3x125GHz full span.

Figure 6.6. (Continued.)

increases network capacity. Flexi-grid technology is based on DWDM techniques in which the fiber low-loss spectrum can be split into multiple spectral windows, each of which can carry independent data streams. This allows for high transmission capacity over a single strand of fiber and multiple channels can be used for different applications or dropped to many locations of different bandwidth/capacity demand. Flexi-grid technology has several advantages over traditional fixed-grid DWDM networks. It allows for more efficient use of the available bandwidth, as multiple

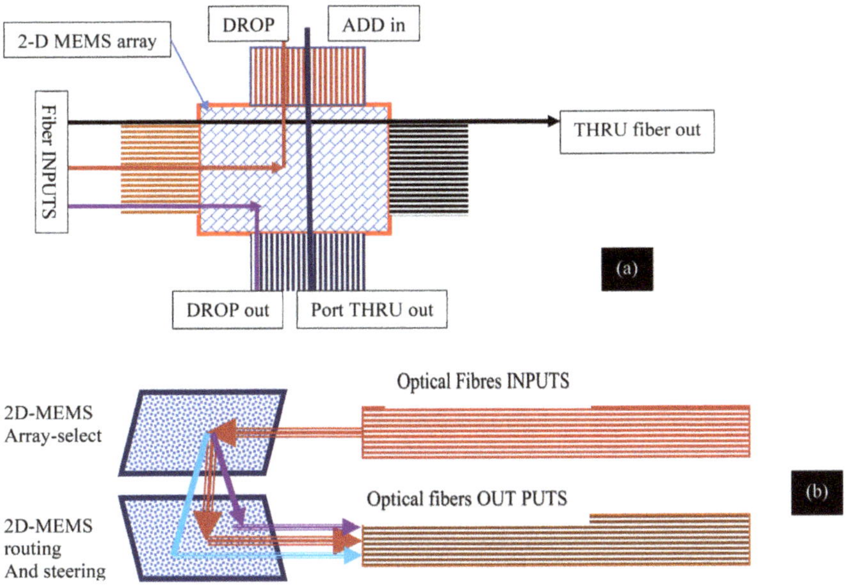

Figure 6.7. Schematic of (a) two-dimensional MEMS structure; and (b) three-dimensional MEMS selection, routing and steering.

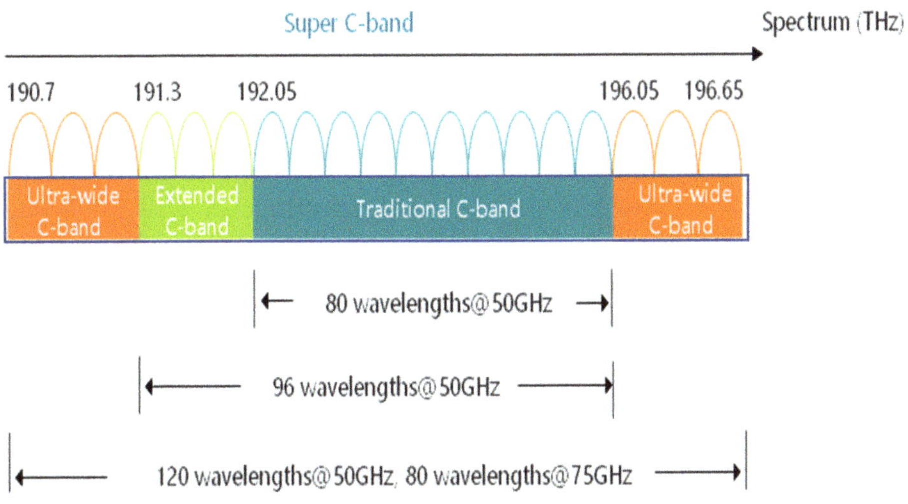

Figure 6.8. Spectrum allocation of super-C channels of a total of 120 of 50 GHz spacing and 80 channels of 75 GHz spacing. For the extended and ultra-wide C-band, it can support 96 channels of 50 GHz spacing. The traditional C-band supports 80 channels of 50 GHz spacing.

channels can be used for different applications. Furthermore, it allows for more energy efficient usage, as the total power can be distributed among the channels. Flexi-grid technology is becoming increasingly popular in optical fiber communication networks as it allows for greater flexibility in the use of available bandwidth.

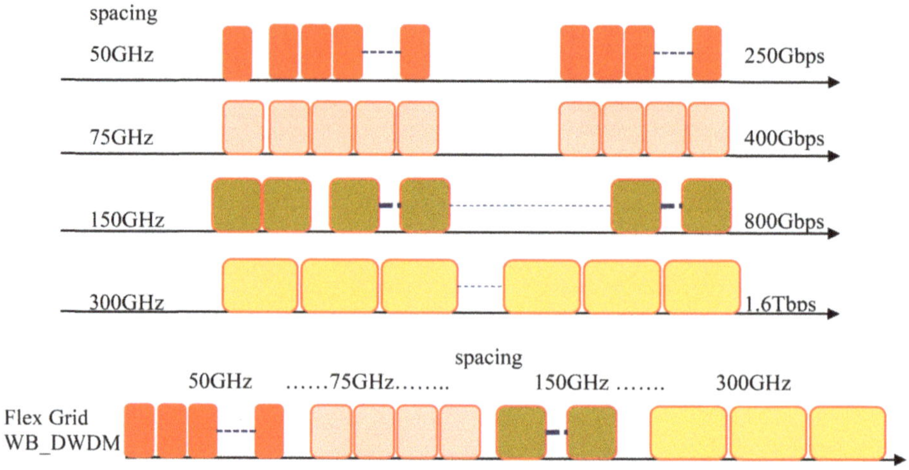

Figure 6.9. Channel bandwidth and grid size and aggregate bit rate per channel.

It is also becoming increasingly used in long-haul networks, as it allows for more efficient use of the available spectrum and power.

The use of super-C-band in optical communications systems is becoming increasingly popular as a cost-effective and reliable way to transmit data over long distances with minimal loss. Super-C-band is a type of optical fiber that operates in the 1550 nanometer (nm) wavelength range. It is a relatively new technology that offers higher bandwidth and improved performance when compared to traditional C-band fibers.

Super-C-band is ideal for applications that require high-speed data transmission capacity over long distances. This type of information band transmission over silica fiber is commonly used in long-distance networks, such as MANs (metropolitan area networks), wide area networks (WANs) and terrestrial as well as submarine networks. It is also used in high-speed DCs and other applications that require transferring and transmitting of agile and massive data.

One of the main advantages of super-C-band is its ability to reduce the amount of signal attenuation that occurs over long distances. This type of fiber is able to transmit data over distances of up to 80 km with minimal loss. This makes it an ideal choice for applications that require high-speed data transmission over long distances. Additionally, this type of fiber is also more resistant to environmental factors, such as temperature and humidity. Another key benefit of super-C-band is its ability to reduce the amount of crosstalk between fibers. Crosstalk is the interference that occurs when signals from one fiber are picked up by another. This type of fiber is designed to reduce the amount of crosstalk, which makes it ideal for applications that require high-bandwidth and low-interference data transmission.

Overall, super-C-band is an excellent choice for applications that require high-speed data transmission over long distances. It offers improved performance, greater bandwidth, and greater resistance to environmental factors when compared to traditional C-band fibers. As the demand for high-speed data transmission continues to grow, this use of this super-C-band is likely to become even more popular in optical networking.

6.3.2 MEMS: operation principles and applications

MEMS can switch different wavelength channels by converting, steering and/or routing optical signals from one or multiple number of optical ports to other ports through which optical channels can be transmitted, added or dropped whose spectral and spatial range can be limited to the transparency of the switches. It has one or more selectable transparent spectral windows in the photonic domain.

In addition to MEMS, there are also other optical switches, including mechanical types, pyro-optic switches, acousto-optic switches, electro-optical switches, magneto-optical switches and liquid crystal optical switches.

6.3.3 Operation principles

Optical MEMS are tiny machines that use light to measure and control physical objects. They are made up of tiny mirrors, lenses, actuators, and other components that are controlled by electrical signals. The operation principles of optical MEMS involve the use of light to manipulate tiny objects. When an electrical signal is applied to an optical MEMS, the mirrors, lenses, and other components move in response to the signal. This movement is used to control the position of the object, such as a laser beam or a light source. The movement can be in any direction, allowing for precise control of the light. The mirrors, lenses, and other components can also be used to focus the light in a specific direction. This is used to create a beam of light that can be used for beam sweeping and thus routing to desired positions.

Optical MEMS are important for optical communications and networking. This section discusses the operational principles of MEMS optical switches, and how they can be used to improve communication systems.

MEMS optical switches are based on MEMS technology. This technology utilizes a combination of mechanical and electrical components to create a switch that can be controlled electronically. The switch is made up of a micro-mirror, a light source, and a detector. When the switch is activated, the micro-mirror reflects light from the light source onto the detector. The detector then sends a signal to the controlling device, which can be a computer or a network switch (figures 6.10 and 6.11).

In generic terms, a MEMS-based optical switch can be either 2D or 3D in structure, the 2D- or 3D-MEMS depending on the switching and routing of lightwaves route in a plane or in three-dimensional space, that is, the mirrors of the switches, can be rotating on a plane axis r in a twisted dynamic. The rotating mirrors of the 2D-MEMS optical switch is monolithically integrated on the silicon substrate via a surface micro-mechanical axis and the mirror rotating is via a surface mechanical force generated by interdigital (IDT) electrodes, then surface wave propagation forces the torsional rotation of the silicon mirror. The collimated light is then connected to the designated output terminal through the rotation control of the micro-mirror. When the micro-mirror is horizontal, the light beam passes and is

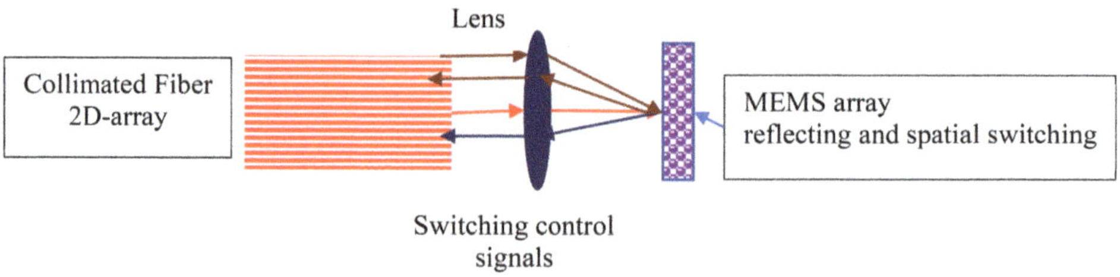

Figure 6.10. Working principle of MEMS optical switch.

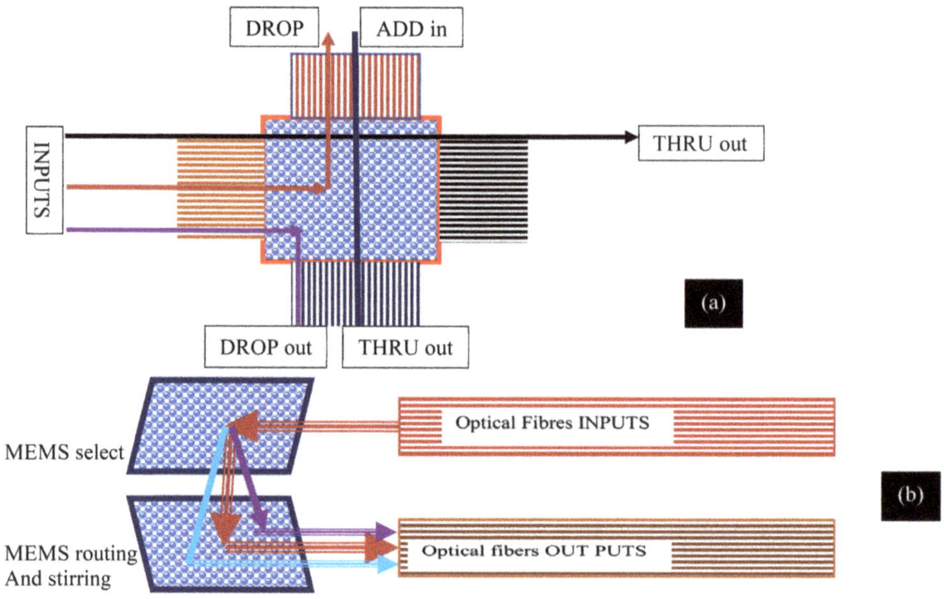

Figure 6.11. (a) 2D planar switching using MEMS matrices; (b) optical switching using 3D twin MEMS planes as controlled mirror deflectors.

reflected when the micro-mirror is switched up, that is, when the micro-mirror is in perpendicular or $\pi/4$ angle with respect to the silicon substrate.

In the 3D-MEMS, the micro-mirror rotates arbitrarily along two axes, then the mirror at different angles can be forced to direct the lightwaves to the desired output. These mirror-arrays appear normally in pairs. Such a 2D mirror can reflect the lightwaves beam to the mirror surface of the second array, and then the lightwaves can be further reflected to the desired output port.

6.3.4 MEMS fibre optic switch

A fiber optic switch (figure 6.12) is a multiport device in which the optical input and output port are arranged on the same face and the electrical control plane is placed on the back face. The optical beam is launched and focused on a MEMS array plane

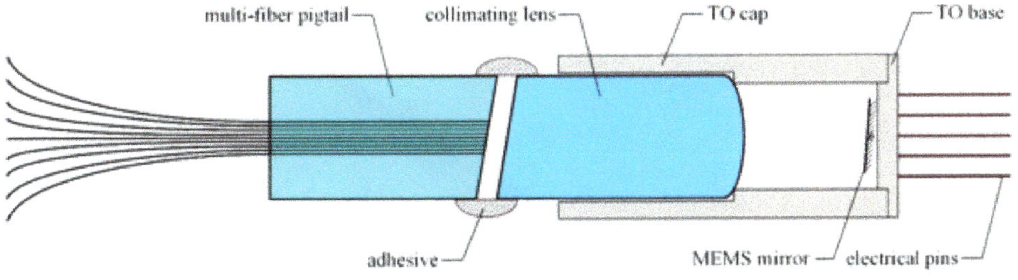

Figure 6.12. Schematic of 1 × N MEMS.

through which the electrical control can steer the MEMS units. The switched and reflected optical beams then enter the output fiber faces and are routed to the desired destination. The port configurations can include 2 × 2, 1 to N ports or N port-to-N port routing.

The disadvantage of this type of MEMS OXC is that it is not possible to select wavelength and switch, thus this switch has only path routing but no wavelength carrier selection. In order to generate wavelength (color) selective switching gratings or wavelength spectral de-multiplexing must be integrated, as we can observe in figure 6.12.

$N \times N$ optical switches are commonly called OXC. A MEMS-based 1 × N port OXC switch is shown in figure 6.12. It consists of a MEMS torsion mirror, a collimating lens and multiple fiber lines. The MEMS mirror is commonly assembled in a TO-base (tin-op package for typical electronic or opto-electronic device) whose cap can be used to assemble collimating lens. Finally, the multi-line optical fiber is actively aligned to the lensed sub-assembly.

6.3.5 MEMS switch and mechanical–optical switch (MOS)

The working principle of MOSs is to redirect optical signals by physically or mechanically moving optical fibers with electronic control or by piezoelectric movement of the lensed fiber directed to the desired output lensed fiber port. There are three types of mechanical-control optical switches: (i) prism-based switching the optical path; (ii) mirror switching type; and (iii) switching by mechanically moving the lensed optical fiber (figure 6.13).

MEMS optical switches are based on the micro-electro-mechanical mechanism described in section 6.2.3, in which micro-mirrors or array mirrors can be micro-mechanically switched to routing the lightwaves to different paths. When exchanging lightwave paths are required, the angle of the MEMS micro-mirror can be micro-mechanically moved via the electrostatic force or acousto-optic or magneto-optic force, then the input lightwaves are routed to the different outputs (figure 6.14).

An MOS offers: (i) low insertion loss; (ii) high isolation; (iii) it is colorless and polarization-less; and (vi) it is not a conductive device. On the other hand, MEMS offers more advantageous features such as: (i) micro-size and high-degree of integration; (ii) high scalability and is polarization-less; and (iii) agile switching.

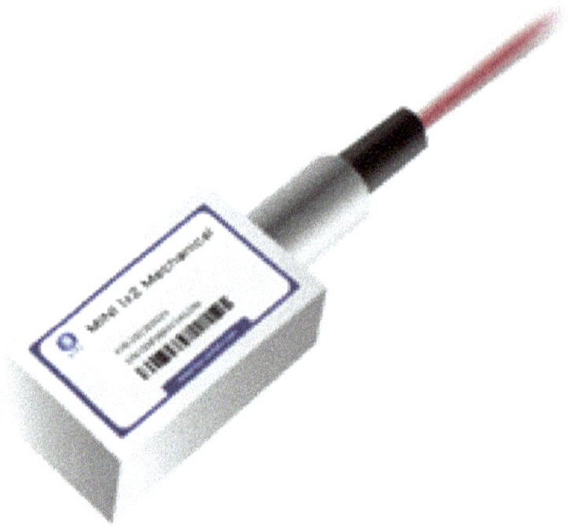

Figure 6.13. Mechanical optical switch.

Figure 6.14. MEMS as optical switch with electrical pins and package optical MEMS.

MEMS are currently the mainstream technology for ultra-high switching capacity optical networking.

6.3.6 MEMS as switching systems optical networks

MEMS optical switching arrays provide a wide range of optical communication networking applications including: network protection, light source control for testing of optical sub-systems, real-time optical monitoring of network performance, forming OXC core sub-systems, forming ROADM, etc.

6.3.6.1 Multicast switching optical switch (MCOS)

MCOS is based on a combination of PLC (planar lightwave circuits) and MEMS technologies. It is a key component of the next generation of ROADM. Its functional unit is composed of *M*x independent splitters and *N*x independent MEMS switches. They can provide N x M interconnections, that is, routing of N upper (or lower) input ports to M output ports, or optical paths.

6.3.6.2 MEMS in intelligent optical distribution

By cascading MEMS and OXCs switches one can form an intelligent optical distribution system to distribute an information data frame in enterprise networks.

6.3.6.3 Switching core of OXC

OXC switches are the core of an all-optical switching system and core lines in DCNs.

6.3.6.4 MEMS for real-time performance monitoring

Integrated with tunable optical filters or an optical time domain reflectometer, combined with monitoring software, and coherent mixing technique of known laser frequency, one can monitor the signal performance of the DWDM channel in a multi-core fiber in the optical cable. It can be used in monitoring optical transmission network, ROADM network, DCI, etc. This is now critical for monitoring of 5G networks in which several 5G users can be reported in time. An integration with artificial intelligent (AI) technique can assist and ease the field engineers to obtain the exact performance and failure of 5G network nodes.

6.3.7 MEMS based optical switch in all-optical networks

A long-haul optical transmission system would incorporate a ROADM through which a wavelength channel can de added or dropped to a desired port of direction NEWS. Such a ROADM is shown in figure 6.15.

6.3.7.1 DWDM optical transmission system and DCN

Figure 6.15 shows a typical optical transmission in which multi-spans of optical amplification stages are employed and optical add/drop multiplexers are used. The data channels are transformed to optical signals via the optical transmitters. These optical channels are then multiplexed in both the polarization modes and

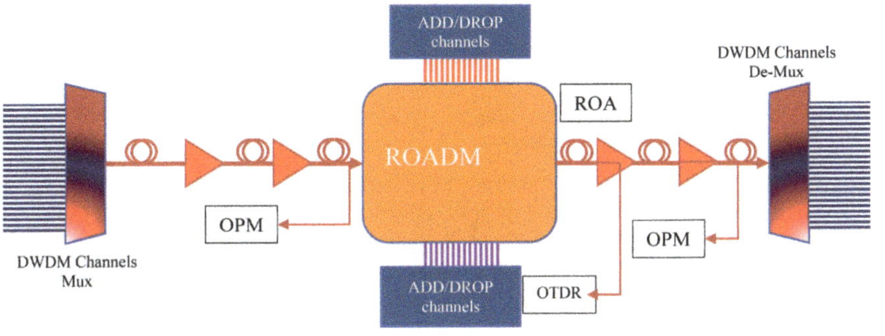

Figure 6.15. Long-haul optical amplified multi-span transmission system incorporating ROADM and optical amplifiers with monitoring facilities. Labels: ROA = Raman optical amplifier, OA = optical amplifier (typical EDFA) ROADM = reconfigurable optical add/drop multiplexer; OTDR = optical time domain reflectometer; OPM = optical power meter; Mux = multiplexer.

wavelength carriers in ITU-T[3] grids. The multiplexed channels are then transmitted through a number of optically amplified spans. Both discrete and distributed applications can be used, erbium doped fiber amplifiers (EDFAs) and Raman optical amplifiers (ROAs). Optical power monitoring (OPM) and reflectometry (OTDR) are used for power monitoring and optical continuity of the fiber systems. A selected wavelength channel can be dropped from the ROADM and then its average power meter and OSNR can be monitored. The optical continuity can be also monitored by using the OTDR.

Figures 6.16 and 6.17 give an overview of the DCN of region/metro and access networks to inter-DC interconnections and internetworking, especially the 5G and 6G access via fiber and wireless networks.

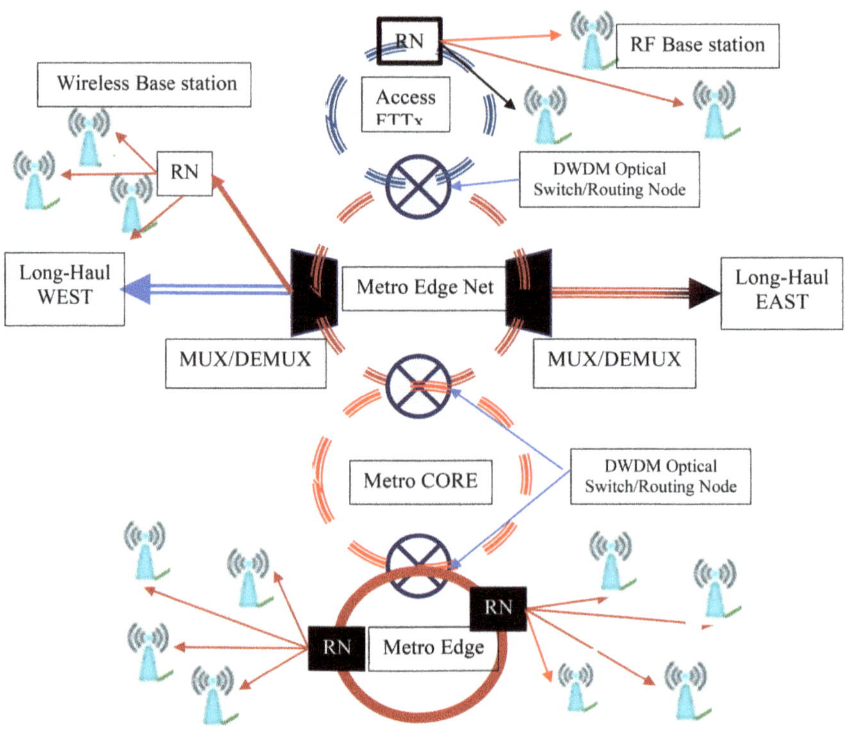

Figure 6.16. Structure of AON integrating ROADM, Mux/Demux and power splitter, remote nodes and 5G/6G wireless base stations into access networks, metro edge, metro core and long-haul networks. Similar structures can be established for DC in which access networks are equivalent to input/output servers, Metro nets equivalent to group of access servers for medium long DC and long-haul equivalent to global DC interconnections. Labels: RN = remote node (optical); PS = power splitter; ROADM = reconfigurable ADD/DROP multiplexer; base station = RF wireless base station; FTTX = fiber to the X. X = home, building, enterprises.

[3] ITU-T = International Télécommunication Union, T = télécommunication section.

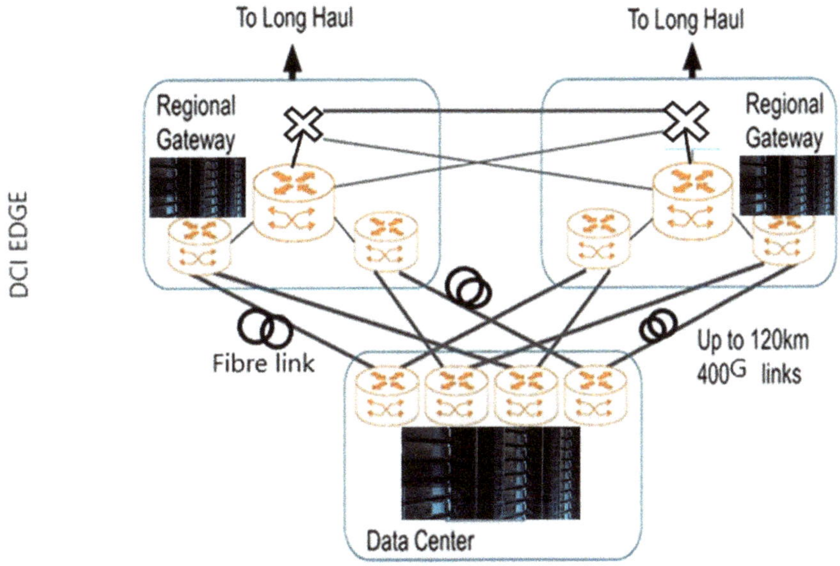

Figure 6.17. Inter DCN with DCI edges and long-haul interconnection networking as well as long-haul and access, metro/regional networks. All fiber links are operating at bit rate > 400 Gbps and reaching much higher when distance about ~2 km. **X** = optical cross-connect system.

6.3.7.2 All-optical networks (AONs)

Optical networks are becoming increasingly popular as a way to transmit data over long distances. AONs employ a combination of optical fibers, lasers, and other components to transmit data plus advanced optical switching and routing sub-systems. Further, they can be composed of several other core components such as ROADM, OXCs, optical network monitoring sub-systems etc.

The all-optical SMF-based communication network (AON) and systems can provide inter-network transmission of information data channels at speeds up to 800 Gbps (gigabits per second) or even at 1.6 Tera-bps. AONs are becoming more common in business, as they provide a cost-effective way to transmit large amounts of data over long distances. They are also becoming more popular in homes, as they are less expensive and easier to install than traditional copper-based networks. AONs are becoming increasingly important as the demand for data transmission increases. They are capable of providing high-speed, secure, and reliable data transmission, making them an ideal choice for businesses and homes alike.

AONs implies the transmission and routing processes of the network all in the optical domain, that is, through optical fiber guided media, optical switches, OXC, ROADMs, WSSs and associate optical devices. Thus, the data transmission process from a source node to an arbitrary destination node can be implemented all in the photonic domain, and the optical/electrical, electrical/optical conversion processes can only be performed at the ingress or egress points of the AON. Therefore, there is no need for intermediate electro-optical (OE) and electric-optical (EO) conversion in the network, thence increasing speed. The main technologies of the AON include optical fiber technology, DWDM (dense WDM), optical switching technology (OST), OXC, passive optical network technology and optical fiber amplification technology. The intensive 5G connections and distribution in a CBA (central business area) are shown in figure 6.18. The 5G signals are composed of allocated 200 MHz bundles to carriers with a total capacity that an reach several tens of Gbps depending on the multi-level modulation M-ary-QAM orthogonal frequency division multiplexing modulation formats.

6.3.8 *N* x *M* arrays of WSSs

Commercially there are many WSSs available for implementation of CDC-ROADM. One such WSS can be shown in figure 6.19 in which there are two common ports which act as a common output of multiplexed channels and thence outputting 34 channels of selected central wavelength and their passbands. The detailed structure of such a WSS can be seen in figure 6.12. The insertion loss per

Figure 6.18. 5G connection in a CBA with interconnection via ROADM to optical networks of metro and long-haul to DCN (DC networks). CBA = central business area.

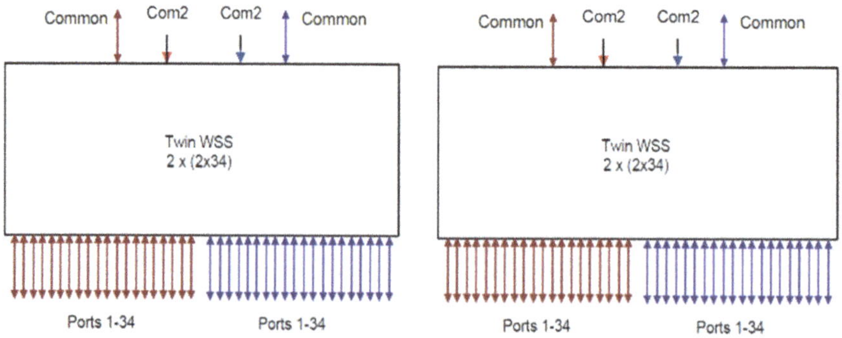

Figure 6.19. A 'twin' set of 2 × 34 WSSs with two common ports per WSS and 34 output ports.

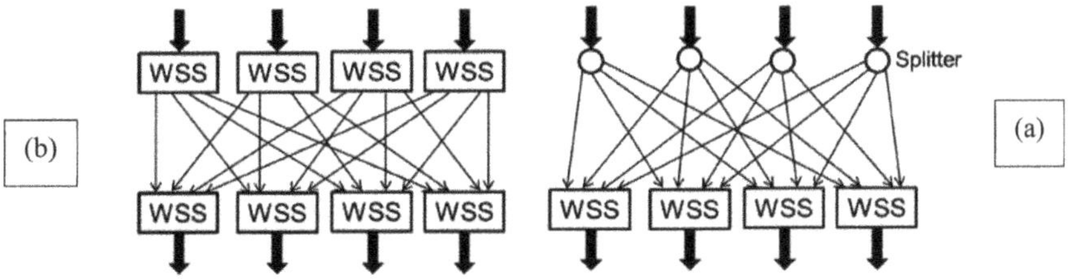

Figure 6.20. Assembly structures of (a) WSSs and multi-port optical splitters; and (b) WSSs–WSSs, to form massive flexible wavelength selective channel arrays.

port can be about 5.7–6.2 dB. This loss is acceptable and much lower than those of array waveguide gratings. Similarly, the common ports can be the multiplexed ports that sum up all the input ports 1–34 channels.

These WSSs can be combined with other WSSs or optical splitters to form massive arrays of a wavelength selective and distribution system, as shown in figure 6.20. Similarly, these WSS arrays can be used for multiple layered ROADMs.

6.4 Non-reconfigurable OADM (optical add-drop multiplexer)

OADMs can be introduced in metro/access networks to meet the bandwidth-hungry networking by private and enterprise users. OADMs are serving as wavelength management with switching and routing of wavelength-dependent lightwaves channels. Selective add and drop of optical channels can be performed to drop or insert information data channels to desired specific networks and inject a demand information channel via such multiplexers. The OADMS offer low insertion loss with low energy use and are reliable and cost-effective as well as scalable, thus much advantageous in all-optical networking. Furthermore, such OADMs can be improved to form reconfigurable OADMs so that flexibility of an AON can be achieved to provide much more economical and flexible agile networking. OADMs are a core element of DWDM and UB-WDM (ultra-broadband wide wavelength division multiplexing) networks. They can selectively drop and insert optical signals from and into an AON.

6.4.1 Configurations and functions of OADM

Generally, an OADM consists of: (i) an optical demultiplexer (demux), (ii) an optical multiplexer (mux), an OXC to reconfigure the optical paths between demux and mux, as well as the optical ports for inserting/adding and dropping of optical channels. The mux is used to couple two or more wavelengths channels into the same optical path. Then the reconfiguration of the optical paths can be implemented by optical switching or cross-connecting to direct the desired wavelength channels to the demux, that is, to drop the channels to the selected ports. The demux separates the multiple wavelength channels to different optical fiber paths (figure 6.21).

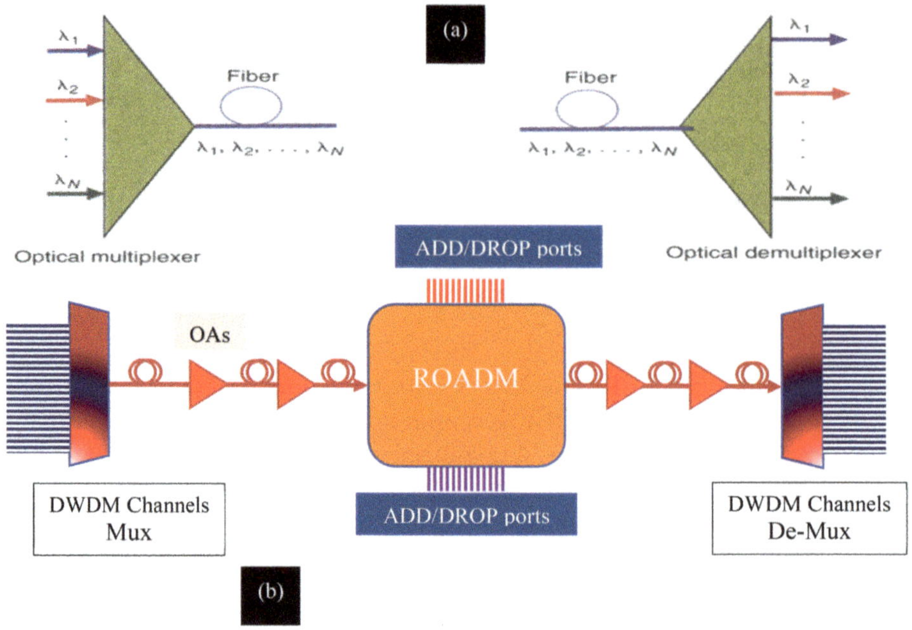

Figure 6.21. Components of (a) a traditional mux and demux multiwavelength (DWDM) transmission, (b) optical transmission with CDC-ROADM. OA = optical amplifiers, circles = fibers.

6.4.2 Optical add/drop filter

6.4.2.1 Optical add/drop filter as a multiplexer for OADM

A three-port tunable optical filter using micro-ring resonator or multilayer thin-film filter with angle input to generate an OADM, which is shown as a device in figure 6.22. Such optical add/drop can use the multi-layer thin film technology to offer high isolation and a sharp roll-off band to ensure that the residual crosstalk is minimal.

A typical spectrum of the three-port OADM filter is illustrated in figure 6.23 showing just two wavelength channels are dropped to two ports and then the transmission port spectra. Such an ADD/DROP multiplexer is commonly employed in OADM incorporating an optical pre-amplifier as shown in figure 6.24. Three optical filtered ports are dropped with three channels CH22, CH40 and CH41, and the rest transmitting through. For multichannel operations, such as C-, super-C-band and L-band add or drop spectra are shown in figure 6.25 via the output optical port P2, P3 and P4.

6.4.3 Configurations of OADM

These three-port filters can be employed in an OADM, as shown in figures 6.23 and 6.24.

There are two basic configurations of an OADM in which they can be distinctly classified by the type of optical filters as by-pass or drop of wavelength channels. The optical add/drop filters can be of two types, the dielectric multiple-layer thin-film filter (TFF) and fiber Bragg grating (FBG) filter. For OADM structures using TFF,

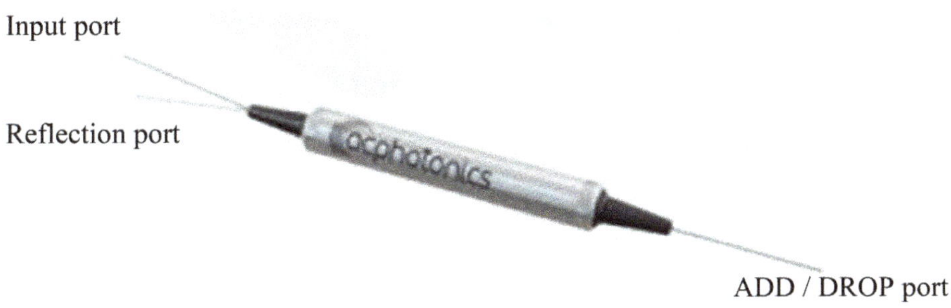

Figure 6.22. An OADM filter with drop/add port and an input port and a reflected port as transmission port (minus the drop channel).

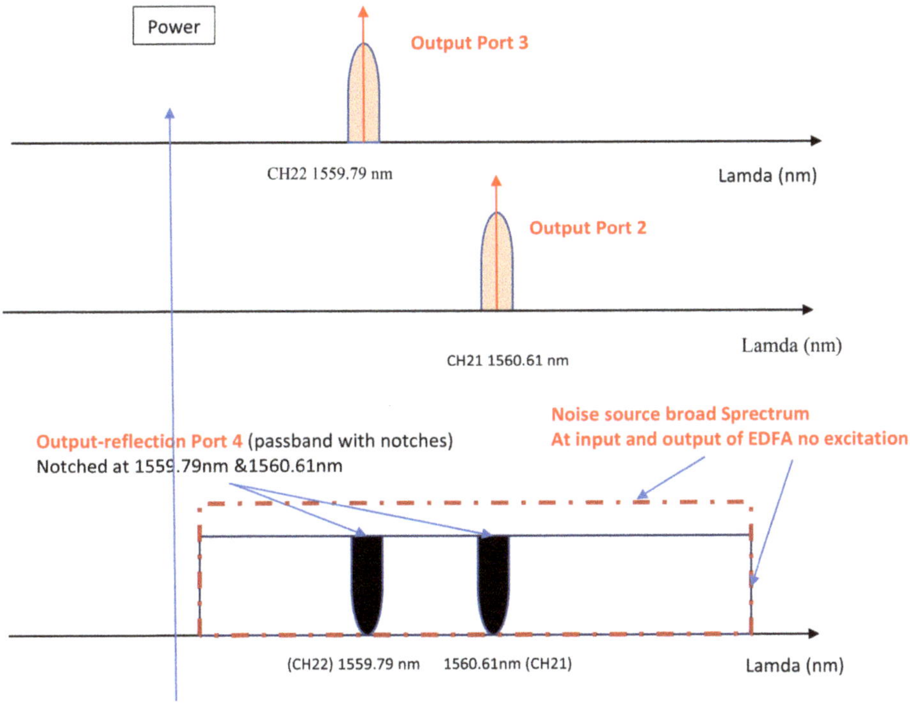

Figure 6.23. Spectrum of PORT 3 output, P2 output and reflection port of filter port of an optical ADD/DROP multiplexer or demultiplexer.

an arbitrary wavelength channel signal is directed and dropped from WDM signals via a narrow band-pass filter (BPF), whereby only the desired signal wavelength is transmitted while others are reflected, as shown in figure 6.26. Simultaneously, an arbitrary wavelength channel can be inserted or added into WDM lines via such an OADM filter, whereby the desired signal wavelength can be transmitted and combined with the reflected wavelength channels (figure 6.26). Similarly, figure 6.27

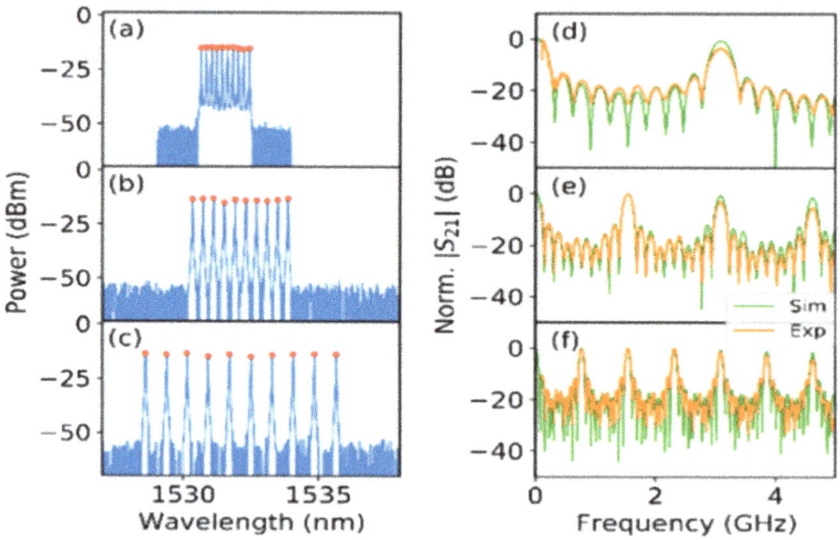

Figure 6.24. Multi-channel ADD/DROP in a DWDM network.

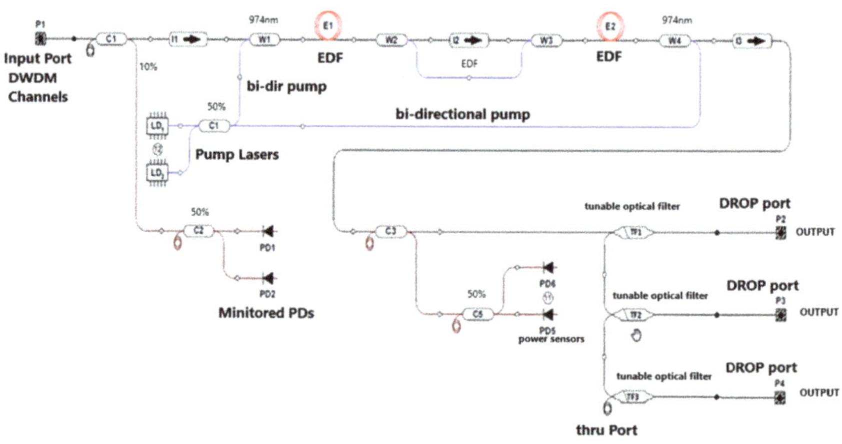

Figure 6.25. Structure schematic of an optically amplified OADM to drop three wavelength channels and a throughput port of all wavelength channels except the three dropped channels. Labels: EDF = erbium doped fiber for amplification E1/E2, PDs = photodetectors.

depicts the OADM using FBG (Fiber Bragg Gratings) as a dual-port reflector of wavelength channels.

While configuring an FBG-based OADM, the WDM channels enter an FBG through an optical circulator through which only one arbitrary wavelength channel is reflected while others are transmitted. The reflected wavelength channels are then dropped to a selected port which is different from the input injection port. In the case of multiplexing an arbitrary wavelength channel, the wavelength channel injected to the circulator is reflected by the FBG, and is inserted into the WDM channels that are transmitted via the other circulator ports (figure 6.27).

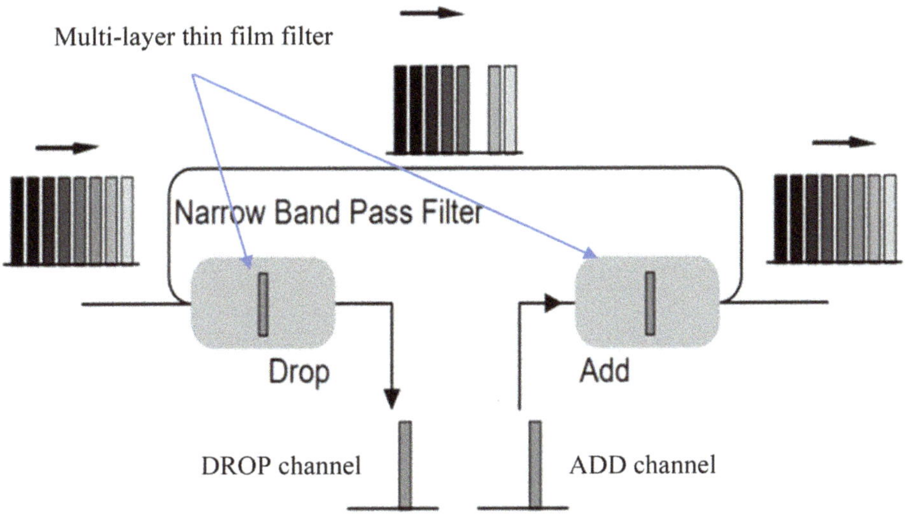

Figure 6.26. Configuration of OADM with TFF (tunable optical multilayer thin film filter or narrow band-pass filter).

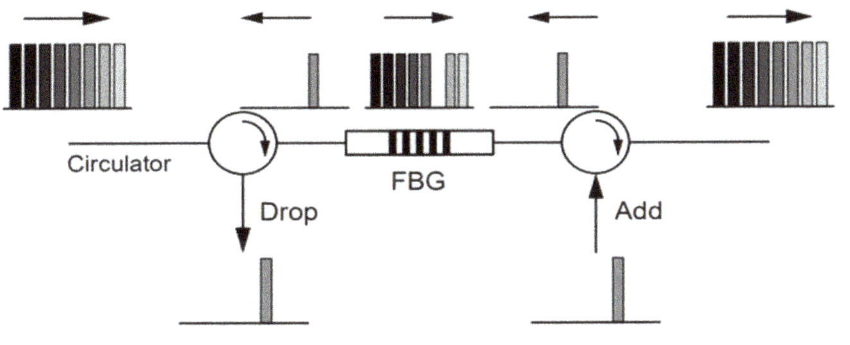

Figure 6.27. Configuration of an OADM with FBGs as reflected filters via circulators.

6.4.4 Main functions of OADM

An OADM can add one or more new wavelength channels to an existing multi-wavelength WDM group of signals, or remove and drop one or more wavelength channels, passing those signals to another optical network path. The OADM can selectively drop a wavelength channel from multiple WDM channels in a fiber, thence separating the signals from the particular group of channels. It can then add the same wavelength in the same flow direction of data channels of different content. OADMs are commonly employed in WDM ring networks and in long-haul transmission systems for more flexible structure and higher capacity in channel distribution and add different channels on transmitting demands. Figure 6.28 illustrates the functionalities of such optically amplified OADM in which optical amplifiers (OA)s are incorporated to compensate for its insertion losses.

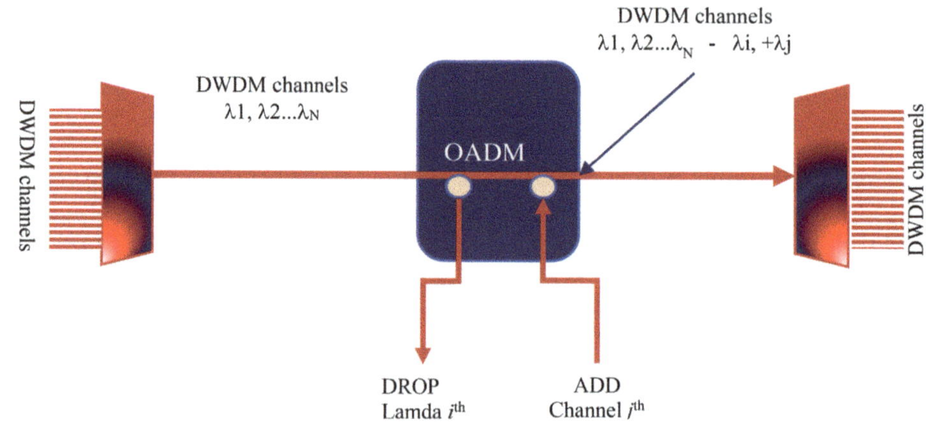

Figure 6.28. The schematic function of OADM. OA = optical amplifier, (pre- and post-)lamdas = wavelength channels. OADM =optical ADD/DROP multiplexer.

6.4.5 Different types of OADMs

OADMs can follow two configurations, the fixed wavelength (FW)-OADM and the ROADM. In an FW-OADM, wavelength channels can be selected and remain the same until intervention. In ROADM, the wavelength channels can be dynamically routed from inputs to outputs of the mux to any inputs of the demux.

6.4.5.1 Fixed-wavelength optical add-drop multiplexers

FW-OADMs are initially used to enhance the delivery of 'express' traffic through optical networks, without going through the OE/EO regeneration. FW-OADMs use fixed wavelength filtering that add/drop a selected wavelength center and passband and let through the remaining wavelengths channels through an optical node. Such wavelength filtering minimizes the attenuation due to de-multiplexing of all DWDM channels in an optical path. The solution is the FW-OADM because the added and dropped wavelength(s) channels are fixed at the time of add/drop filter installation on the optical path through a node. No additional filters can be possibly added without interrupting express wavelengths traveling through the node.

6.4.5.2 Reconfigurable optical add-drop multiplexers

ROADMs provide flexibility in re-routing selectively optical wavelength channels, bypassing faulty connections, allowing minimal optical channel service disruption. A ROADM can adapt and possibly be upgraded to different optical network configurations of WDM channel grids, that is, flexible grids optical networking. ROADM uses a WSS to provide such wavelength selective flexibility. WSSs can have a plane dimensional cross-connect and provide quick service networking, remote cross-connection and WDM mesh networking. A ROADM allows ingress and egress functionalities of a single wavelength or group of wavelength channels from and to optical ports. In ROADM node, there is no need to conduct OE and EO conversion and routing via conventional electronic switches such as some flexible OADMs (FOADMs) do. A ROADM can configure, as required, without affecting traffic, selected wavelength channels, as shown in figure 6.29.

6.4.5.3 Application of OADM

In long-haul transmission systems, the main targets are the high capacity and the transmission distance. In metro/access networks, however, low-cost and system flexibility are highly important. ROADM can meet both demands. The main area of OADM employment is in a MAN (metropolitan area network), with their high flexibility, ease in upgrading and amplification to compensate for channel power losses. OADM nodes also allow different wavelength channel mux at different nodes. OADM can also be associated with OXCs to provide a much higher degree of networking, as shown in figure 6.30.

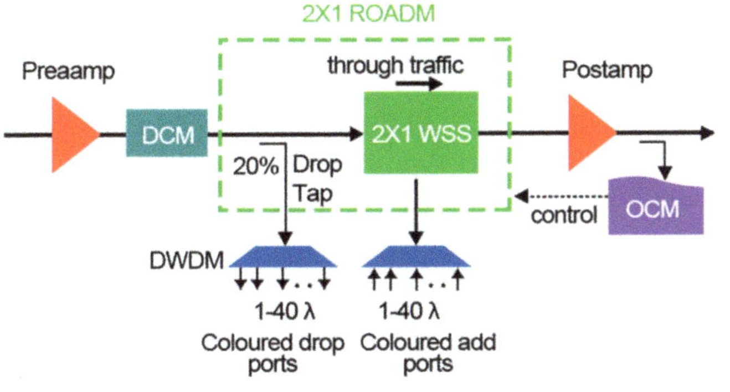

Figure 6.29. The principle of reconfigurable OADM. OCM optical channel monitoring. DCM = dispersion compensator module. WSS = wavelength selective switch. Post-amp, pre-amp = optical amplifiers.

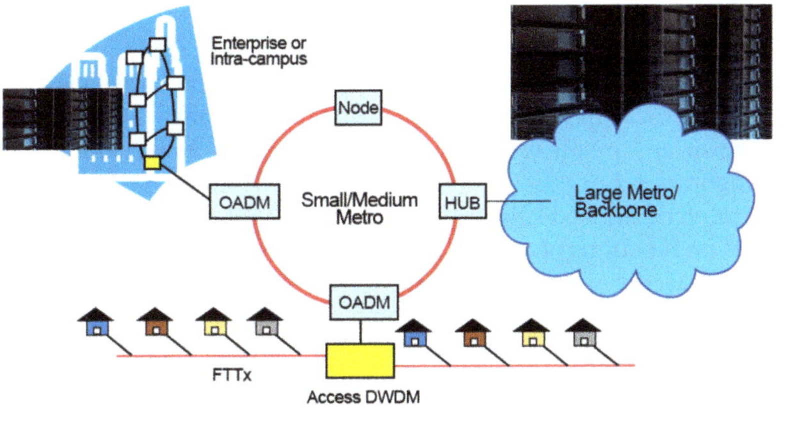

Figure 6.30. Application of OADM. FTTx = fiber to the 'X' – X = home, building, premises etc.

6.5 Space NEWS-ROADM photonic routing node

6.5.1 Demands and configurations

Multi-dimensional ROADM (MD-ROADM) networking has recently been implemented and has revolutionized the management and design of the structures of optical networks. MD-ROADMs are a powerful tool that can be used to increase the capacity, flexibility, and scalability of AONs. An MD-ROADM is a type of optical multiplexer capable of switching and routing light signals across *multiple* optical paths that allows for a greater degree of flexibility in the design of an optical network, especially allowing for more efficient use of available optical resources. Furthermore, MD-ROADMs can be used to optimize the performance of an optical network by allowing for more efficient utilization of bandwidth and better management of network resources.

The most common type of MD-ROADM is the WSS. A WSS is a type of optical switch that can switch light signals across *multiple* wavelengths. This allows for a more efficient use of optical resources, as the WSS can be used to divide the optical signal into multiple paths or channels, each carrying different wavelengths of lightwaves, thus allowing greater flexibility in the design of an optical network, as each wavelength can be used to carry different types of traffic.

The advanced type of MD-ROADM is the colorless, directionless, and contention-less CDC-ROADM. This type of ROADM is capable of switching light signals across multiple optical paths without the need for wavelength-specific components.

CDC-ROADMs are reconfigurable optical networking devices used to route optical signals in different dimensions, in a network. They are designed to reduce contention and maximize network capacity by allowing for the transmission of multiple signals on the same wavelength. CDC-ROADMs also enable WDM to increase the capacity of the network by combining multiple signals into a single wavelength. They are also capable of providing multiplexing, switching, and routing of optical signals/channels. Additionally, they can be used to reconfigure optical paths in an optical network, making them ideal for dynamic and agile networks.

6.5.1.1 Geometrical structures of multi-directional ROADM

ROADMs have been widely deployed in long-haul and metro WDMs and internal and inter-DC networks in recent years. They enable flexibility in injection or adding and dropping of any or all multiplexed channels at the wavelength level without manual configuration or O–E–O (optical to electrical and then converted to optical) switches. In mesh or meshed ring optical networks, MD-ROADMs can also provide cross-connection functions of multiplexed signals among different paths, so they can also be termed as the wavelength optical cross-connects (WOXCs). The present MD-ROADMs can be constructed by simply interconnecting individual single-degree ROADMs. However, such a node cannot provide full flexibility for add/drop and switching. This section investigates the requirements for the next generation MD-ROADM nodes and their modes of operation.

An MD-ROADM is a type of optical switch that is used for wavelength routing in optical networks. It is a key component of optical networks, as it allows for the efficient management of the optical spectrum and allows for the addition and removal of wavelengths from a particular fiber. The main advantage of multi-dimensional ROADMs is that they can be used to create larger, more complex networks. This is because they can be used to add or drop multiple wavelengths from a single fiber, thus increasing the number of connections that can be made within a network. Additionally, MD-ROADMs can be used to create networks with multiple nodes, allowing for more efficient data transmission.

MD-ROADMs are also more cost-effective than traditional ROADMs, as they require fewer components and are more easily installed and maintained. Furthermore, they are more reliable, as they are less susceptible to failure due to their advanced technology. Multi-dimensional ROADMs are used in a variety of applications, including telecommunications, DCs, and research high performance computing networks. In telecommunications, they are used to create high-speed networks with fewer components and more reliable connections. In DCs, they are used to create high-capacity networks with greater flexibility. And in research networks, they are used to create networks with greater scalability and flexibility.

Colorless indicates that add/drop ports are not wavelength specific. Directionless features enable any transponder to be connected to any degree, that is, multiple shuffling of optical ports. In this section we describe different multi-degree ROADM node architectures that offer full colorless and directionless features.

Figure 6.31 shows the structure of a ROADM conventional node of degree 1,2, …,*N*. The node consists of WSSs in each input set of ports and output set of ports. Under WSS operations, there are through ports and routing ports. The through ports can lead wavelength channels through the switch and reserve the channel wavelength as well as their bandwidth for other devices. On the other hand, the routed channels can be directed to predetermined wavelength connections as indicated by the cross-connection path in figure 6.31.

The MD-ROADM nodes exhibit two major limitations due to the following. (i) Firstly, the channels to be configured and switched are wavelength-dependent, that is, each optical transponder can correspond to a fixed wavelength channel, because the wavelength corresponding to each demultiplexer output port can be predetermined at the manufacturing phase. Furthermore, even the transmitter of the transponder can be tunable across the entire operational band of the optical guided medium. Therefore, the operational spectral bands of all transponders must be predetermined, thus they are not colorless. (ii) Secondly, a further limitation of MD-ROADM, is that the add/drop operation is not directionless but strongly dependent on the channel flow direction. This prevents the routing of ADD channels, thence there is no flexibility of the nodes to be reconfigured to be integrated with other ROADMs, they must be manually reconfigured. For an MD-ROADM to offer directionless features it must have the properties that 'for any channel dropped at the local ROADM node, the corresponding ADD channel can transport to any output port, regardless of the position of coming input port'. The muxes and demuxes are employed for multiplexing and de-multiplexing channels of different wavelengths

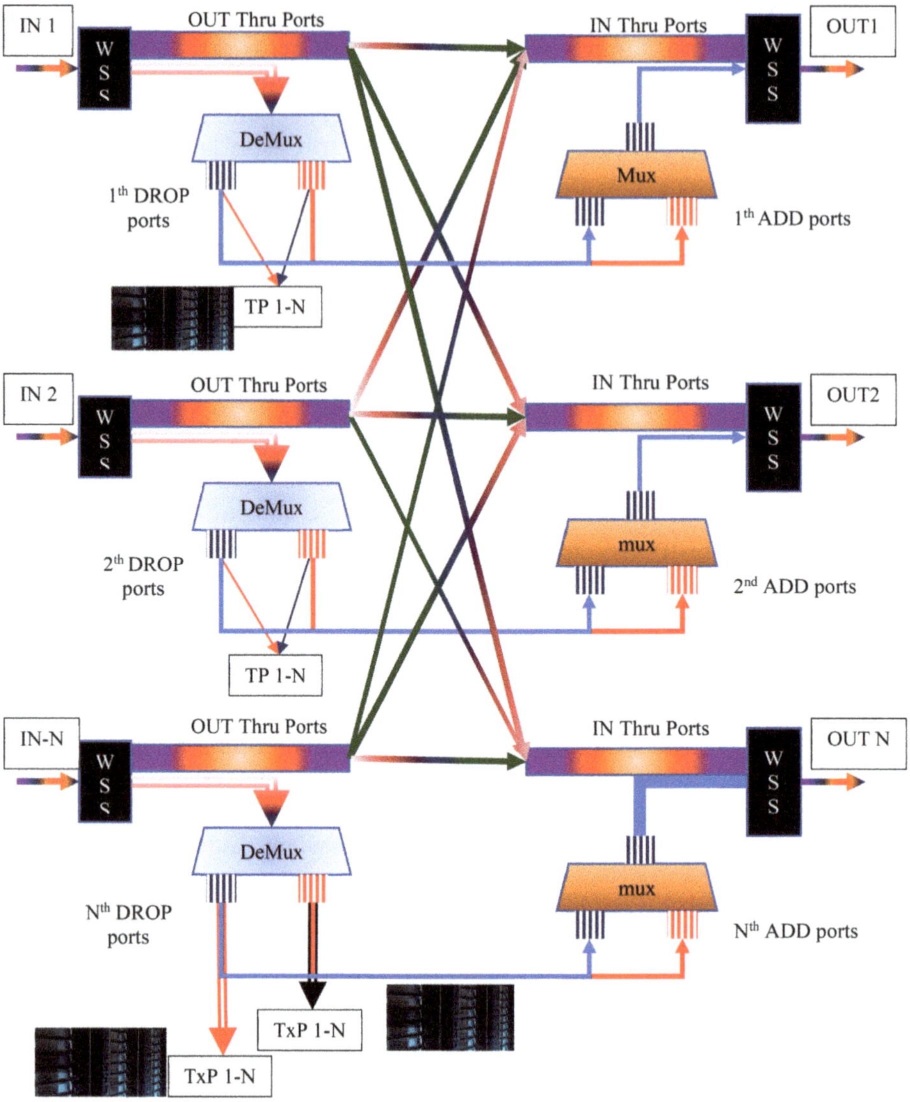

Figure 6.31. Structures of conventional ROADM node. TxP = transponder.

(assigned in spectral grids) and the ADD or DROP to predetermined ports. The muxes and demuxes with the N Ports, are set to 'drop to' or 'add from' the transponders (TxPs). It is clear that the wavelength channels in this case, must be predetermined and not flexible in selecting of wavelength or assignment of any spectral channel groups to be routed. The directionless feature permits much more efficient sharing of TxPs in a node among different paths, and significantly improves flexibility of protection schemes.

Figure 6.32 shows the structure of CDC-ROADM nodes of similar degrees 1 to N. The main aim is to enable pre-filtering and software-defined service provisioning

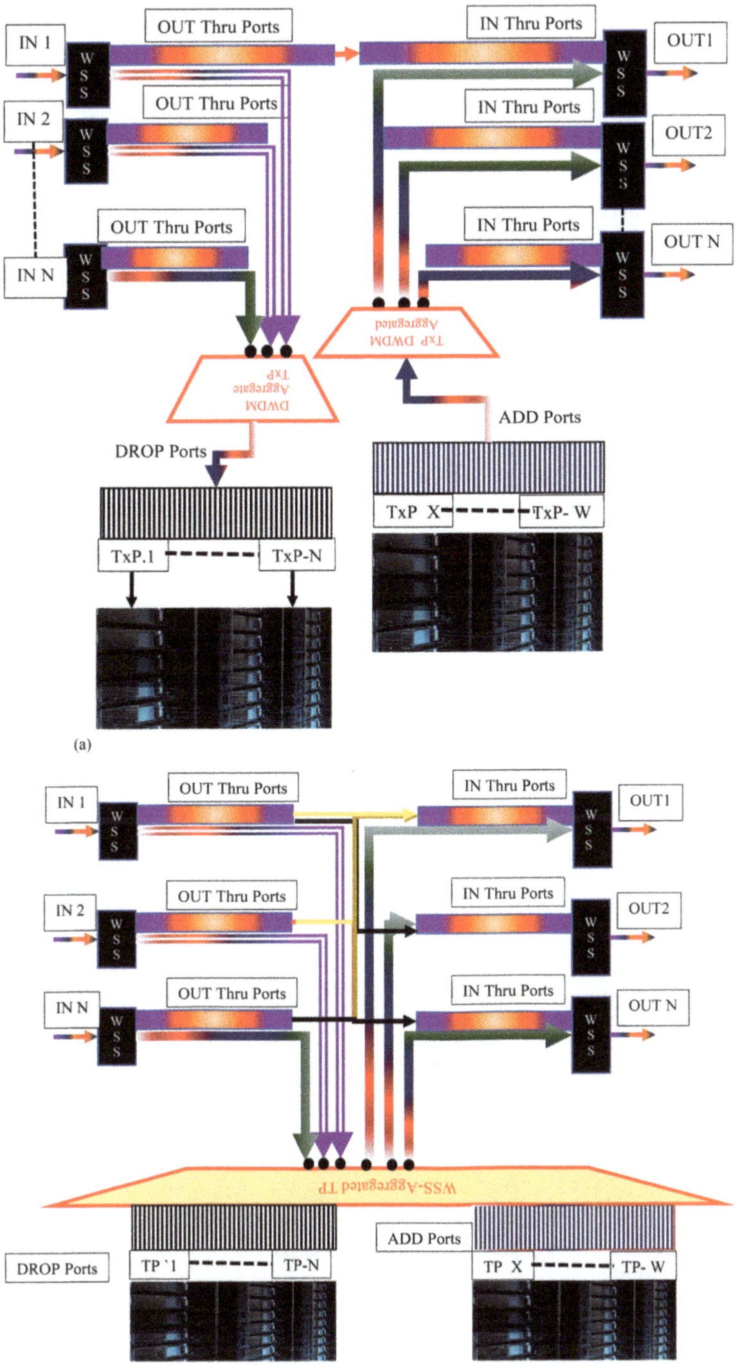

Figure 6.32. (a) Conventional *N*-degree ROADM node, with (b) schematic structure of a hybrid conventional ROADM and CDC-ROADM node. WSS = wavelength selective switch. TP = transponder.

such that the ROADM node is not fixed to a predetermined wavelength, thence eliminating the need for manual operation. The TxP can be aggregated to two muxers and demuxers and all routing for ADD and DROP. The WSSs can switch the aggregated channels to the DWDM muxers and demuxers. The wavelength channels are aggregated together via a massive wavelength muxer or WSSs and groups of selective wavelengths can be assigned to channels and they can be dropped or added to the desired destinations. Using WSSs in ROADMs, one can select desired channels of specific wavelengths to route in or out of the CDC-ROADM and direct them to TxPs.

Figure 6.33 shows the interconnections of CDC-ROADM nodes to an OXC device so as to generate an MD-ROADM node. A number of input and output ports of OXC are reserved for shared TxPs so that exchanged and switched channel data can be performed via this OXC. Therefore, the MD-ROADM can offer a much higher degree of ADD/DROP with extending wavelength operations, hence the term multi-dimension. The structure of an MD-ROADM can be extended to many configurations and the use of WSSs and ROADMs as well as TxPs can be reconfigured to evolve the MD-ROADM to generic systems.

6.5.1.2 Mathematical applications in tensor or indexing problems

Due to complexity of the MD-ROADM, tensor analyses can be employed for indexing, and then computerized indexing of input ports and output ports as well as programming the OXC ports. The transmittances and attenuation of the through ports and switched ports can also be integrated and automatic control via MCU (microcontroller unit) can be used to perform these functions.

In the next section we address these automatic expansions and interconnect these MD-ROADMs for large-scale systems, especially for applications in DCs. Under the high demands of ultra-dense data transfer and servicing online, such massive switching and routing systems are necessary. Furthermore, the optical

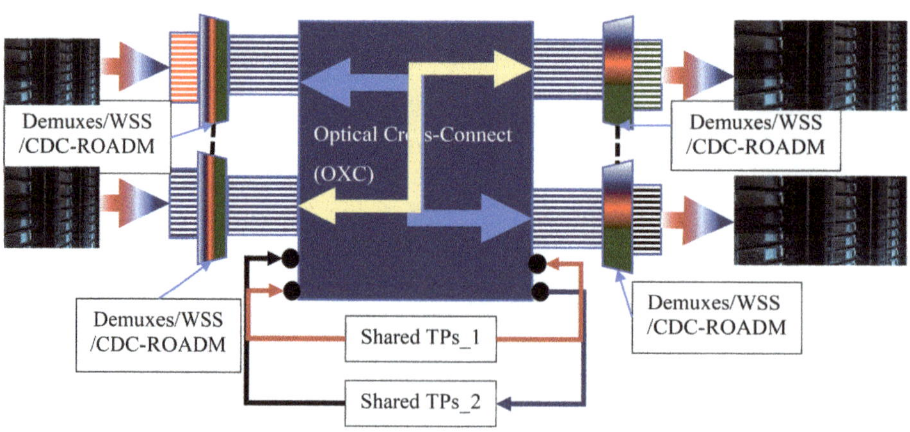

Figure 6.33. CDC-ROADM mode with shared TxP and OXC to interconnection and switching between ROADMs and muxes and demuxes as well WSSs. Several MD-ROADMS can be evolved from this generic structure. OXC = POC (photonic/optical cross-connect).

interconnection and switching in the optical ports will reduce the power consumption, hence there is less demand on power/energy supplies.

The mathematical representation of ROADM must be able to address the indexing of the following features: flexible add/drop of wavelengths, *N*-degree ROADM and flex-grid ready, rates of support 10G/100G/200G and 400G wavelengths, up to 120 (50 GHz spacing) super-C-band add/drop wavelengths, WDM spacing—50 GHz or 100 GHz, A-to-Z provisioning of wavelengths and protection.

6.5.2 Scalable ROADM nodes using WSSs plus slave-OXC (OXC_S)

When it comes to network infrastructure, scalability is key. This is especially true in the case of ROADMs. ROADMs are used to provide flexibility and scalability to optical networks, allowing for the addition and removal of wavelengths without disrupting the existing traffic. In order to ensure scalability, ROADMs must be designed with the ability to expand and contract as needed. This means that the a ROADM should be able to handle an increased number of wavelengths without needing to be completely replaced. It should also be able to adjust its configuration to accommodate new wavelengths or changes in existing wavelengths.

Scalable ROADMs are essential for creating an efficient and reliable network infrastructure. They provide the flexibility needed to quickly add or remove wavelengths as needed, while still maintaining the existing traffic. This allows networks to quickly respond to changes in demand or traffic patterns.

Scalable ROADMs also provide cost savings by reducing the need for expensive upgrades or replacements. By allowing for the easy addition and removal of wavelengths, ROADMs can help keep networks running smoothly and efficiently, while still providing the flexibility needed to respond to changes in demand. Finally, scalable ROADMs provide greater security and reliability. By allowing for the addition and removal of wavelengths, ROADMs can help protect against malicious attacks or service outages. This helps ensure that networks remain secure and reliable, even in the face of unexpected changes or disruptions.

Overall, scalable ROADMs are essential for creating a reliable and efficient network infrastructure. They provide the flexibility and scalability needed to quickly add or remove wavelengths as needed, while still maintaining the existing traffic. This helps ensure that networks remain secure and reliable, even in the face of unexpected changes or disruptions.

This section describes and gives structures of a number of scalable ROADM nodes. Firstly, ROADM node by integrating WSSs and Tx muxes and demuxes to Rx with optical interconnect. Secondly, an OXC acts as the central part to exchange and connect wavelength channels with great flexibility to input lightwaves channels and out to desired output ports with flexible wavelength assignment. Thirdly, non-blocking MxN switches interconnect to form a non-blocking central part and accept muxed or demuxed channels to desired outputs with selection of wavelengths.

Regarding muxes and demuxes, WSSs are becoming increasingly popular for their ability to serve as both a mux and demux. A mux is a device that enables multiple input signals to be transmitted over a single output channel. In other words, it allows multiple signals to be sent and received simultaneously. A demux is the opposite, allowing a single input signal to be divided into multiple output signals. WSSs are ideal for multiplexing and de-multiplexing because they can handle multiple simultaneous connections. This means that a single WSS can be used to send and receive multiple streams of data at once. This is especially useful when dealing with high-traffic applications, such as video streaming, as it enables the server to handle multiple requests at once. WSSs also offer low latency and high throughput, making them ideal for applications that require fast response times. Additionally, they are secure, as they use the latest encryption protocols to protect data from malicious actors. Overall, WSSs are an ideal choice for applications that require multiplexing and de-multiplexing, as they can handle multiple simultaneous connections, offer low latency and high throughput, and are secure. If you're looking for a reliable and efficient way to send and receive multiple streams of data at once, then a WSS is definitely worth considering. The integration of such WSSs and optical cross-connect can form a ROADM with a high degree of flexibility and scalability.

6.5.2.1 ROADM using WSSs

ROADMS can be structured using WSSs, OXC and aggregate optical switches MxN of non-blocking features, as shown in figures 6.34(a)–(c), respectively. When WSSs are used they can act as muxes and for switching of assigned input and output ports, as shown in section 6.3.8. That is, the ports of the WSS at the input ports are multiplexed to output to the common port and then connect to WSSs at the output or cross-connect via an OXC matrix, as shown in figure 6.34(b). The multiplexed channels are switched to desired output ports of the WSS, WSS 1o to WSS 4o. The number of switched output ports can be higher by using a larger number of WSSs at the input and output. In figure 6.34(a), the optical splitters can be used to shuffle the input ports to outputs which are then employed as input ports to the WSSs so that a larger number of input channels can be designed. The equivalent structure of figure 6.34(a) can be shown to be the OXC and WDM mux and demux, as shown in figure 6.34(b). The main different features of these two structures are dependent on the total insertion losses (ILs) between the input and output ports. For WSSes typical IL is 5 dB between the ports at input to output, then there are additional losses in connection of other elements. While the coupling losses for the WDM couplers are in the order of <10 dB depending on the number of the ports, the WSSs are operating in free space and MEMS, so the losses are much lower. So, the higher the port number the more losses for the WDM couplers as compared to those of the WSSs. Thus, the selection of these structures is to be optimized in the use of these two types of ROADMs. Alternatively, ROADMs can be designed using MxN optical switches as shown in figure 6.34(c). These switches are non-blocking by using an intermediate switching matrix. The microcontroller unit can control the switching between the ports. This type of non-blocking ROADM can be best for a small number of ports. The assignment of the ports can be formed by a control matrix by using tensor analysis or the index of complex systems or crystal structures.

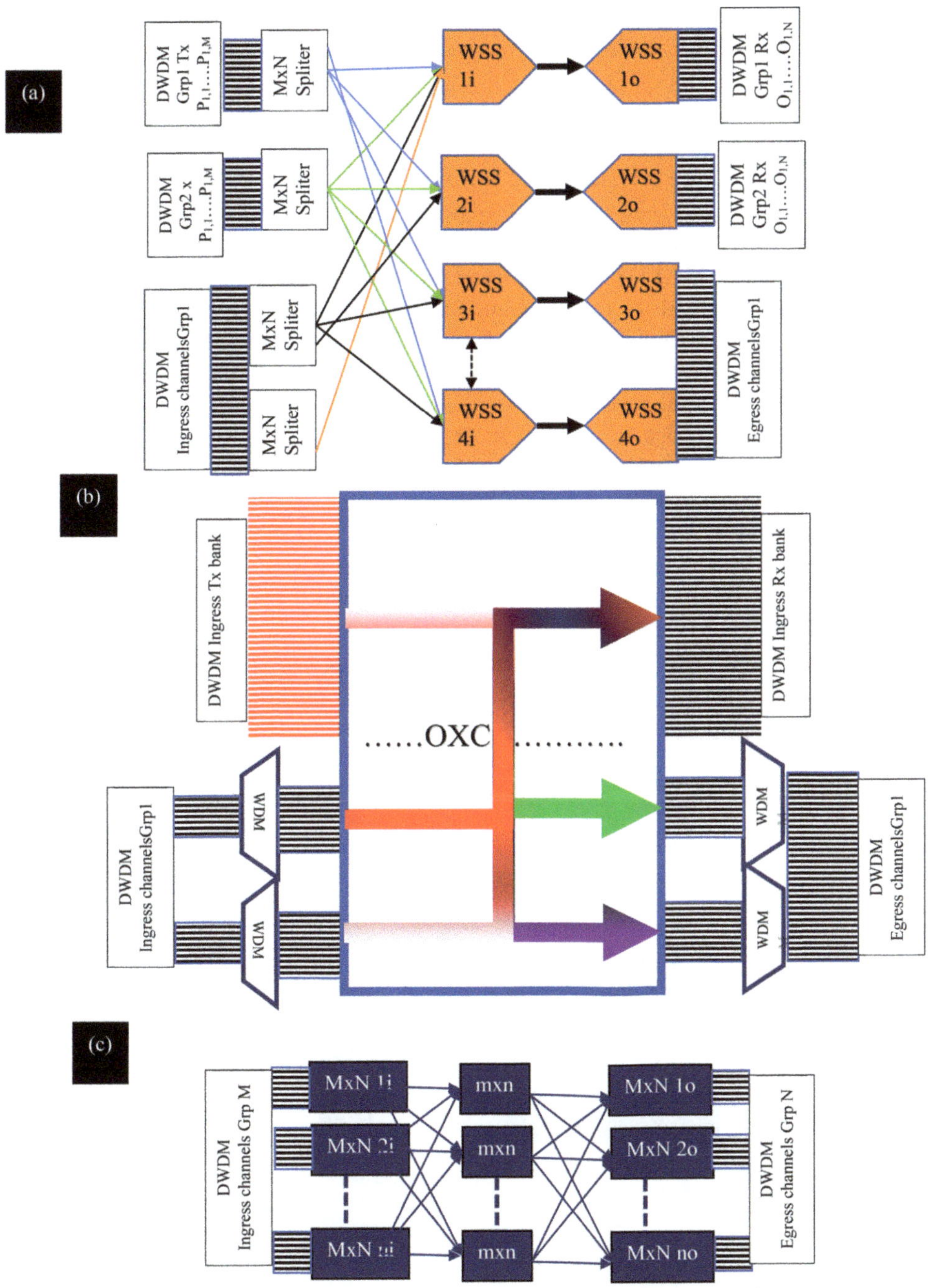

Figure 6.34. Schematics of (a) WSS-based ROADM; (b) 3D OXC switch-based ROADM; (c) non-blocking switching mesh network-based ROADM. Note: wide signals connecting WSS to WSS to indicate the aggregate channels of DWDM. Labels: $P_{i,j}$ = input port of *i*th device and port *j*th, $O_{k,l}$ = output port of *k*th device and port *l*th.

The output ports can be fed into the optical receivers if they are the terminal ports, as shown in figure 6.34(b), or to the 'egress' ports to continue to other sub-systems of the networks of DCs.

If the insertion losses of the OXC and WSSs are high, hence degrading the OSNR, then one needs to use OAs to compensate for these losses. The amplifiers are not shown in these structures. OAs would normally have the gain of about 30 dB with the input signal level of −20 dBm to 0 dBm and output saturated level at 10 dBm. Therefore, the power budget and dispersion budget for the optical transmission links associated with the ROADM-based optical networking must be considered and OAs or dispersion compensators inserted accordingly. The power budget technique is given briefly in appendix C of this book. The OSNR measurement and definitions are given in the appendix, section 6.11 of this chapter. More details can be found in appendix D of this book, especially in the measurement of in-band noises of the OSNR so that in practice this measured OSNR can reflect the true values of the OSNR in an optically amplified ROADM-based optical transmission.

6.5.2.2 Scalable ROADM

A scalable ROAM node is shown in figure 6.35(a) in which two types of OXCs are employed, one is the master OXC for communication through the ROADMs and then there is the slave-OXC, OXC_S, which then further switches to and interconnects with many selected wavelength channels. The scale of such ROADMs can be scalable by increasing either the number of planar ports of the master OXC or

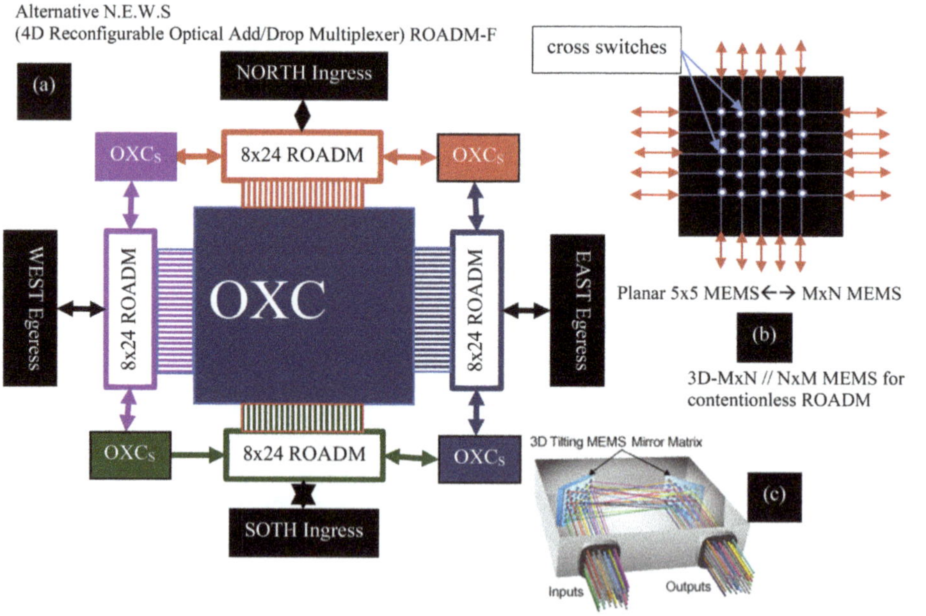

Figure 6.35. Scalable space ROADM by WSSs + OXC. (a) Structure of optical routing and add/drop of wavelength channels; (b) planar 5×5 MEMS interconnection/switching for optical cross-connect; and (c) 3D-MEMS routing and switching.

OXC_S. There are four ROAMS located as slaves in NEWS directions though which any drop wavelength channels from a ROADM can be then add to the next and then switched to predetermined output ports of the master OXC.

Therefore, for any wavelength channel, they can be routed to any desired ports from there to a TxP. We note that the ROADMs can be best if they are CDC-ROADM or MD-ROADM as described in section 6.5.1.1.

Figure 6.35(b) shows the structure of a planar OXC using integrated photonic technology in which the guided waves are switched via a switching guided optical coupler and possible wavelength filtering depending on the coupling length and strength of the coupling region on the wavelength. Further, in figure 6.35(c) a 3D-MEMS switch is illustrated in which two planes of MEMS can be controlled and switched to an appropriate port via a free-space region. This free-space coupling is via the GRIN lens collimator and the insertion loss is less than <0.1 dB per coupling. We note that a multimode planar coupler can also be used in figure 6.35(b) to distribute the channels and be switched or filtered by the interferences through the MMI (multi-mode interferometer) section.

An extensive scaled NEWS-ROADM is shown in figure 6.36 when additional space division muxes or demuxes are integrated with a scalable NEWS-ROADM. Figure 6.37 shows also further extension of the NEWS-ROADM with flexible CDC-ROADM, which generates a massive optical router and switches for ultra-high capacity systems in internal DC communications and services of ultra-fast Internet channels. $N \times M$ WSSs indicates N input ports switched and filtered to M output ports. Selections of wavelengths and filter bandwidth can also be possible. The interconnection buses are communicating with direct connections between WSSs. The 4×1 or 1×4 CDC-F WDM muxes and demuxes are used to distribute or

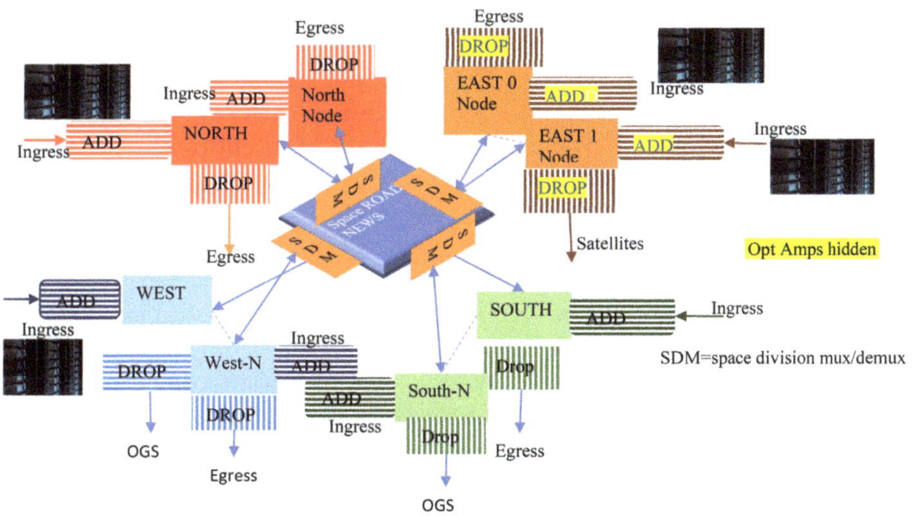

Figure 6.36. Schematic diagram: North–East–West–South (NEWS)-ROADM (CDC-Flex). Labels: SDM = space division mux or demux; egress = DROPPED to output destinations; ingress = ADDED to ports for switching and routing. N = node; OGS = optical ground station.

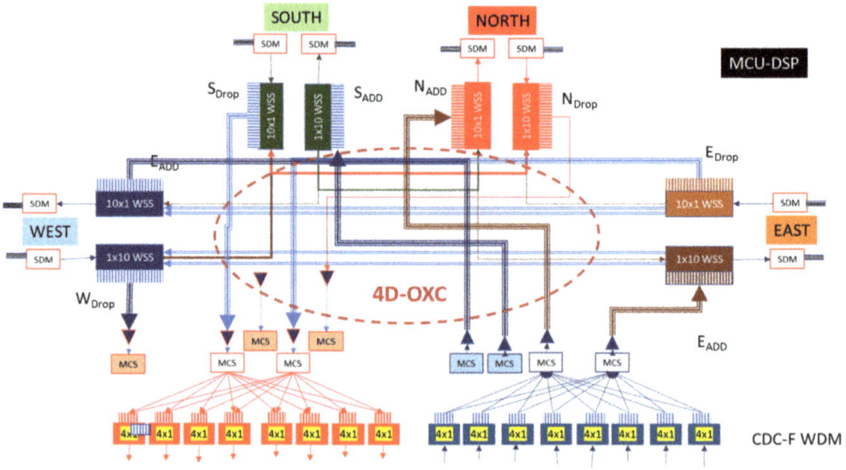

Figure 6.37. NEWS space optical routing add/drop mux node. Labels: SDM = spatial division mux/demux; WSS = wavelength selective switch; OXC = optical cross-connect; CDC-F = contention-less, directionless, colorless-flexible grid; S = satellite; MCS = multi-cast switches; N = north; E = east; W = west; S = south; MCU = micro-control unit.

combine the wavelength channels for routing and drop as well as dropping to appropriate destinations. The central integrated ROAA is a NEWS-ROADM type and then via the SDM (spatial division mux/demux) to split the channels in ingress and egress ports. The MCU is responsible for control of the setting of switching and routing. The speed is not fast but stable to setting the ports in and out routing, so a medium speed MCU is enough to conduct or manage these functions.

Figure 6.38(a) depicts the intersatellite constellations in which each satellite would use an optical routing system, the CDC-ROADM, to distribute and add different channels from different satellite orbits. Figure 6.38(b) shows more detailed structure of a NEWS-ROADM with WSSs as ROADM and switches to distribute wireless channels for terrestrial as well as space RF distribution of 5G and 6G channels to be used for satellites to optical ground stations. OAs are essential to compensate for attenuation of channels when passing so many stages of optical routing and switching. Optical power amplifiers are also used to boost the average power of the channels to other satellites or to ground stations. This forms an all-optical routing node in a satellite deep space communication network.

The ADD/DROP Ports of each ROADM are assigned in four directions north–east–west–south (NEWS) so that the assignment matrix for switching and routing can be assigned with an additional dimension, the wavelength dimension.

One of the potential applications of such ROADM networking is the switching and routing of a space laser node, as shown in figure 6.37, for space laser networks for a high capacity space computing networking. Such a ROADM node consists of OAs as preamplifiers to amplify the low power input from the telescope that is receiving at low level after propagating over a long distance and with high loss as there is no mid-stage amplification, as in a fiber-based transmission system. The

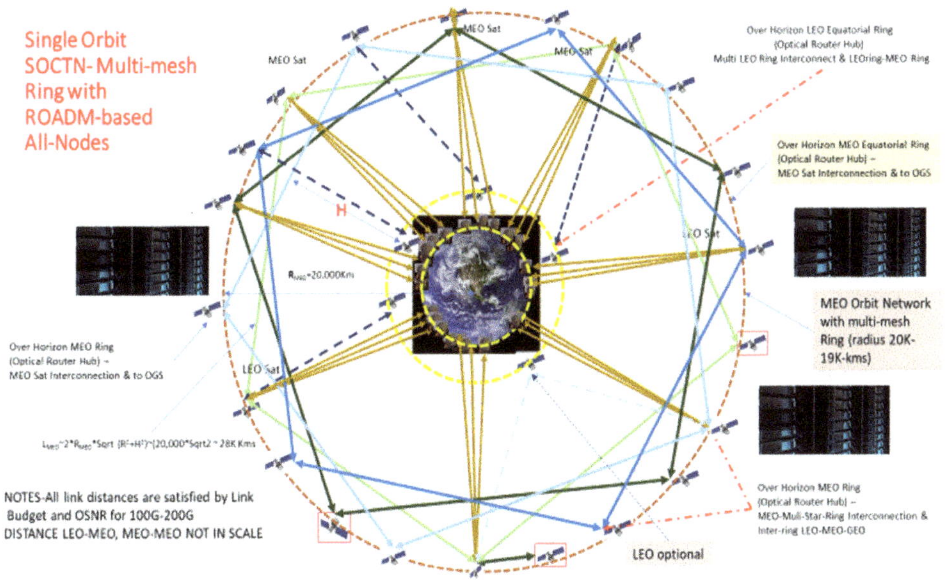

Figure 6.38

(a) Inter-orbital constellations and optical transmission paths for SOCS networking. Shown are GEO, LEO and MEO satellite constellations.

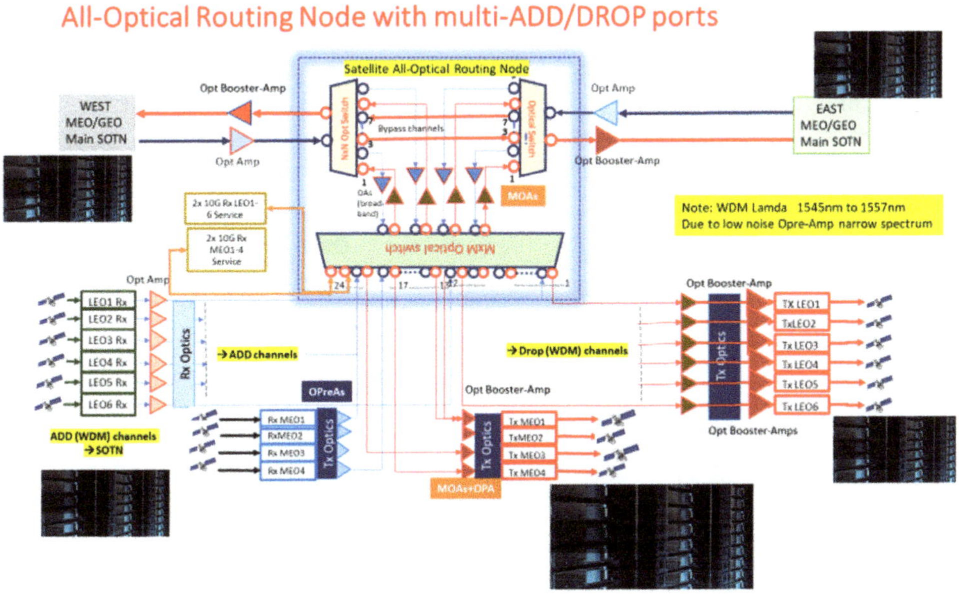

Figure 6.38

(b) All optical routing Space Node

Figure 6.38. (a) Schematic of an intersatellite constellation of space communication network; (b) optical routing in an optical node using CDC-ROADM in intersatellite transport networking. TX = transmitter; Rx = coherent receiver; MEO = medium orbit; GEO = geostationary orbit (satellite); Amp = optical amplifier; SOCS = space optical communication systems.

drop channels can be assigned to an optical ground station or to connect to other satellite nodes. The ROADMs are operating with optical switching so there is a huge saving of power as the optical switches consume much less energy than their electronic counterparts. Saving energy or power in a space satellite is important as this can save space and solar energy absorption panels, thence saving weight and simplifying the launching technique.

6.5.3 Optical network: architecture on demand

6.5.3.1 Non-blocking CDC-ROADM networking with architecture on demand

Software defined networking offers significant advantages in the flexibility for forming architecture of networks on demands (AoD). This section addresses this type of network, the flexible AoD networking technique.

As an example, considering figure 6.39(a), which shows the switching matrix for the networking of figure 6.39(b) in which the diagonal row is 'zero' that means that no self-switching node is possible but to external nodes. This is six-degrees of freedom for communication and switching with add/drop capability, and with 20 demands/requests. The network configuration uses bi-directional switching and routing nodes for DCN optical routing and reconfigurable networking. The nodes can operate as ADD/DROP nodes with reconfigurable routing as indicated with SA1 to SA6. Note the number of ports of [1×*N*] splitters incorporated with the ports of WSSs. The non-diagonal rows and columns with the cell numerals address the ports of the transfer nodes. The nodes exhibit the wavelength channel ports, as indicated in figure 6.39(c). The nodes are ROADM-based and inter-communications and switching with add/drop capability can be implemented with the matrix of (a).

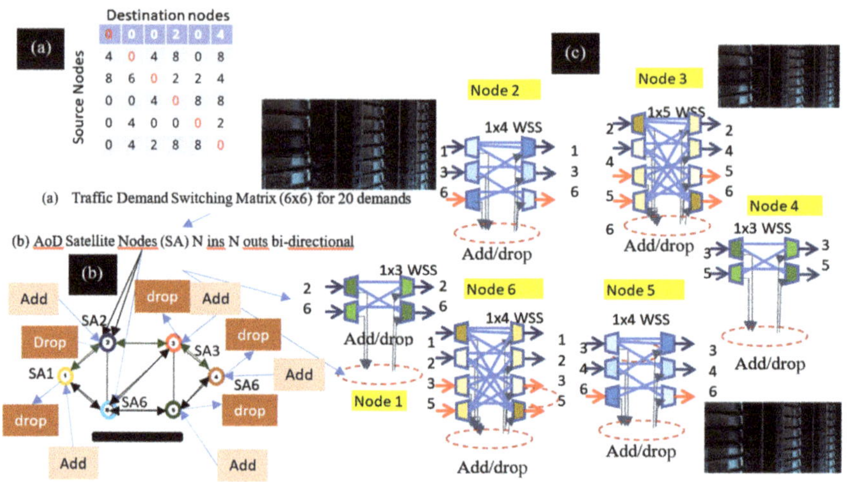

Figure 6.39. Optical network architecture on demand: example on a six-node-CDC-ROADM with 20 demands/reswitching matrix; ROADM satellite nodes quests, bi-directional. Node = satellite and/or DC optical routing node in inter-switching system nodes. (a) Traffic demand switching matrix; (b) networks with nodes and ADD/DROP nodes; (c) non-blocking switching matrix of six nodes with ADD/DROP Ports. AoD = architecture on demand.

The nodes can be satellite nodes in free deep space constellation laser networking or DC server rack nodes so as to transmit desired wavelength channels to other communication networks, thence to clients via the access networks of internet service providers. The ROADMs are thus CDC-ROADMs as required for maximum flexibility.

The switching matrix given in figure 6.39(a), follows a tensor assignment of the ports' input and outputs. The setting of the matrix can be conducted using MCUs. The port index and its corresponding bandwidth, as well as the switching or routing nodes, can be assigned without much difficulty.

Thus, we can see that for deep space inter-constellation of satellites the non-blocking CDC-ROADM interconnection with add/drop capability is highly important. Similarly, this feature is also highly critical for DCs operating at ultra-high capacity.

These non-blocking CDC-ROADM nodes form the bases for DC interconnecting to other DCs or metropolitan and access networks. This is to be addressed in the next section.

6.5.3.2 Global DC internetworking and intercommunication access

A global Internet is equivalent to networking of networks or data-centre-networks. This new emerging networking is quite different from traditional telecommunication networks in which the layers of communications protocols are simplified so as to serve the ultra-high speed and ultra-high capacity communication. Figure 6.40 shows a schematic of the networking of networks.

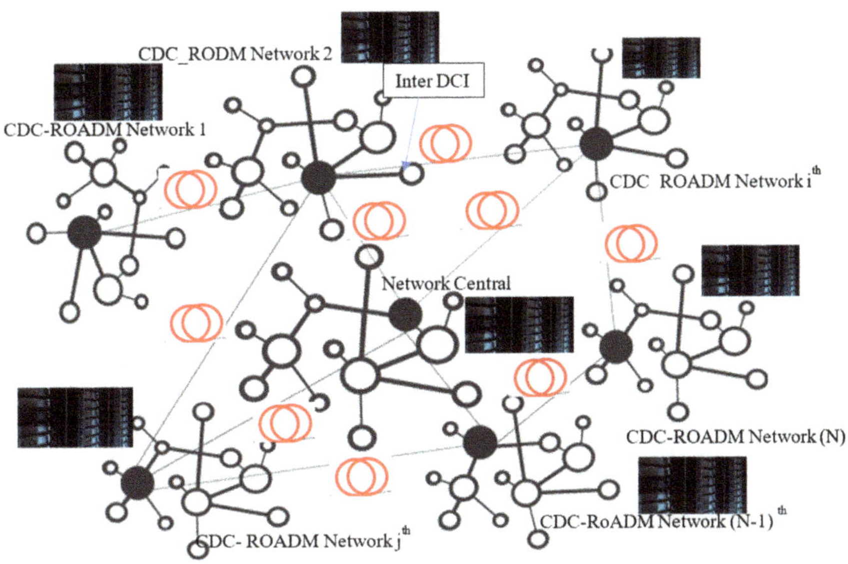

Figure 6.40. Global Internet is equivalent to network of network or DCN. Each network is a DC-based network. Optically amplified multi-span fiber transmission systems and ROADM-based routing not shown.

There is access to more networks directly with lower delay times and fewer optical hops by getting a high-speed IP transit connection international Internet backbone, thence DCN, especially for voice and video channels.

Global internetworking can support voices over IPv6 and IPv4 (Internet protocol versions 6 and 4) over the same connection at no additional charge. Once an IP transit connection is obtained from DCs of DCN, they are already set to add IPv6 support for access.

6.6 Concluding remarks

We have presented the photonic technologies and novel devices for reconfiguring optical and photonic devices that can route, switch and add or drop optical channels of different wavelength carriers, the C-band, super-C-band and L-bands. The routing and switching are color-independent, directionless and contention-less, allowing great flexibility and agility in DCN, as well as global networking of optical networks or DC networks.

There is no doubt that optical networking will be based on networking of DCs and not traditional telecom networks. The all-optical/photonic CDC-ROADM technologies will make a significant contribution to such networking. This chapter has addressed those technologies. However, the photonic technologies will keep evolving and further novel ROADMs will be emerging and thus the evolution of optical networking in DCN and OSDN will be critical for flexibility in configuring new networks of DCs, both inter and internal, DCI and DCN.

6.7 Appendix: Fabrication processes of typical optical MEMS

Surface micromachining technology for MEMS has contributed significantly to the development of miniaturized optical devices in integrated and free-space optics, especially in optical networking [7]. The unique MEMS characteristics are in systems integration, and operation of micro-optical systems. The precision mechanics of MEMS by microfabrication techniques and its functionalities make possible a wide variety of micro-mechanical movable optical elements in optical networking sub-systems. This section briefly describes the fabrication of such devices.

The fabrication steps of a bandwidth tunable optical filter using MEMS technology can follow the steps below:

(i) Fabrication to integrate silicon-based MEMS actuators can be based on micro-machine technology in silica on insulator (SOI). The suspended silica optical waveguides can be actuated by a Si electrostatic-comb drive actuator, e.g., with a maximum displacement of 8 μm at 50 V bias, as shown in figure 6.41.

(ii) Optical tunable-bandwidth filters can be highly useful in optical communication systems and networks, especially in optical signal processing. A micro-disk filter with a tuning range from 3 to 80 GHz can be implemented using Si photonic MEMS technology that offers higher performance filtering performance. This device can be employed for 6G/7G (6th or 7th generation) wireless communications between servers of DC networks. But

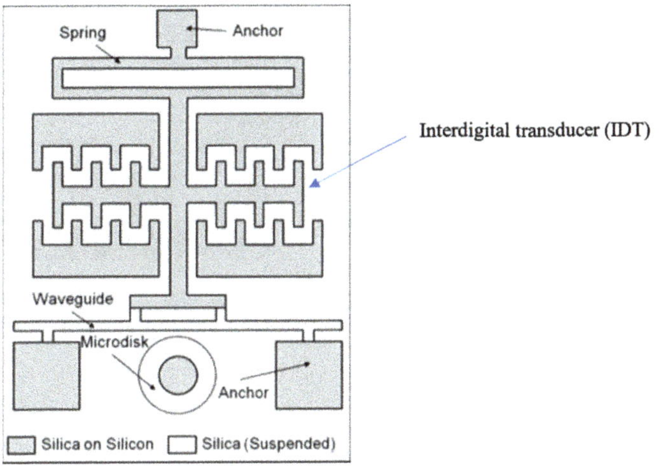

Figure 6.41. Schematic of a MEMS structure with IDT integrated over an optical waveguide for tunable-bandwidth optical filtering. Reproduced with permission from Ming Wu Lab/University of Berkeley.

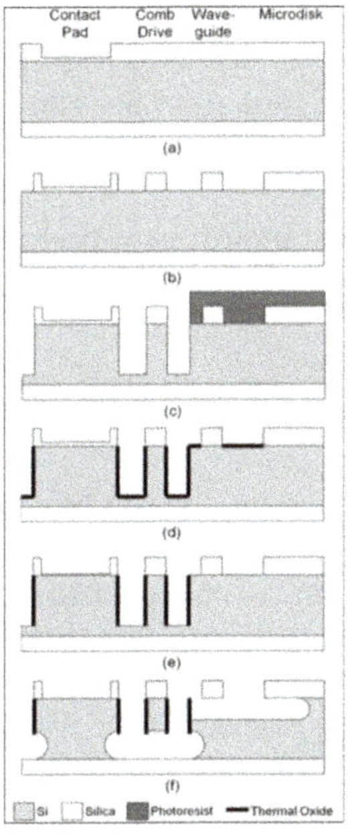

(a) Starting with SOI substrate wafer. Deposit p+-doped silica. Etch partially MEMS contact pads; Etch device layer.

b) Etching device structure layer.

c) Deep Si-etching with photoresist protection layer over the optical parts.

d) Removing PR and oxidization.

(e) Removal of base oxide layer by reactive ion etching.

(f) dry etching . Gas reflowing to enhance smoothness

Figure 6.42. Fabrication process on SOI wafer with details from SOI wafer to MEMS in steps (a) to (f) to create a MEMS structure. Reproduced with permission from Ming Wu Lab/University of Berkeley.

the tuning can only be accomplished with an additional bulk micro-positioning stage. This sub-section describes the fabrication process that integrates a silicon-based MEMS comb drive actuator with silica optical components, as shown in figures 6.41 and 6.42.

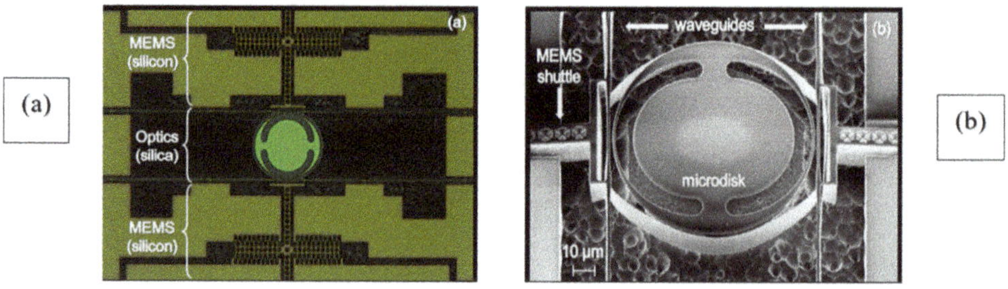

Figure 6.43. (a) Fabricated device composed of integrated silica optics and silicon MEMS. (b) MEMS scanning micro-graph. Reproduced with permission from Ming Wu Lab/University of Berkeley..

An integration of silicon MEMS actuators with silica optical component can produce a tunable-bandwidth filter. By employing an *electrostatic comb* one can drive to change the coupling of the filter by moving the silica waveguide with respect to the micro-disk, hence rotating the silica mirror (figure 6.43).

6.8 Appendix: Form factors modules in DCN

Differences between QSFP-DD and QSFP+/QSFP28/QSFP56/OSFP/CFP8/COBO. Compatible modules for DCN are shown in figure 6.44. These modules can be found in many manufacturer websites and are illustrated here only for readers to search in relevant websites.

The evolution over the last two decades of the transmission rate per module is shown in figure 6.45.

6.9 DCO (digital coherent optics) for DC interconnection and networking

DCs and long-haul transmission systems are the backbone of the modern information world, and their interconnection and networking capabilities are essential for providing reliable services and servicing applications of massive data and ultra-high transmission rates. As the demand for faster and more efficient data transmission increases, so does the need for advanced technologies to support it. One such technology is commercial DCO (digital coherent optics) or coherent transmission and reception technology with integrated photonic technologies, which is becoming increasingly popular for DC interconnection and networking.

DCO is a coherent type of optical transmission technology that uses DSP to encode and decode data to transmit coherently the polarized channels of data and coherent reception by mixing with an LO laser, especially the incorporation of digital signal processors in both the transmitter and receiver sub-systems so that corrections of distortion and their compensation techniques can be applied. This allows for high-speed transmission, and ultra-high capacity of data over long distances with minimal distortion. It is ideal for DC interconnection and networking because of its high bandwidth and low latency.

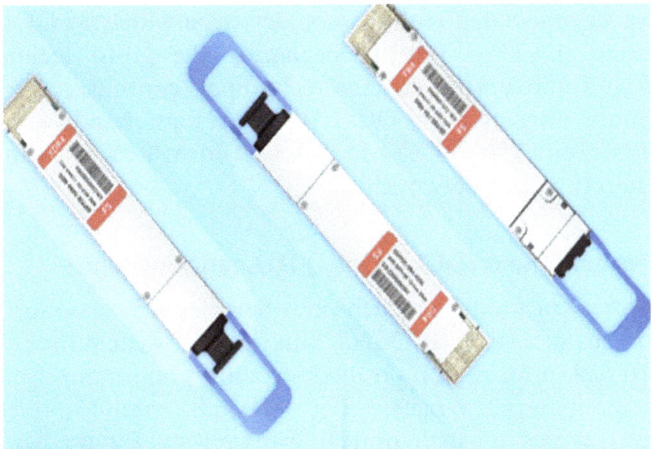

Figure 6.44. Compatible modules for DCN; FS400BASE-DR4/XDR4/FR4 QSFP-DD with manufacturers' Identity number.

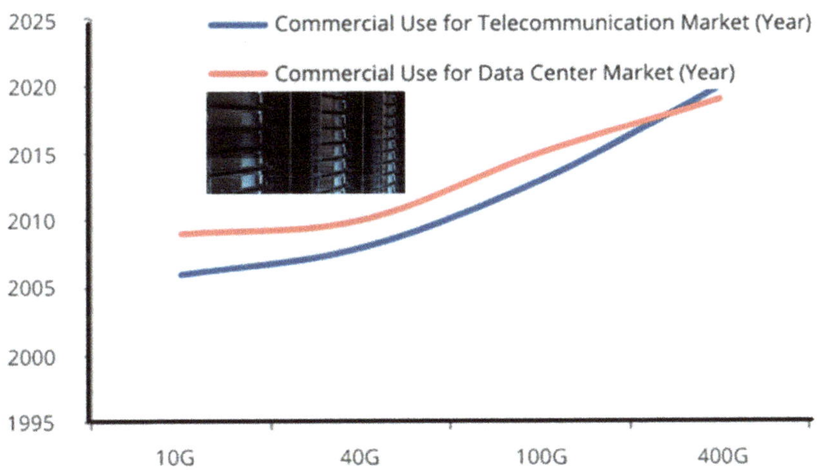

Figure 6.45. Evolution of bit rates per wavelength carrier in commercial transmission systems between 1995 and 2025.

DCO also offers a number of advantages over traditional optical transmission technologies. For example, it is more reliable and efficient than traditional optical transmission techniques, and it is also more secure. Additionally, DCO is more cost-efficient than other technologies, which makes it a highly attractive option for DCs.

DCO is also highly scalable, which makes it ideal for DCs that need to expand their capacity or add new services. This scalability also makes it easier for DCs to upgrade their networks as their needs change. In addition to its advantages, DCO also has some drawbacks. For example, it is more expensive than traditional technologies, and it requires specialized integrated photonic and electronic hardware and software to operate. Additionally, DCO is now no longer new, so there is still extensive experience associated with it to be deployed widely in DCs and long-haul optical networking. DCO is quickly becoming the go-to technology for DC interconnection and networking. As the technology continues to improve, it will become even more attractive to DCs. With its high bandwidth, low latency, scalability, and cost-efficiency, DCO in different form factors is an ideal solution for DC interconnection and networking.

6.9.1 Digital coherent optics (DCO) QSFP-DD optical modules

There exists a variety of modules which offer flexible and cost-effective options for all types of interfaces such as CISCO Inc., Huawei Ltd, Acacia Inc., etc. Readers are recommended to seek information on their structures and main technical specifications on these companies' websites. This appendix outlines only the working principles and structures of such optical transceivers. Figure 6.46(a) shows the complexity of the functional block diagram of a DCO module with the transmitter section and the reception section with DSP dedicated to both sections. Figure 6.46(b) depicts structure of the optical modulation and coherent reception of the module.

In this section, we briefly explore the concept of polarization diversity multiplexing, Quadrature Amplitude Modulation (QAM) and coherent reception. We look at the advantages of these techniques and how they are used in modern

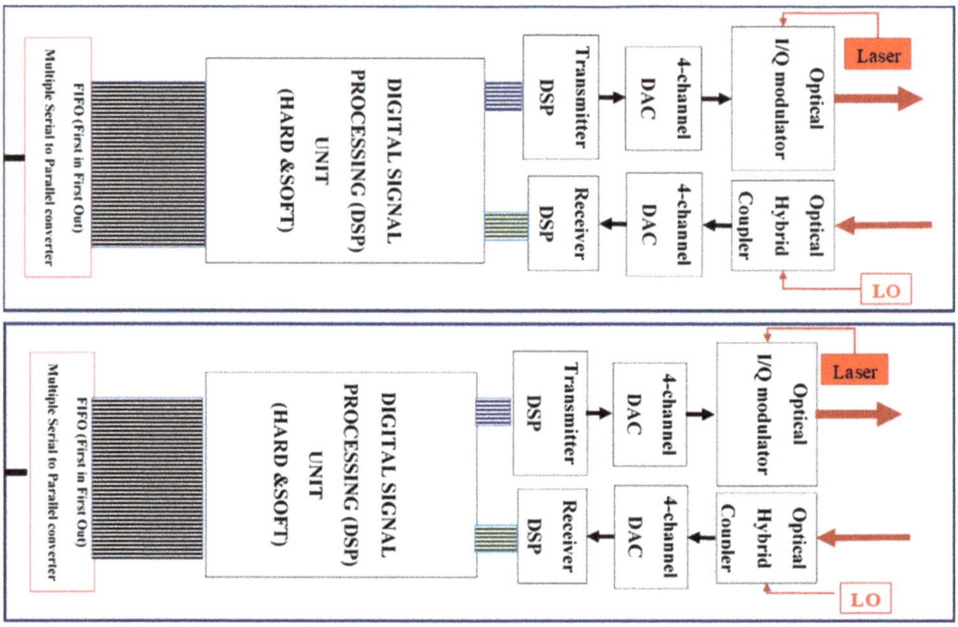

Figure 6.46. (a) OTRx module block diagram by a manufacturer of the transceiver [8] which can be seen in (b); (b) schematic detailed diagram of the transmitter and coherent receiver sections for IQ modulation and polarization de-multiplexing coherent balanced reception. LO = local oscillator; Pol = polarization; H = horizontal; V = vertical; I =in-phase; Q = quadrature phase; QAM = quadrature amplitude modulation.

communication systems, especially in DCO modules for ultra-high data rate transmission, the PDM-QWAM modulation of higher level modulation formats.

Polarization diversity multiplexing is a technique used to improve the performance of communication systems by using multiple antennae with different polarizations. This technique is used to increase the signal-to-noise ratio of the signal, as well as to reduce the effects of fading and interference. By combining the signals from multiple antennae, the system is able to take advantage of the diversity in the signal path and improve its performance.

QAM is a digital modulation technique that is used to transmit digital data. In QAM, the digital data is encoded onto a carrier signal using two amplitude-modulated signals that are 90 degrees out of phase with each other. This technique is used to increase the data rate of the system, as well as to reduce the effects of interference.

Coherent reception is a technique used to improve the performance of communication systems by using multiple receivers with different polarizations. This technique is used to increase the SNR of the signal, as well as to reduce the effects of fading and interference. By combining the signals from multiple receivers, the system is able to take advantage of the diversity in the signal path and improve its performance.

In conclusion, polarization diversity multiplexing (PDM), QAM modulation and coherent reception are all techniques that are used to improve the performance of communication systems. Thus, the resultant total capacity for a PDM-QAM is (number of effective bits/symbol) × (number of polarization beams), for example PDM-64-QAM = 2 × 8 × symbols/bit of the modulation. So, for 50 GSymbols s^{-1} PDM-64-QAM the total bit rate is 8 × 2 × 50 Gbps = 800 Gbps. These techniques are used to increase the SNR of the signal, as well as to reduce the effects of fading and interference. By combining the signals from multiple antennae, receivers or modulators, the system is able to take advantage of the diversity in the signal path and improve its performance as well as increase the transmission capacity.

6.9.2 DCO OTRx (optical transceiver) modules

Nowadays compact modules with ultra-high-speed transmission are commonly used in DCs and networking. The digital coherent optic modules of the transceivers under PDM-QAM modulation formats are used. Figure 6.47 shows the evolution of such DCO transceivers over the last decade. Integrated photonic technological and integrated ASICs or processors are now employed to provide such compact subsystems. QSFP and CFP2 (centum form factor pluggable) formats are commonly employed.

6.10 Optical module requirements for DCI

The internal interconnection of the DCs (DCI) accounts for a large proportion of the overall traffic distribution of the DC. The typical requirements for optical modules are shown in table 6.1, and there are development trends towards high speed, low power consumption, low cost, intelligence, etc.

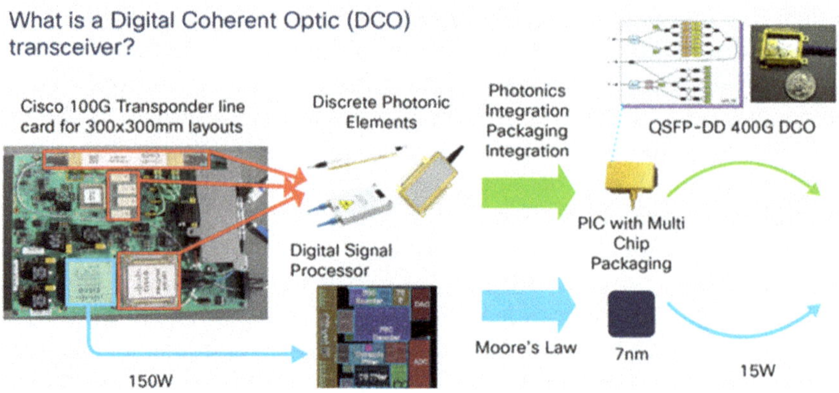

Figure 6.47. Evolution from discrete components to integrated coherent pluggable QSFP DCO.

Optical modules are an essential component of DCs, providing the necessary internal interconnection for communication between servers and other components. In order to ensure that DCs are running optimally, it is important to understand the requirements of optical modules for internal interconnection.

First and foremost, optical modules must be designed to meet the specific requirements of the DC. This includes the type of DC, the type of equipment used, and the type of data being transmitted. The module must be able to handle the data rate, bandwidth, and latency requirements of the DC. Additionally, the module must be able to support the physical environment of the DC, including temperature, humidity, and other environmental factors.

The optical module must also be designed to provide reliable and secure communication between the components. This includes ensuring that the module is able to provide the necessary encryption and authentication protocols to ensure that the data being transmitted is secure. Additionally, the module must be able to provide the necessary error-correction protocols to ensure that data is transmitted accurately and without errors.

Finally, the optical module must be designed to be cost-effective. This includes ensuring that the module is able to provide the necessary performance at an affordable cost. Additionally, the module must be designed to be energy efficient, ensuring that the DC is able to run as efficiently as possible.

By understanding the requirements of optical modules for internal interconnection in DCs, DC administrators can ensure that their DCs are running optimally and securely. By selecting the right optical module, DCs can ensure that their data is being transmitted accurately and securely, while also reducing costs and increasing efficiency.

6.10.1 Trend towards ultra-high speed

The internal interconnection within Amazon, Google, Microsoft, Facebook, and other North American ultra-large DCs began commercial deployment of 400 Gb s^{-1}

Table 6.1. Optical module requirements of internal DCI.

Types of optical modules		Form factor	Optical interface rate *N*x Gbps	Electrical interface rate (Gbps)	Transmission distance (m)	No. of fibers	Power consumption
100 Gbps	VR	QSFP28	100	100	30/50 m	1	<3.5 W
	SR4		4×25	4×25	70/100 m	4	
	PSM4		4×25	4×25	500 m	4	
	CWDM4		4×25	4×25	2 km	1	
	LR4		4×25	4×25	10 km	1	
200 Gbps	VR2	QSFP56	2×100	2×100	30/50 m	2	<6.5 W
	SR4		4×50	4×50	70/100 m	4	
	FR4		4×50	4×50	2 km	1	
	LR4		4×50	4×50	10 km	1	
400 Gbps	VR4	QSFP-DD/OSFP	4×100	4×100	30/50 m	4	<12.0 W
	SR8		8×50	8×50	100 m	8	
	SR4.2		8×50	8×50	100 m	4	
	DR4		4×100	4×100	500 m	4	
	FR4		4×100	4×100	2 km	1	
	LR4		4×100	4×100	10 km	1	
800 Gbps	VR8	QSFP-DD800/	8×100	8×100	30/50 m	8	16W
	PSM8	OSFP/QSFP224/	8×100	8×100	70/100 m	8	
	DR8	CPO	8×100	8×100	500 m	8	
	DR4		4×200	8×100	500 m	4	
	2×FR4		8×100	8×100	2 km	2	
	FR4		4×200	8×100	2 km	1	

optical modules from 2019–2020. Domestic DCs are gradually transitioning from 100 Gb s^{-1} to 400 Gb s^{-1} transceivers, and the scale deployment is expected to be realized in 2022. As shown in figure 6.48, the throughput of DC switching chips is expected to reach higher than 51.2 Tb s^{-1} in 2025 and 102.4 Tb s^{-1} after 2025. Higher rates of 800 Gb s^{-1} and 1.6 Tb s^{-1} will become important choices to realize high-bandwidth data exchange (figure 6.49).

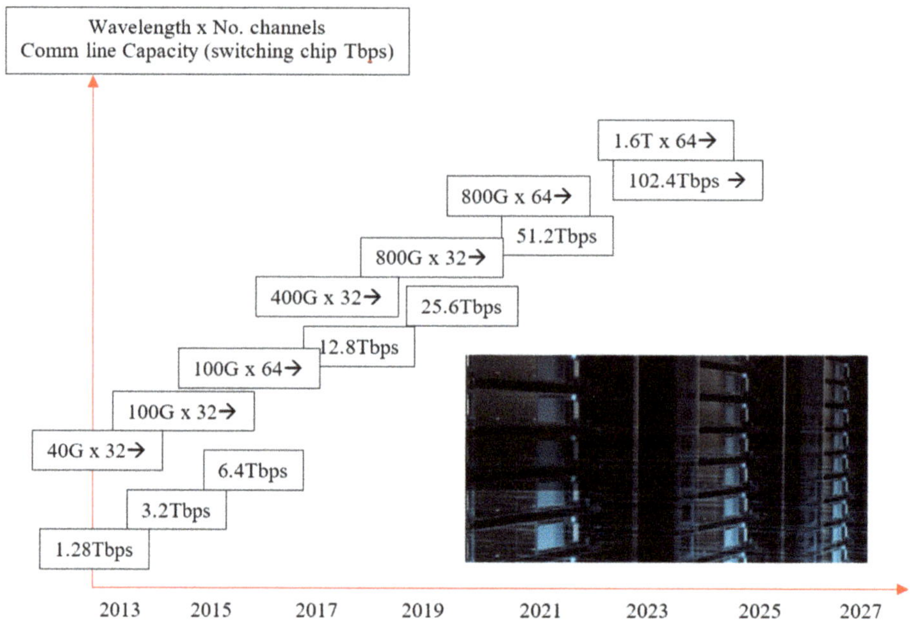

Figure 6.48. Bi-annual developing growth of DC switch integrated chip (ASIC—application specific integrated circuit) throughput.

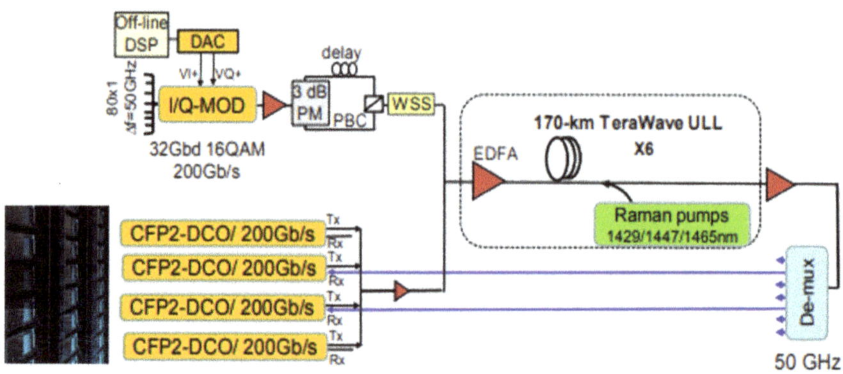

Figure 6.49. Optical transmission of 16 Tera-bps using CFP2 modules of 200G.16000/200 = 80 channels, carrying 200 Gbps per channel, grid channel spacing 200 GHz, bi-directional and interleaving between forward and backward transmission.

6.10.2 The trend towards low-consumption

As the capacity of switching chips continues to increase, the power consumption of optical modules has begun to exceed that of switching chips, becoming a key factor in network solutions. The early power consumption of 400 Gb s^{-1} optical modules is 10–12 W, and the long-term power consumption is expected to be 8–10 W; the power consumption of 800 Gb s^{-1} optical modules is about 16 W. In addition, the industry expects to reduce the power consumption and cost of interconnecting serializer/deserializer by encapsulating the optical engine and the switching chip, and CPO (co-packaged optics) technology encapsulates electronic chips and optical engines together, which has become a research hotspot in the industry.

6.10.3 The trend towards low cost

There are massive interconnection requirements in DCs, and low cost is one of the main driving forces for the continuous development of optical module technology solutions. First, the access cables in scenario one show a trend of diversification. Some solutions reduce the interconnection distance by adjusting the cabinet layout, and using lower-cost direct-attached copper cables instead of optical cables; second, with the stable operating environment and fast replacement of DC optical modules, the industry is actively exploring solutions to lower costs by reducing requirements for temperature and long-term reliability and so on; thirdly, as the speed continues to increase, the sinking trend of coherent solutions is obvious, and non-coherent solutions are also striving to expand to long distances. The two schemes 'meet' in some application scenarios, and the proportion of demand for different schemes in the 'meeting' scenes will be closely related to factors including cost.

6.10.4 The trend towards intelligence

Optical network operators and manufacturers have begun to pay attention to the enhancement of the operation and maintenance capabilities and quality improvement of optical modules. The health monitoring of optical modules and early warning of faults are realized through artificial intelligence, machine learning, and big data, which puts forward new requirements for the functional characteristics and specifications of optical transceivers.

6.11 Appendix: OSNR and measurement in ROADM optical networks[4]

Optical transceivers, whose structures are shown in figure 6.51, employ dual polarization multiple-level (M-ary) quadrature amplitude modulation (DP-M-ary-QAM), and are commonly employed as the transmitter and receivers in many coherent optical communication systems. They consist of a high-speed optical

[4] This appendix adapts the information provided by Viavi Inc. to clients and summarized here on measurement techniques, issues and solution provided by ultra-high resolution OSA of Viavi Inc.

modulator, the optical Mach–Zehnder modulators in parallel, to provide the in-phase and quadrature modulated lightwaves which come from an input source laser.

The DSP-based optical receiver is shown in figure 6.51 which consists of an optical hybrid coupler that mixes the LO laser and the polarized components of the signals in the balanced photodetector pair and electronic preamplifiers. The signals at output of the electronic amplifiers are then sampled by an analog-to-digital converter. The digital streams are fed into a DSP and processed by a number of algorithms to produce various performance parameters. The receiver sensitivity in dBm and OSNR is highly critical for determination of the BER of the receiver, thence the power budget required for the transmission system. As a result, the noise at the channel wavelength is not the same as the noise to the left and right of the channel. In this case, relying on a standard, out-of-band OSNR measurement can give misleading ideas about network performance. This appendix thus describes with definitions and techniques how to determine in-band OSNR in the frequency or wavelength domain, that is easy to obtain from the optical signal and noises spectra obtained from an optical spectrum analyzer. The noise of the OA that the channels are passing through is embedded in the signal band. This is the technique defined in the standard by IEC61280-2-9 [9]. IEC61280 provides a parameter definition and a test method for obtaining OSNR using apparatus that measures the optical spectrum at a multichannel interface. Because noise measurement is made on an optical spectrum analyzer (OSA), the measured noise does not include source relative intensity noise (RIN) or receiver noise.

The central part of a ROADM is the WSS, which is used to select which wavelength channels to be routed. At the input, wavelength channels can come from different optical nodes and thus have different optical noises. Since any wavelength channel can be switched to any corresponding port, as the wavelengths enter the ROADM, they are thence de-multiplexed. Because the optical filters are wider than the DWDM signal band, some parts of the residual noise go through, as shown in figure 6.50.

The spectrum of two DWDM optical channels after passing through a WSS can be observed, as shown in figure 6.50. This depends on the passband spectra of the channels of the WSS. The passband of a WSS is normally similar to a super-Gaussian distribution of order 5–12. Thus, the signal spectra and that of the WSS are not matched, but sharper in the roll-off band than those of the signal bands and thus the in-band noise spectrum can be wider or narrower and at an appreciable level. So, if the standard IEC is to be followed then there may be a significant error when the out-of-band noises only are considered.

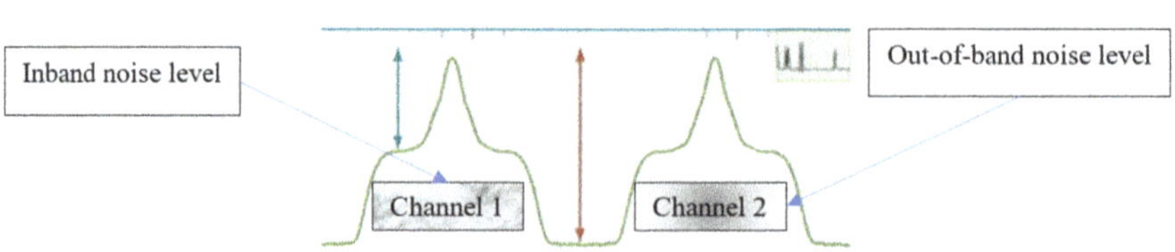

Figure 6.50. Spectrum of two DWDM channels after transmitting through a WSS.

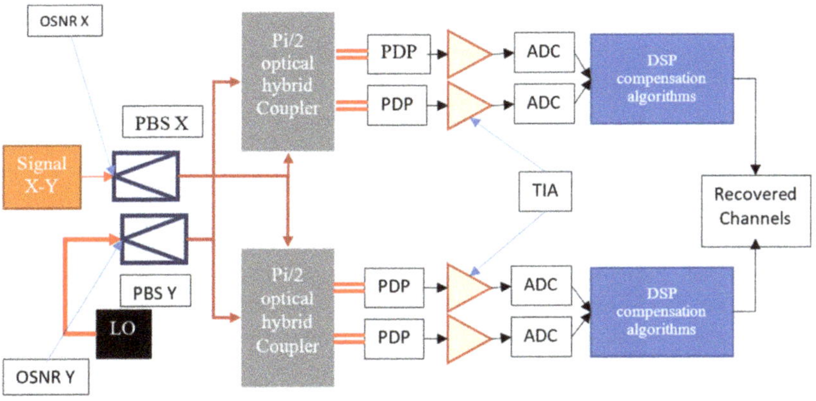

Figure 6.51. Optical receiver structure for coherent reception of PDM-QAM signals. Labels: TIA = transimpedance amplifier; LO = local oscillator; ADC = analog-to-digital converter; PBS = polarization beam splitter; DSP = digital signal processor.

The standard IEC61280-2-9 Fiber Optic Communication Subsystem Test Procedures Part 2-9: Digital Systems Optical Signal-to-Noise Ratio Measurement for DWDM Systems [10] defines OSNR measurement as the average between the left and right OSNRs, which are measured as the difference in power between the peak power and the noise at half the distance between the peaks by interpolation. Unfortunately, this description is no longer appropriate for ROADM-based networks.

It is critical to determine the close to exact value of OSNR so as to determine the performance of an optical communication system. In the DC optical links and networking of ultra-high capacity with channels packed closely to ensure high capacity. To determine the total average signal power of an optical modulated channel the optical power per resolution window of the signal spectrum must be measured. Similarly, for the noises. In optically amplified links under ultra-high speed and closely packed the spectrum may overlapped. So, the definition of OSNR and experimental method for measurement must be established, especially during the test and manufacturing of the optical link. The OSNR monitoring technique can be found in Viavi.com website in which the original development was from JDSU.com.

References

[1] Shannon C E 1949 Communication in the presence of noise *Proc. Inst. Radio Eng.* **37** 10–21

[2] Binh L N 2017 *Advanced Digital Optical Communications (Optics and Photonics)* (Boca Raton, FL: CRC Press)

[3] Petersen K E 1977 Micromechanical light modulator array fabricated on silicon *Appl. Phys. Lett.* **31** 521–3

[4] Petersen K E 1980 Silicon torsional scanning mirror *IBM J. Res. Develop.* **24** 631–7

[5] Kikuchi K 2008 Coherent transmission systems *ECOC 2018, 34th European Conf. on Optical Communication* pp 1–39

[6] Colbourne P D, McLaughlin S, Murley C, Guadet S and Burke D 2018 Contentionless twin 8 × 24 WSS with low insertion loss *2018 Optical Fiber Communications Conf. and Exposition (OFC) (San Diego, CA)* 1–3
[7] Le H T *et al* 2021 MEMS inductor fabrication and emerging applications in power electronics and neurotechnologies *Microsyst. Nanoeng.* **7** 59
[8] Extracted from figure 6.1, of AC1200 Single-Carrier Product Specification Rev0.2 by Acacia Inc. 7 of 86 pages
[9] International Electrotechnical Commission Standards Document IEC 61280-2-9, 'Digital systems—optical signal-to-noise ratio measurements for dense wavelength-division multiplexed systems,' 2009. https://standards.globalspec.com/std/966286/iec-61280-2-9
[10] Fibre optic communication subsystem test procedures—part 2-9: digital systems—optical signal-to-noise ratio measurement for dense wavelength-division multiplexed systems, https://webstore.ansi.org/preview-pages/IEC/preview_iec61280-2-9%7Bed2.0%7Db.pdf

IOP Publishing

Le Nguyen Binh

Chapter 7

Power-limited (quantum) and bandwidth-limited (ultra-high-capacity) network transmission

The transmission technologies presented in this book for data center networking, especially in association with reconfiguration optical add/drop multiplexers (ROADMs) have led to networks of capacity reaching multi-Tbps and even Exabits s^{-1} in the near future.

Further, the security aspects of the transmission and switching/routing of massive data communications require security transmission technique. There are several secure transmission techniques such as the traditional coding, the chaotic states, and quantum communications. Quantum states are the most secure communication technique in which the keys are transmitted at quantum levels in either discrete or continuous variables depending on the reception of the bits in sampling bits by bits in discrete photonic detection. When the total energy per bit is detected as a whole, then the quantum reception is termed as continuous variable quantum key distribution (CV-QKD). The keys are formed by the sequence of the secret bits whose energy is at the level of a photon or even much lower, say 30 dB below this single-photon state. Thence the quantum keys are either at single-photon states or at a single-photon quantum level that is either at the noise level or well below that, the continuous variable quantum key distribution (CV-QKD).

This concluding chapter addresses the capacity and bandwidth-limited regions of the well-known curve given by Shannon in 1947 [3], that is, the extreme limits of either scenario. Therefore, in this chapter we describe the transmission of quantum bits (Qubits) for continuous variable quantum key distribution (CV-QKD) as well as ultra-high capacities of multi-Tera-bits through the C-band, high transmission atmospheric, and intersatellite optical channels.

In the bandwidth-limited cases, the optical amplification methods, booster power amplifiers, and highly sensitive input optical preamplifiers are demonstrated to

doi:10.1088/978-0-7503-5849-1ch7

provide the compensation of ultra-high attenuation of terrestrial installed fibers or in DCI and DCN networks. The amplifiers are accumulated over several transmission spans and then at the stage where the noises reach a level at which the OSNR cannot satisfy the sensitivity of the receiver. The signals must be regenerated for further transmission.

On the other hand, the digital signal processing (DSP) algorithms in coherent receivers for ultra-high capacity systems and ultra-low power quantum bits are shown to play the critical roles in system performance. DSP techniques extensively developed over the last decade for ultra-high speed coherent optical communications, have boosted the ability to detect signal levels at a pre-FEC BER (bit error rate) of only 2e−2 to achieve an optical-to-signal ratio (OSNR) reaching 10–13 for 100 GSy s^{-1} dual polarization quadrature amplitude modulated signals with DSP compensating for dispersion and phase rotation for matching local oscillator (LO) laser frequency and that of the carrier. Therefore, the transmission rate and distance can be significantly improved, reaching to 1.6 Tera-bps with symbol rate of 224 GSy s^{-1}.

In this chapter we specifically address the two extreme cases of transmission that would affect both the secure communication and ultra-high capacity transmission. That is the transmission system operating in the case of power-limited and bandwidth-limited scenarios. By applying a Nyquist equivalent sampling theorem to the power-limited case, it would considerably improve the quantum BER (QBER). We can base on a combination of the Nyquist sampling theorem and the Heisenberg uncertainty principle to achieve an acceptable error rate that allows the forward error correction technique to boost it to an error-free transmission, when the signal power is reaching a level well below, say −30 dB, that of a single photon's energy.

Further, in the case of band-limited transmission, the Nyquist pulse shaping method and probability shaping of a constellation can be used to maximize the Tera-bits transmission via space channels, with a small amount of reduction of the effective bit rate.

7.1 Introduction

Given that optical and photonic technologies have progressed tremendously in fiber optic communication terrestrial systems over the last six decades reaching more than higher than 50 Tbps (tera-bits s^{-1}) over the entire C- and L-bands of silica single-mode fibres (SMFs), the standard SMF, ITU-T standard G.652 type. Advanced modulation and multiplexing techniques incorporating DSP with ultra-high sampling rates have allowed the symbol rates reaching >200 GSymbols s^{-1} (GSy s^{-1}). Thus, combining higher order modulation formats and coherent reception techniques, the bit rate using a single wavelength can reach over tens of (tera-bits s^{-1}) Tbps in terrestrial and submarine systems, even over the intersatellite multi-hop optical links.

Only recently, the transmission of such ultra-high capacity over terrestrial and space channels, especially over multi-stage optically amplified or hops inter-orbital satellite constellations have been considered [1, 2]. Simultaneously the secure transmission of secret keys over quantum level of optical bits is critical for coding and decoding at absolute security over the same channels by optical technology.

This chapter, thus, briefly presents recent works on the coherent optical transmission technology to transport both the ultra-high capacity channels and moderate bit rates for low power at quantum level, the QKD over the space inter-orbital satellite constellations with multiple hops and all-optical routing and switching in data center inter and networking, DCI and DCN, as described in previous chapters (chapters 1–6), and at optical satellite nodes of free-space and partially deep-space laser communication networks.

It can be shown that with signals at quantum level or power-limited signaling, a sampling 'theorem' that combines the Shannon channel capacity (theorem) and the Heisenberg uncertainty principle with quantum noise fluctuation the recovery of an SNR of 10^{-N} is possible with $N > 2$ for a proportionate sampling factor.

This chapter is organized as follows: in the next section we give an overview on the channel capacity under bandwidth-limited scenarios and power-limited case and their relations with the sampling rate of the systems. Quantum noise fluctuations and the sampling for channel capacity to reach an SNR of −30 dB are given. Section 7.3 gives the technological development of multi-Tbps transmission capacity over SOCS (space optical communication systems) and terrestrial or submarine or DCs or DCN, with initial results obtained for modules to provide bit rates of 200 Gbps per wavelength under digital coherent optics for an optical transceiver. Section 7.4 gives insights into the design of an quantum optical transceiver and QKD protocol so that absolute security can be obtained.

7.2 Shannon 'channel' and bit rate

7.2.1 Transmission systems

Figure 7.1(b) shows a generic optical transmission link over a space channel, while figure 7.1(a) is a fiber-based multi-span optically amplified link in which there are two optical amplifiers per stage, one is for preamplification and one for boosting power to transmit the channels to the next stage. These amplifiers in a fiber-based transmission system are required to compensate for high insertion loss of dispersion-compensated fiber (DCF) which is required in the fiber transmission line to compensate for the dispersion of the preceding fiber line. The core of such DCF has high propagation loss. The lightwaves carrier can be of the C-band (150 nm as a typical wavelength) and 1060 nm spectral windows. The optical carrier frequencies can be selected based on two factors:

- The transparency of the space propagation media, that is, the refractive index of the medium, in vacuum or in atmospheric free space. For an optical fiber-based system, the refractive index of the guided media is confined by the guiding condition or boundary condition of the fiber, thus the group velocity and hence the dispersion is a function of the wavelength. This is due to different refractive index and the propagation vector of the guided waves changes, thus there are differences in the propagation time of the channels of different wavelengths.
- The spectral windows through which the laser beam can be transmitted with minimum attenuation.

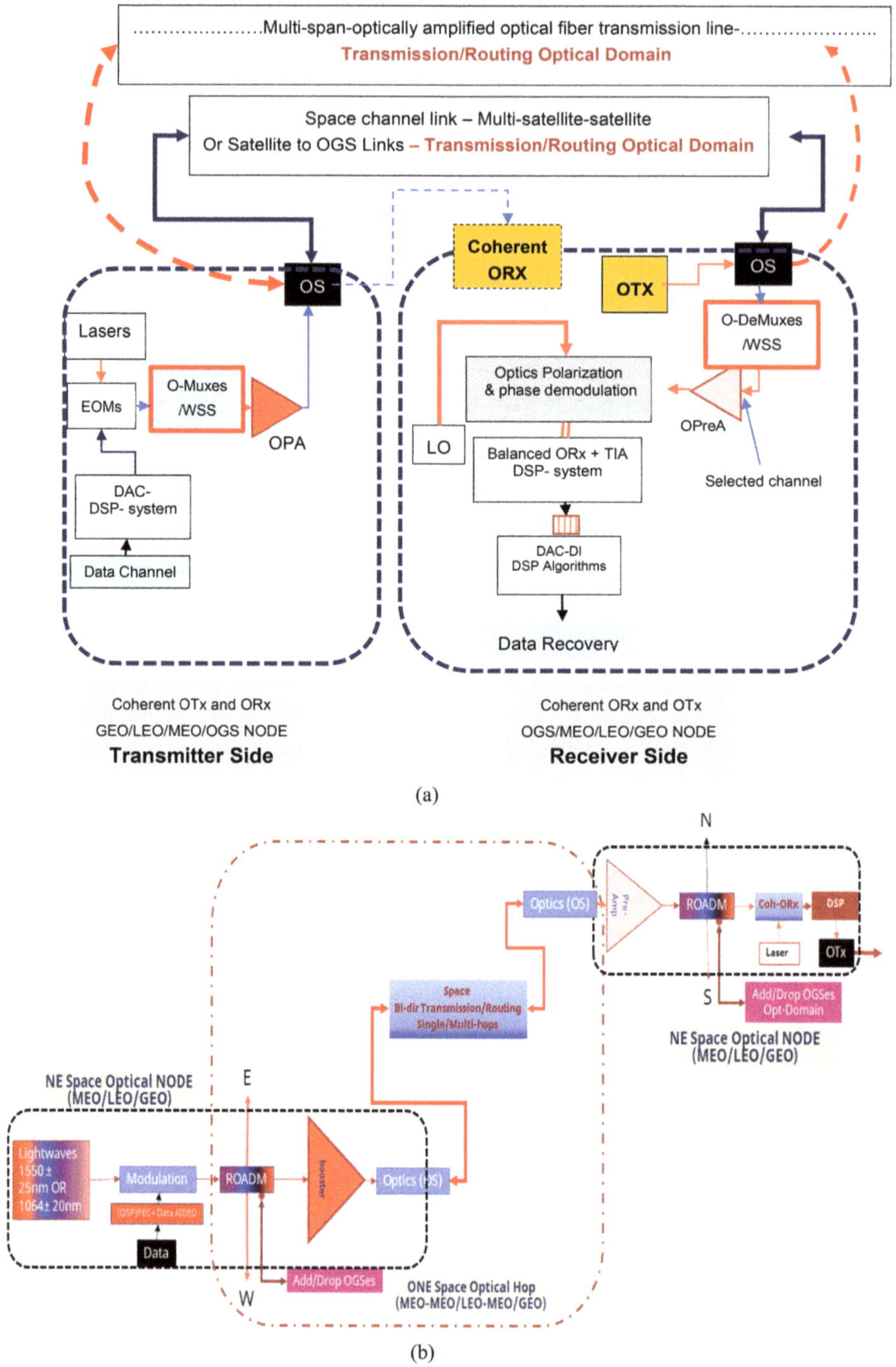

Figure 7.1. (a) A generic multi-stage optically amplified optical fiber link using coherent reception techniques and multi-level advanced modulation format. A ROADM is integrated at the NE space optical node. (b) A generic schematic of space optical communication system (SOCS) link and transport transceiver. Labels: Tx = transmitter; BPF = bandpass filtering, DSP = digital signal processor; BW = bandwidth;

ROADM = reconfigurable optical add/drop multiplexer; S = south; N = north; E = east, W = west; RxOS = optical system at receiving end; TxOS = optical system at transmitting end; LEO/MEO/GEO = low/medium/geo orbits; Coh-ORx = coherent optical recption; DSP = digital signal processing. DAC = digital-to-analog converter; OS = optical system (optical head); NE = North-East; LEO = low earth orbit, MEO = medium earth orbit, GEO = geostationary earth orbit.

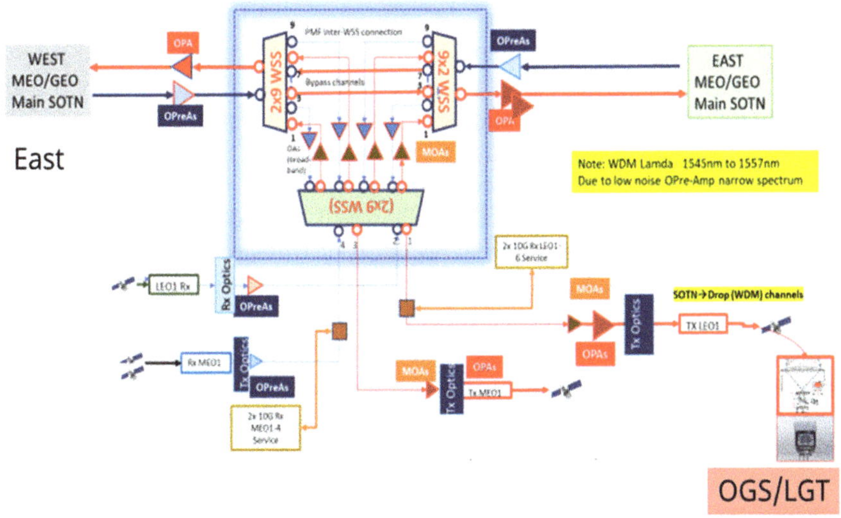

Figure 7.2. Schematic of a ROADM node employing 2 × 9 WSS for east–west transmission through add/drop ports. Labels: OGS optical ground station of optical nodes to be dropped with selected wavelength channels. OPA = optical booster or power amplifier; OpreA = optical amplifier at the input stage to amplify the weak signals. LGT = light ground terminal.

- The availability of optical amplification in these spectral widows so that amplification can be used to compensate for the signal attenuation over long-distance transmission in space. The space transmission distance can reach 8000 km if the communication is between MEO (medium earth orbit) with a loss of approximately 270 km using the inverse square distance law of attenuation factor of the laser beam.

As mentioned above, it is similar to a terrestrial long-haul transmission system, except that the optical amplification in this system is done over 80 km with compensation of chromatic dispersion. In space there is no mid-stage amplification, naturally, thus the loss is very high and amplification of low noise is critical.

The routing of optical channels of different carrier wavelengths can be implemented using a WSS (wavelength selective switch), as shown in figure 7.2. This is also reconfigurable due to the properties of WSS as described in chapter 6. Thus, the name ROADM (reconfigurable add/drop multiplexer). These ROADMs are also extensively employed in DCI and DCN.

The channel bandwidth is considered to include the bandwidth of the sampler. The sampling rate can be much higher than the bit rate of the information data sequence, hence this increases the transport link bandwidth.

7.2.2 Capacity, sensitivity and limits

Now taking to the limit of the $\frac{E_b}{N_0}\rightarrow$ quantum limit or ~1→ 0 or ≪ 1 (l'Hospital's rule) i.e., $\frac{e^{\eta}-1}{\eta}$ leading to 2*ln(2)→0.6931 (or −1.592 dB). Thence plotting energy per bit against η, as shown in figure 7.3. The energy per bit and noise power per bit is used to ensure that the scaling of bandwidth and noise can be done separately so that one can simplify the comparisons of different modulation and transmission methods. The channel capacity given by (equation 7.1) can be plotted as shown in figure 7.3, in particular, the region in the circle of figure 7.3, in which the signal energy approximately approaches the noise levels and even much lower (>30 dB). It is thus necessary to extend this power-limited case to the limit by applying the sampling rate theorem and the uncertainty principles for extremely narrow time window for an ultra-low energy particle. This will be addressed in section 7.4.

Equation (7.1) gives the Shannon expression of transmission capacity versus the ratio of the energy per bit and noise spectral density contained in this bit period. In the channel capacity given by Shannon [3] we denote C as the bit rate, BW as the system bandwidth, E_b is he energy per bit, N_0 is the total noise energy per bit, η is the ratio between the transport bit rate and the system bandwidth.

$$\text{Then} \rightarrow \quad \frac{C}{\text{BW}} = \eta = \log_2\left[1 + \frac{E_b}{N_0}\eta\right] \rightarrow \frac{E_b}{N_0} = \frac{e^{\eta}-1}{\eta}; \tag{7.1}$$

in which the SNR is very low and even at a level where the signal energy per bit is well below the noise level. The region marked with the dotted circle is where the

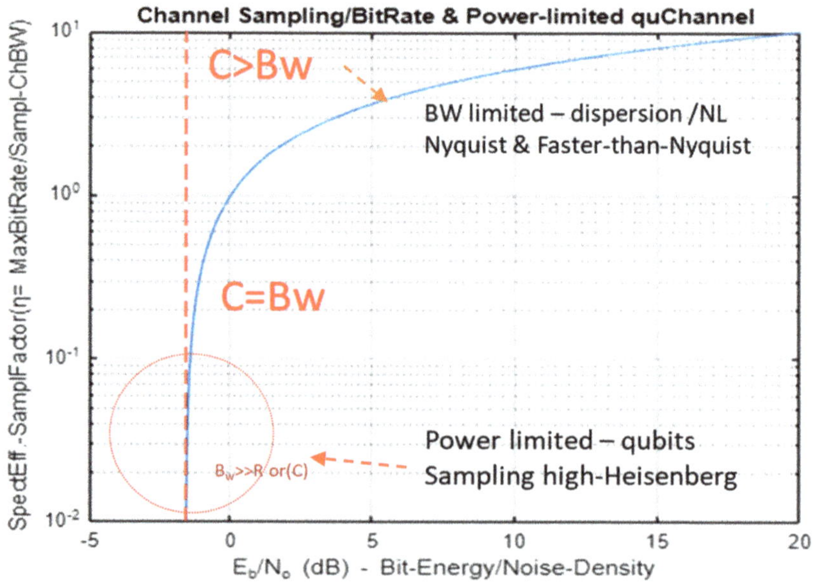

Figure 7.3. Channel capacity (C) by sampling factor/channel bandwidth versus signal energy per bit over noise power per bit [3]. Labels: R = bit rate, Bw = channel bandwidth (or ChBw), C = capacity or bit rate under transmission.

signals are limited by the average power that can be well below the noise level or the OSNR can be <0.

However, in the above equation (equation (7.1)) the noises are considered to be Gaussian, that is, not quite correct when considering the noises have fallen into the quantum level.

So, in the limit of BW/C (bandwidth over capacity) reaching a large ratio (e.g., 10–1000) E_b/N_0 results in an SNR of 0.691 or −1.599 dB, that is, the noise power > signal average power in the bit period. The ratio E_b/N_0 can be defined as the power efficiency or equal to energy density (per bit)/noise density. The energy/noise per bit can be used here so that this amount is independent of the channel bandwidth that can be scaled appropriately for specific transmission systems. Therefore, the sampling rate can be much greater than transmission bit rate, thence it is highly possible to recover the signal. This also satisfies the Nyquist sampling theorem, that is, for the case of noiseless systems.

In the scenario when the channel energy or power is at the quantum level, the sampling rate can be increased as indicated in the small region indicated by the red circle of figure 7.3, that is, when in time the window becomes very narrow. Thus, the Heisenberg's uncertainty principles must be applied as given in equation (7.2)

$$\Delta E \Delta t \geqslant h/2\pi \tag{7.2}$$

where ΔE is the signal energy contained in the time window Δt, and h is the Planck constant. So how do we set the noise and quantum signal to obtain absolute secure transmission? This is considered under noise modelling in section 7.4.

7.2.3 Bandwidth-limited system

By inspecting the regions of figure 7.3 in which the OSNR is much greater than 1 dB. For the limit of the bit rate or channel capacity under a Nyquist sampling rate there needs to be sufficient power to obtain the maximum bit rate that can be transmitted over the channel. This is the region in which Tbps can be achieved by firstly obtaining the maximum symbol rate under the available bandwidth of the SOCS sub-systems. Such multiplexing and higher order modulation formats are shown in figure 7.4, and the flow of signals in figure 7.5 shows a generic schematic of the bandwidth-limited transmission system. The modulation is coherent and complex M-ary-QAM (quadrature amplitude) (as described in chapter 1), or quadrature amplitude modulation with M-constellation points. Furthermore, polarization multiplexing diversity is also employed. Thus, the advanced modulation technique is PDM-m-QAM or M-ary-QAM pol-mux schemes, m is the number of constellation points.

Thence, by multiplexing of the optical carriers and polarization diversity one can increase the aggregate capacity over several Tbps. Section 14 expands these aspects so that under the transmittance characteristics of the intersatellite links the maximum Tbps can be achievable. Figure 7.6 illustrates the multiplexing of wavelength channels and corresponding bandwidth with bit rates under dual polarization M-ary-QAM modulation formats. Therefore, to reach an aggregate Tbps we need to multiplex, for example, 5 wavelength-carriers with 200 Gbps per

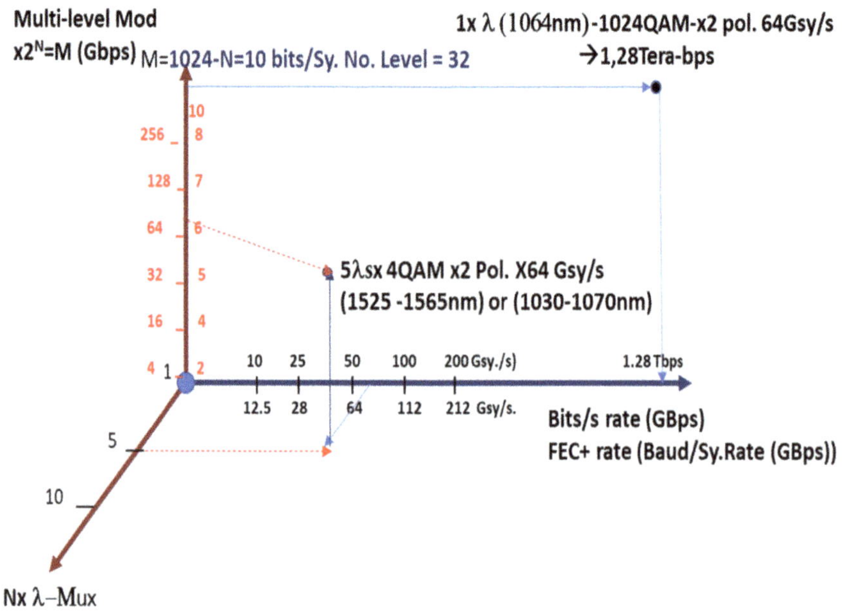

Figure 7.4. Optical multiplexing and higher order modulation formats to reach aggregate-capacity to multiple Tbps. λ = wavelength; GSy/s = Giga-symbols s^{-1}; QAM = quadrature amplitude modulation; Mux = multiplexing.

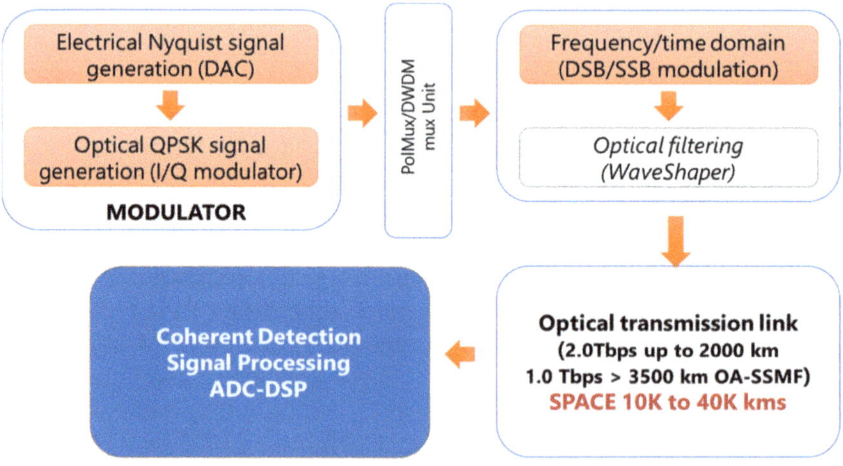

Figure 7.5. Generic schematics of optically multiplexing Tera-bits s^{-1} transmission system.

carrier etc. We note that as the level of modulation increases, the required OSNR is increased accordingly, as shown in figure 7.4.

Figure 7.5 gives a generic schematic of an optically multiplexing tera-bits s^{-1} transmission system. The individual wavelength channel bit rate can be 100G, 200G or up to 1600 Gbps, as indicated in figure 7.4. The different carrier channels are

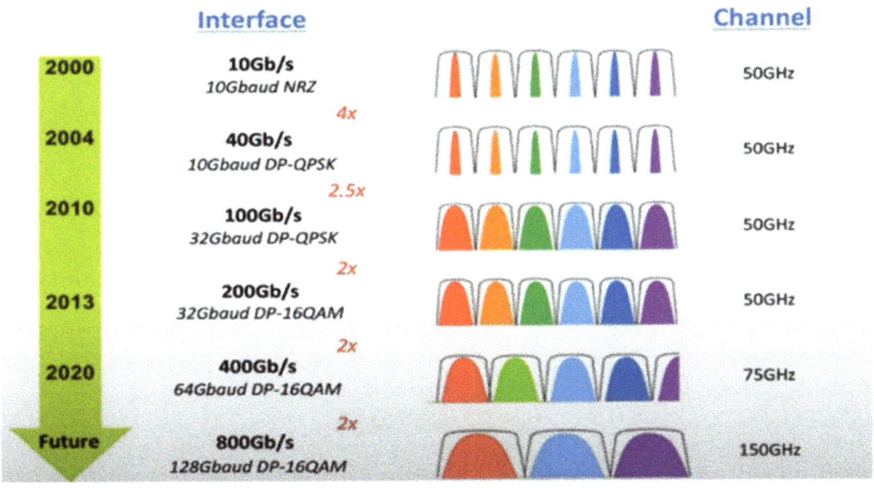

Figure 7.6. (Left) wavelength division multiplexing and required double sideband bandwidth in an optical spectral window, (right) transmittance property of atmospheric medium as a function of optical wavelength.

multiplexed via a multiplexer which can also be a WSS, as given in chapter 6. Polarization modes can multiplexed as well so that if two orthogonal polarizations can be multiplexed then the total transmission capacity can be doubled. The signal passbands of the channels are given in figure 7.6(right side) to determine the spectral window over which the carriers are transmitted so that Tbps capacity can be obtained. It is noted that the capacity can be increased if a higher order of modulation is used, thence the number of bits/symbols can be increased, hence the capacity. Figure 7.4 shows such a three-dimensional combination of the parameters, the modulation format of bits/symbol, the multi-level, then the baud rate or symbol rate and wavelength carrier (frequency).

Eventually the multiplexed information pipe is entering the demultiplexer and individual channels are separated and detected by optical receivers which can incorporate the digitaliser and the DSP. An optical front processor in front of the photodetection can also be used to separate the polarized channels and then phase shift in the optical domain to resolve the in-phase and quadrature components.

7.2.4 Power-limited system

The power-limited case occurs when the signal average power is very low, at the quantum level, or much lower, say about −30 to −50 dB below the quantum level of the energy of a photon in a bit period. This is the region in which the SNR $\ll$ 0 dB, as indicated in figure 7.3. With further expansion of the region of power limited with the increasing of the sampling rate and applying the Heisenberg uncertainty principles with quantum noise fluctuation, we can obtain the modified channel capacity for this case, as shown in figure 7.7. It is noted that the higher the sampling rate, the higher the channel bandwidth and the lower the signal average power that

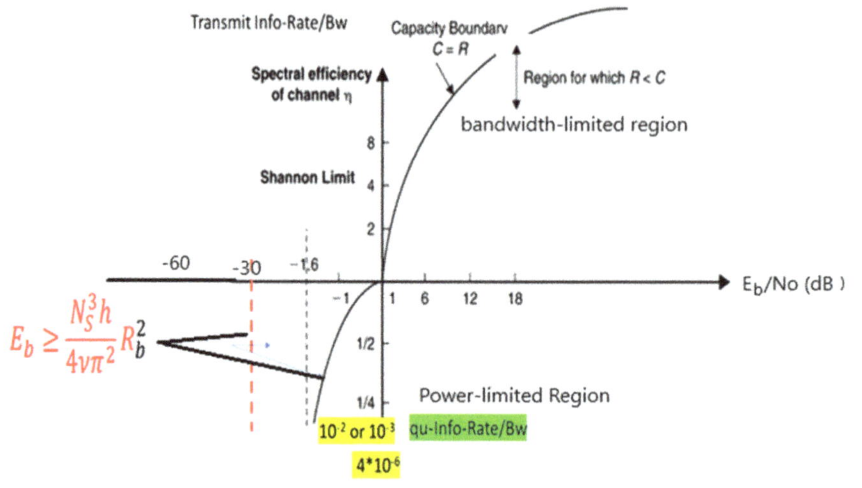

Figure 7.7. Modified channel capacity versus the SNR. E_b= energy per bit; R_b = bit rate; N_s = sampling rate scaling factor; h = Planck constant; qu = quantum, C = capacity; R = bit rate; BW = bandwidth.

can be detected with respect to the noise power level, by a nonlinear scaling by a cubic factor.

7.3 Multi-tera bits s^{-1} transmission

7.3.1 Modulation and multiplexing

Reaching capacity of Tbps can be achieved by polarization multiplexing and wavelength multiplexing in the optical domain. Further increase of the aggregate bit rate can be achieved by employing multi-level modulation in the complex plane.

For example, for 256-QAM or 256 constellation points in the complex plane, the number of bits per symbol is 8. Thus for five wavelength channels and dual polarization multiplexing we can obtain 80 × symbol rate. So for a 64 GSy s^{-1} rate the aggregate capacity is 5.12 tera-bits s^{-1}.

Naturally the increased number of bits per symbol must satisfy the required SNR so that error-free can be achieved. Under contemporary FEC (forward error coding) techniques, such as soft-FEC or hard-FEC, a BER of 2.e−2 can be achieved with an SNR of only 3.5 dB for QPSK (quadrature phase shift keying) modulation format. In figure 7.8 we also show the region in which the QKD system can operate. This region is the encircled area that can only be inspected in detail in figure 7.3, that is when quantum noise fluctuation and Heisenberg uncertainty principle can be applicable. Further improvement in receiver sensitivity can be achieved by statistical modulation. Such a modulation scheme and effective bits per symbol for 16-QAM modulation is shown in figure 7.9. For statistical modulation, the constellation points are arranged in such a way that it is less than the Euclidean distance as far as possible so that the overlapping is minimized, thus reducing <BER. Under such a probability feature, there are some constellation points that would be missing and

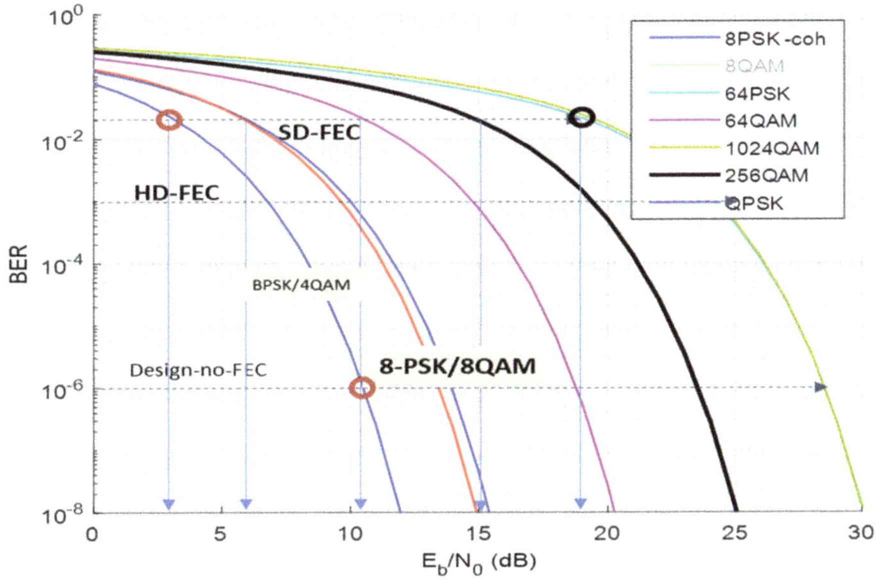

Figure 7.8. BER versus SNR (E_b/N_0) for different modulation formats. Labels: HD-FEC = hard decision-forward error coding; SD-FEC = soft decision FEC.

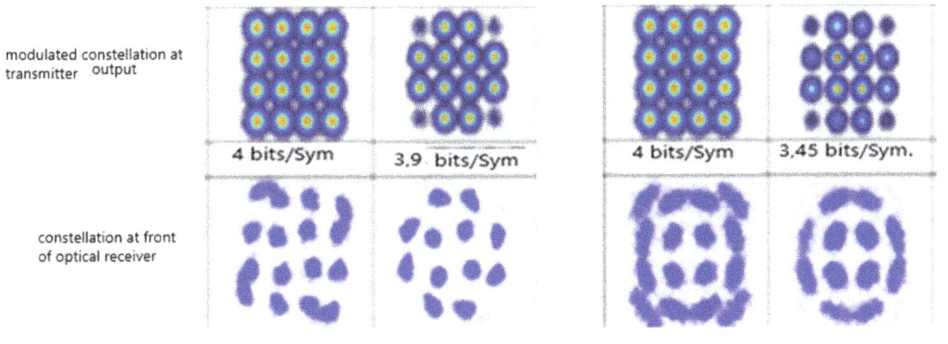

Figure 7.9. Constellations and effective bits/symbol for 16-QAM under statistical modulation.

thus the effective number of bits per symbol can be reduced, thus a minor reduction of the effective number of bits per symbol, thence the aggregate bit rate.

Figure 7.10 shows the schematic structure of a high-speed transceiver which consists of two sections, the optical transmitter, the coherent reception and electronic DSP. DSP serves both the transmitter and receiver sections, to generate pre-distortion, hence to compensate for any distortion of the transmission media or post-compensation for and distortion of the transmission. Besides these sections, there is an optical hybrid sub-section for processing optical signals, polarization demultiplexing, quadrature phase shift, and LO mixing with signals prior to the electronic photodetection. Thence the QAM modulator is for both polarizations using integrated optic Si-photonics.

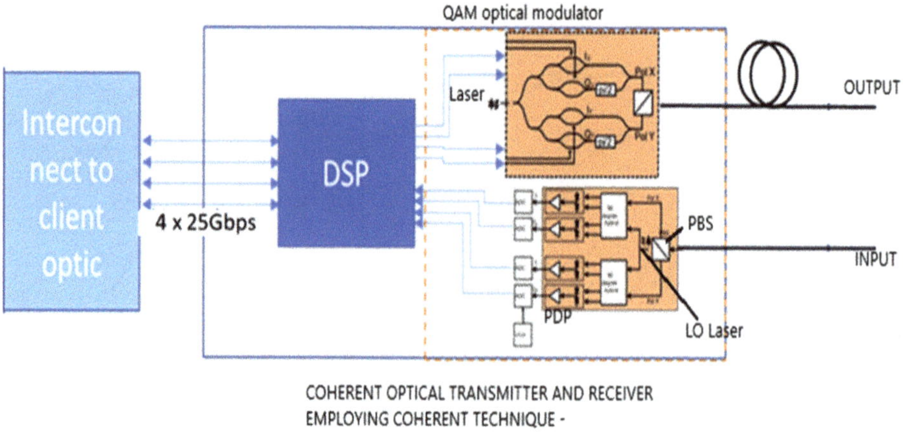

Figure 7.10. Typical structure of an integrated DP-QPSK optical transceiver to obtain 100G, 200G or 400 Gbps. DSP = digital signal processor; ICR = integrated coherent receiver; CFP2-DCO = compact form factor 2 digital coherent optics. QAM = quadrature amplitude modulation; PBS = polarization beam splitter; LO = local oscillator; Pol = polarization; X = horizontal, Y = vertical; PDP = photodetector pair.

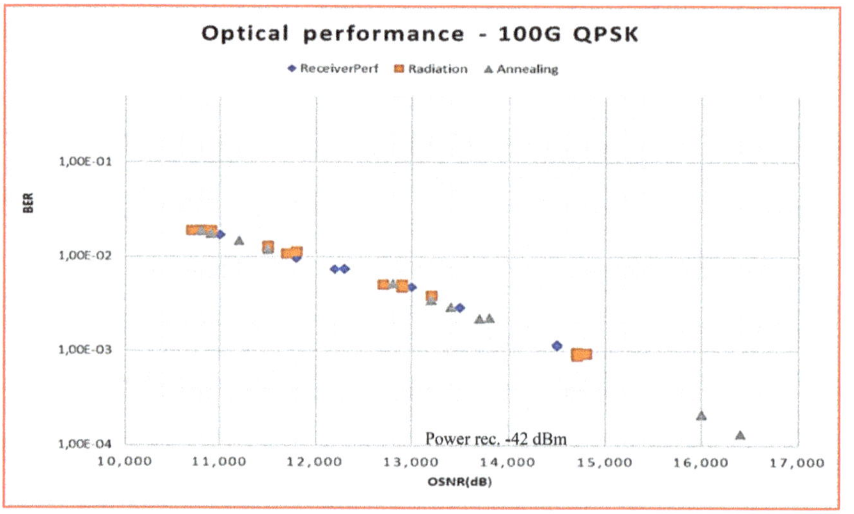

Figure 7.11. Measure receiver sensitivity as a function of optical SNR or DP-QPSK receiver. Quantum level of 5 photons/bit at −42 dBm for 100 Gbits s^{-1}.

The measured performance of a 100G DP-QPSK transceiver is shown in figure 7.11 in which the BER is plotted against the OSNR with specific average power at the receiver. The integrated optical and electronic module is evaluated for space transmission, which employs DP-QPSK modulation format with integrated DSPs and lasers for both optical carrier and local oscillator. 5 photons/bit can be achieved at −42 dBm for 100 Gbits s^{-1} with the symbol rate of 25 GSy s^{-1}, that is reaching the quantum limit of the coherent receiver. The local oscillator power is injected at 5 dBm per polarization so as to boost the sensitivity. The integrated optical transceiver module is shown in figure 7.10 in which the upper section shows the IQ optical modulator, and the lower section shows the optical hybrid to split the polarized beams and orthogonal phase shift of the in-phase and quadrature channels, thence photodetection, electronic amplification and digital processing.

7.3.2 Optical amplification

In free-space long-haul laser communications, due to the long transmission distance, the total loss of the signals can reach 80–90 dB after the gains of the optical systems due to telescope optics at the transmitter and reception systems. Such high attenuation factor must be compensated by using a booster optical amplifier at the transmitter to launch the laser beam into space and an optical preamplifier in front of the coherent receiver. The optical power at the booster amplifier can reach 10–20 W (40–43 dBm) and the gain of the optical preamplifier can reach 50 dB with low sensitive input level down to −5 dBm. In addition, the coherent reception can boost the signal gain to 20–30 dB with a moderate average power of the LO laser.

In long-haul fiber transmission, the loss is compensated by a mid-span optical amplifier whose span distance is typically varied from 80 km to 120 km. The dispersion is compensated by a dispersion compensating fiber length, thus higher loss and thence compensated by a mid-band amplification device. The number of spans can reach more than 200 provided that the OSNR can be as high as 30–35 dB.

7.3.3 Impairments and mitigation over space and long-haul transmission channels

Space optical communication systems and long-haul installed fiber systems can face a few impairments as follows:

- Distortion of the optical wave-fronts due to optics systems which are required to launch optical beam through the free-space channel from the transmitter and reception at the receiver. These wave front errors (WFEs) can be directly related to the error vector magnitude in the digital processed data at the receiver. A digital pre-distortion at the transmitter can be used to mitigate this distortion effect.
- Frequency shift of the optical carriers due to the Doppler shift (DS) effects under different rotation speeds of the satellite constellation of different orbits. For example, the DS can be −7.5 to +7.5 GHz twice a day between the satellites in GEO and LEO rotating at speed 7 km s^{-1} and 3 km s^{-1}, respectively. These DS can be mitigated by tuning the frequency of the LO laser so that homodyne detection can be achieved without the DS effect.
- Crosstalk and in-band crosstalk due to nonlinear effects in fibers which follow the optical booster amplifiers, even with a short length of a few meters. The self-phase modulation or cross-phase modulation effects play a critical role in this crosstalk and intermodulation effects. Such crosstalk should be limited to less than 1 dB by using large effective area (spot size) fiber.

- Polarization rotation via propagation of free-space beam. This can be mitigated via an adaptive optics system and control signal from the DSP system.
- Dispersion and random polarization dispersion and compensation techniques under DSP processing power.

These impairments can be mitigated by a number of advance techniques using DSP algorithms. These are referred to in [4].

7.4 Quantum key distribution

7.4.1 QKD system over space channels and quantum signals

Consider the following parameters of a QKD transmission system over space channel: quantum key rate = 1–100 Mbits s^{-1}, hence using a random key generator of rate less than 10 Mbits s^{-1}. Note: at 1550 nm, over 100 Mbps rate one-bit period contains several thousands of the light-wave periods, thence the capacity C = max key rate = (Bw/2).1/N_s; where N_s = sampling ratio = sampling rate or the key rate; Bw = sampler bandwidth (e.g., 100 GSa s^{-1} → B_w = 50 GHz); so, if the key rate = 1 Mbps, then N_s = 100e9/1e6 = 100 000. So, if C = 100 Mbps then N = 1000. At this rate it is still well below the Shannon limit.

The sampling window Δt must satisfy Heisenberg uncertainty principle and the quantum state for secure transmission with minimum eavesdropping—or the BER would be increasing significantly when the photonic signal energy is reduced when the channel is under an eavesdropping. An illustration of the time window and energy per temporal window is shown in figure 7.12.

Under this narrow temporal slit, we must then apply the Heisenberg uncertainty principle. Thence, the energy contained in one sampled window, can be estimated as: $\Delta E = E_b/N_s \geqslant (h/2\pi)(T_0/N_S) \rightarrow E_b/(N_S^2.\ R_b) \rightarrow E_b \geqslant (h/2\pi)(R_{b.}N_s^2)$. The averaged values are assumed.

However, under noise fluctuation, expected value of finding a quantum state → the expected value $\langle E \rangle$ → fluctuation |.> = Coherent states: $|\,\alpha\rangle = e^{-\frac{|\alpha|^2}{2}}\sum_{n=0}^{\infty}\frac{\alpha^n}{\sqrt{n!}}\,|\,\alpha\rangle$; α = fluctuation; n = photon number; $\bar{n} = |\,\alpha\,|^2$ = average photon number.

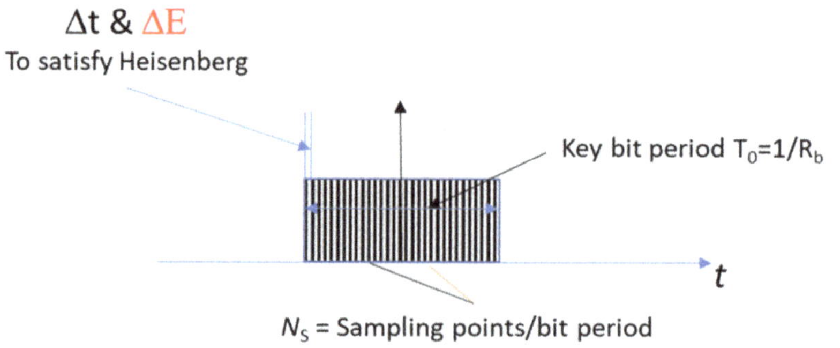

Figure 7.12. Illustration of narrow time window and energy per sampled window.

Hence, using Heisenberg uncertainty → energy fluctuation in one sample window: $\sigma_E = \mathrm{var}(\Delta E) = h/2\pi\ \omega\ \mathrm{sqrt}\,(\langle n\rangle) = h/2\pi\ \omega\ \mathrm{sqrt}\,(E_b\,/(N_S\ h\omega/2\pi))) \geqslant (h/2\pi)/(T_0/N_S) \rightarrow \omega/R_b\ \mathrm{sqrt}(E_b\,/(h/2\pi\ \omega)))/\ N_S^{(3/2)} \geqslant 1$; with n = number of photons in one sampled window. Thence we can obtain $E_b \geqslant \frac{N_S^3 h}{4\nu\pi^2} R_b^2$; where N_s = sampling number in a bit period; R_b = information/key rate; ν = optical frequency; h = Planck's constant.

We can illustrate this with two numerical examples, as follows:

Example 1: R_b = 10 Mbps and taking 1 photon per symbol → N_S = 250 000 (or 250K × R_b) → fulfilling the uncertainty relation → Heisenberg limit. $E_b = 1.2 \times 10^{-19}$ Jor one photon = $1.27\,766 \times 10^{-19}$ J or about one photon in a bit @250 GSa s^{-1}.

Example 2: R_b = 10 Mbps then energy per bit E_b over 1 s of 10 Mbps key rate and sampling factor N_s = 250 or sampling rate = 2.5 GSa s^{-1}, of 1550 nm → E_b = 1.25e −28 J or about 1e−9 of one photon. Thence, if N_s = 2500 or 25 GSa s^{-1} then energy per bit = 1e−6 of one photon at R_b = 10 Mbps.

Given that one photon energy = $h\nu = 1.053e^{-34} \times 3e8/1.55e^{-6} = 2.03e^{-19}$ J, thence the number of photons per bit must be ⩾ 0.94 photons to satisfy Heisenberg's 10M key rate 1550 nm.

For coherent reception and QPSK modulation format with the transceiver structure given in figure 7.10, typical constellation for QPSK at FEC BER is at a receiver sensitivity of −42 dBm. The receiver noises and all noise sources at the transmitter as well as nonlinear noises crosstalk in the quantum receivers must be minimized so that they are at the quantum level so that the quantum signal can breach minimum level or embedded at or below quantum level. Figures 7.13(a) and (b) show the constellations received under these two transmission conditions. The constellation indicates that the overlapped sampled points can fall into either 'one' or 'zero' and the determination processing algorithm should be appropriately developed. As we have analyzed in section 7.2, the sampling rate of the received analog quantum level should be high enough so that the distinction of the level can be distinguished, thus satisfying the Heisenberg uncertainty principle and the Shannon capacity for the OSNR of the quantum signals.

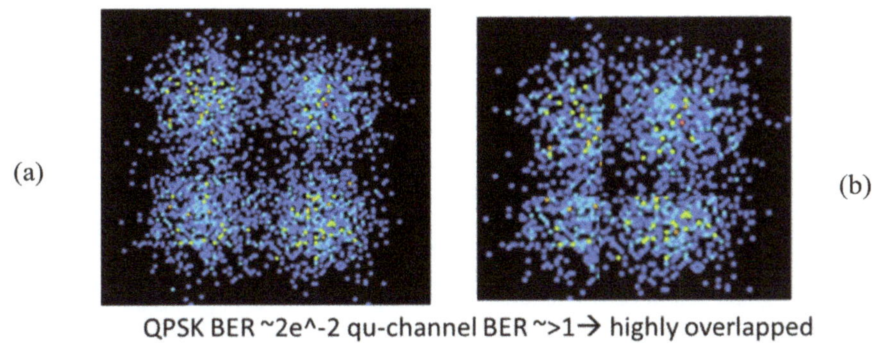

Figure 7.13. Constellation of (a) QPSK under 5 photons/bit received average power; (b) quantum receiver at signal power much lower than noise average power.

7.4.2 QKD transmission system

7.4.2.1 Transmitter, receiver sensitivity, sampling, and protocol

A QKD transmission using coherent reception techniques which are similar to the Tbps transmission system given in figure 7.14. The intersatellite distance can reach 40 000 to 50 000 km over which a total attenuation factor can be 80–90 dB. If the QKD signal is at quantum level or around −80 dBm at 10 Mbits s^{-1} then no receiver can detect such quantum signals at the receiving end of the space channel. This is the major challenge for QKD over space channels, especially when there are multi-hop satellite links, as shown in figure 7.2. The main sub-section of the QKD systems given in figure 7.14. Figure 7.15 shows a typical optical implementation of the QKD transmission in which a narrow linewidth laser is used for transporting the quantum bits and modulated by 'Alice' with the quantum bits which are not known by 'Bob' at the receiving end. Bob1 and Bob2 are used to generate signals which are modulating two OPMs with a phase shift so as to detect the phase modulated signals transmitted by 'Alice', using either constructive or destructive interference with the coherent reception with a LO laser.

The optical demodulated optical signals are mixed with the LO laser in the photodetectors and electronically amplified, then sampled and digitally processed in the DSP. The transmission and reception signals by 'Alice' and 'Bob' are then decoded via the protocol given in the next section when either polarization or phase are used. Thence a sequence of secure bits is decided for further coding of the tera-bits channels.

7.4.2.2 Protocols for the case of polarization coding/modulation and polarization decoding/demod

Alice generates a random bit string and a random binary sequence of coding bases. Assuming now that the quantum information be transmitted by either vertical (|)–horizontal (rectilinear —) or ±45° diagonal (/) polarization of single photons.

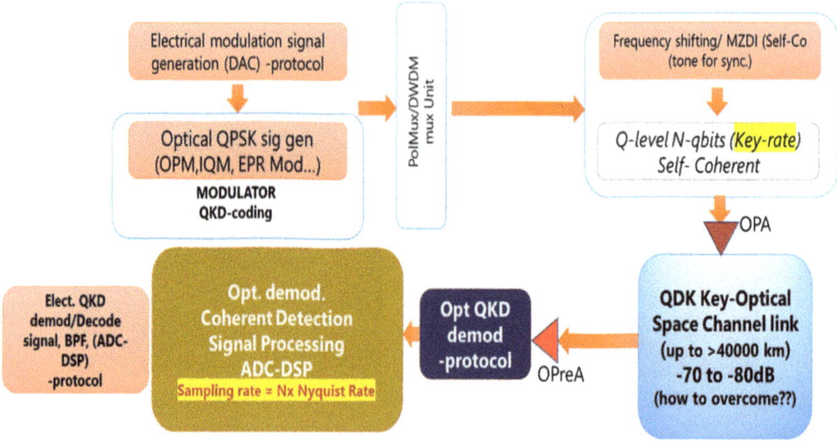

Figure 7.14. Generic optical transceiver for complex quantum signaling for QKD.

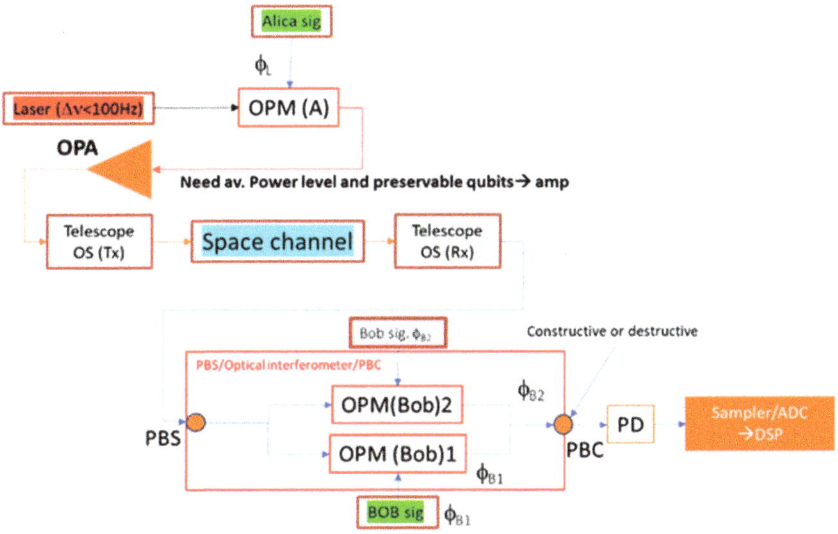

Figure 7.15. Schematic of optical transmission over space for QKD. Labels: OPM = optical phase modulator; PBS = polarization beam splitter; PD = photodetector; OS = optical system; Bob and Alice positions can be interchanged; OS = optical system—coupling optics, ϕ = phase; sig = signal; A = Alice; PBS/C = polarization beam splitter/combiner.

A '0' = '|/—' rectilinear basis and a '1' = '/' = diagonal basis. Further, a '0' in the bit string would be represented by horizontal or—45° polarization, depending on the basis, and a '1' would be represented by vertical | or +45° polarization.

Alice codes a stream of photons randomly as dictated by the sequences she generates and the previous definitions. Alica then transmits the stream of photons to Bob. Alice retains both of her random sequences.

As Bob receives the photons, he randomly decides the basis in which he is going to measure their polarization. Essentially, he also generates a random binary sequence of coding bases. It is crucial that Bob's sequence is entirely independent of Alice's. Bob then measures the polarization of the photons he receives according to his basis.

In the case when Bob has selected the same basis as Alice he will correctly measure the bit that Alice (transmitting party) has generated. When Bob selects a different basis to Alice he may or may not obtain the same bit, as the polarization he measures will be purely random.

A ±45° diagonal polarization has equal components of horizontal and vertical polarization and hence has equal probability of being detected as a horizontally or vertically polarized photon in a rectilinear basis.

It is similar for a rectilinearly polarized photon measured in a diagonal basis. So, when Bob (as receiving party) has selected the opposite basis to Alice he obtains no meaningful data. The maximum meaningful data Bob can obtain is half the random sequence that Alice transmits—those photons for which he and Alice (transmitting party) used the same basis—although this will be somewhat lower due to losses incurred during transmission and detector inefficiencies. The realization of such a quantum protocol system can be shown in figure 7.16 or figures 7.17 and 7.18, respectively.

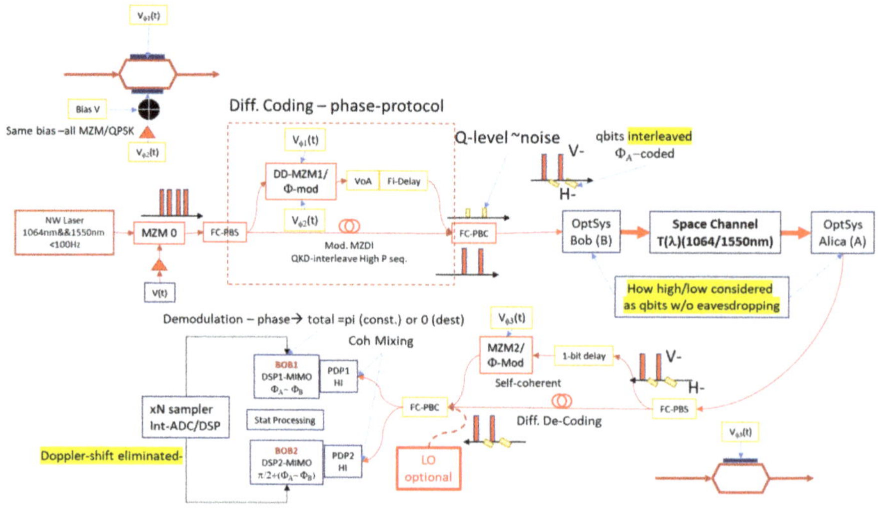

Figure 7.16. Realization of a CV-QKD transmission employing self-coherent reception techniques with option of coherent reception using an external LO laser. Modulation technique = polarization switching and self-coherent or coherent reception technique.

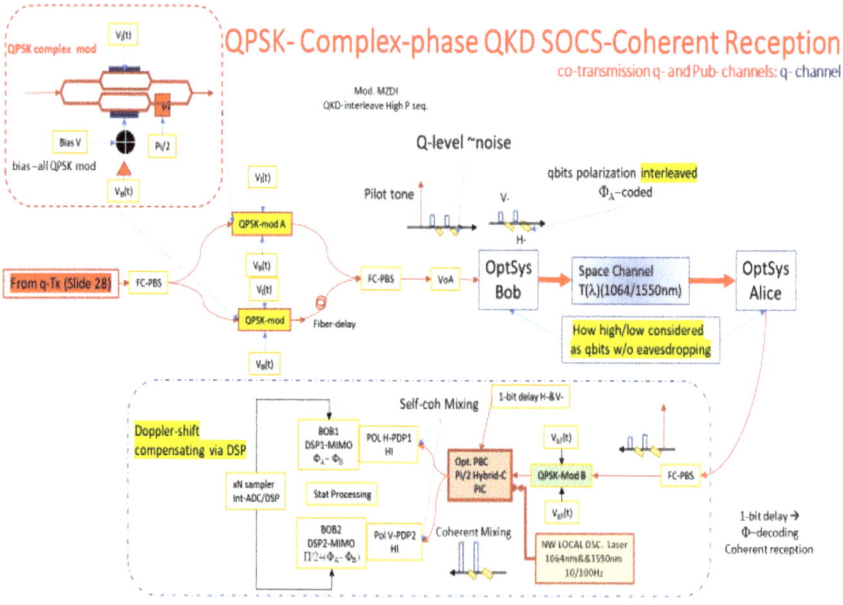

Figure 7.17. QKD transmission system employing complex multi-level modulation QAM of QPSK or m-QAM (e.g. 16-QAM). Reception techniques: coherent. Operating wavelength = C-band 1550 nm spectral window or 1060 nm spectral window. SOCS = space optical communication system.

The self-coherent reception requires two consecutive bits to represent one phase state. So, the differential mode is required so that the phase of the two consecutive bits can be beating with each other so that one can act as the 'LO' and the other as

QKD transmission													
Alica random bits	1	0	0	1	1	1	0	0	1	1	0	0	0
Bob random bits	1	0	1	1	0	1	1	0	1	0	1	1	0
		0		1		1		0	1				0
Bob informs Alica		Yes	no	Y	No	Y	No	Y	Yes				Yes
Raw Key (if no Evesdrop)		0		1		1		0	1				0

Figure 7.18. BB92 protocol ($LiNBO_3$ OPM/IQM –ER → 30 dB), OPM = optical phase modulation; ER = extinction ratio; IQM = in-phase quadrature modulator.

signals to be detected. In this case the high-power pulse can act as the reference and synchronization pulse that can mark the location of the weak quantum state pulse. The polarization states can be added to the modulation of the phases of the sequence so that the capacity increases, hence half the sampling rate, as shown in figure 7.16.

For coherent reception, the LO is injected to mix with the received signals in a balanced photodetection pair (PDP). Then comes the linear amplification stage and then the sampling and processing by DSP in which a number of processing algorithms are applied to recover the phase and matching of the LO and the carrier, plus compensation of a number of impairments of the transmission media.

Figure 7.15 gives a typical block schematic of optical transmission over space using continuous variable QKD (CV-QKD) using optical phase modulation and detection with two parallel receivers which can distinguish both the phase states. This increases the detection of the quantum states and allows a higher degree of determination of the 'grey' states or the region of the overlapped constellations, as given in figure 7.13. The notation of the components of the system are: OPM is the optical phase modulator, PBS is the polarization beam splitter, PD is the photodetector, OS is the optical system including the optics to launch an optical beam into free space, Bob and Alice positions can be interchanged; ϕ = phase, sig is signal. A = Alica, PBS/C demotes the polarization beam splitter/combiner.

Figure 7.16 shows a realizable CV-QKD transmission system employing self-coherent reception techniques with option of coherent reception using an external LO laser. Self-coherent indicates that the original signal can be delayed by one bit and then detected by a reverse delay of the bit so that the beam itself can be combined and detected with the maximum interference or destruction interference. Differential coding must be implemented so that maximum chance of detection of the quantum bit can be achieved. The modulation technique is by polarization switching and self-coherent or coherent reception technique. Common homodyne or intra-dyne coherent reception can be implemented by using an LO laser whose frequency equals that of the signal carrier. Otherwise, within a certain limit, a phase recovery technique can be used to get a homodyne reception.

On the other hand, the complex modulation QPSK modulation can be used with coherent detection, as shown in figure 7.17. Multi-level QAM can also be used. The phase detection can happen throughout the duration of the symbol period. This is true due to the in-phase and quadrature components forming the phase angled vector by their amplitude and a $\pi/2$ phase shift between them. The QAM modulators and receiver can be referenced to those given in chapters 1 and 2. The sampling and detection can be referenced in section 7.2.

The rest of the BB84 protocol is implemented over a public communication channel, with any eavesdropper, Eve, free to listen to, but not alter, transmissions between Alice and Bob. Therefore, the classical communication channel used to determine which photons are received and which are detected with the correct basis, is that Bob tells Alice which bits he detected and the basis he used to measure them, but not the result he obtained from the measurement, and Alice responds by telling Bob which bits he used the correct basis for. Both Alice and Bob discard all other bits, and assuming the transmissions are not disturbed, Alice and Bob should now share a random sequence of bits that are identical. To test for Eve's presence on the quantum transmission link, Alice and Bob now share a random subset of this sequence publicly to determine any anomalies. If excessive anomalies exist, the sequence is discarded, otherwise the sequence is retained and used as the key for the one-time pad encryption technique. Figure 7.18 tabulates the above protocol steps.

7.4.2.3 Receiver noises and sensitivity

In order for the transmitting bits to reach the absolute level, the receiver noises must be at the quantum level so that any eavesdropping can be detected by Bob who can then discarding the bits. Thus the 'Bob' receiver must reach the absolute minimum level so that the QKD system can be assured of the absolute secure level. In addition, the sampling system must be at the maximum rate so that the DSP can recover the QKD key bits whose levels can be several times below the noise level, as described in the sections on power-limited system and QKD system over space channels and quantum signals. There is an equivalent small-signal model of the noise sources as referred to the input port of optical receivers including the total equivalent spectral density of the electronic amplifiers that follows the photodetection as referred to the input, the dark current noise, the laser relative intensity noise, and finally, the LO oscillator shot noise if coherent reception is used or due to the high-power pulse in the self-coherent case.

The reception sub-system for CV-QKD must provide the lowest noise level by the electronic amplification, the optical noises in transmission system as presented in figure 7.19. Coherent receivers are incorporating digital signal processors with the samplers and the ADC so that the sampling rate can be multiple times the key rate. This makes it possible to operate the system in the power-limited region of figure 7.7.

Photonic transceivers utilizing photonic integrated circuits in silicon photonics and indium phosphide technology of electro-absorption modulation lasers (EMLs), are presented as well as all-optical routing over space networking sections. Both technologies are applied for CV-QKD and tera-bps transmission systems, respectively.

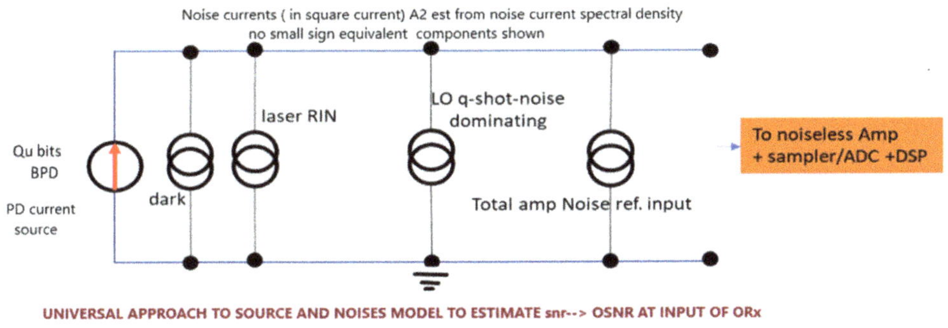

Figure 7.19. Equivalent noise model of an optical receiver for evaluation of noises and quantum level to be determined.

7.5 Remarks

This chapter has addressed the transmission technology for ultra-high-capacity channels and quantum bits over long-haul terrestrial and submarine transmission, and/or the optical transmission over ultra-long-distance inter-orbital constellations of satellites or the web of optical networks, the networking of networks.

The channel capacity and the OSNR are given for the cases of bandwidth-limited and power-limited transmission systems with the principal role of the sampling rate so that under the quantum signal reception, the signal average power that can be much lower than the noise level can be recovered with minimum error. Thence a sequence of secure keys for coding. ON the other hand, the bandwidth-limited case, the sampling must be sufficient high so that the recovery can be commonly recovered with a minimum of doubling factor of the signal bandwidth. Capacity can be increased by multi-level modulation technique and frequency multiplexing. The two cases are in contrast with each other showing the limits of the extreme cases. In DCI and DCN, both cases are important in terms of increasing capacity of information and security of the transmitting and receiving as well as routing of information.

A simple summary by an illustration of multi-Tbps and QKD transmission optical systems, is given for systems that can be valid to both the DCN and DCI and for free-space optical communications of a satellites web or space high performance computing networks.

References

[1] HydRON optical communication for broadband in space https://esa.int/ESA_Multimedia/Images/2021/07/HydRON_optical_communication_for_broadband_in_space

[2] Binh L N, Fechtmann C, Koegler M, Hof A and Horwath J 2022 Transmission of multi-Tera-bits/sec to quantum bits over Space intra-and inter-orbital links *Proc. ICSO Croatia* September 2022

[3] Shannon C E 1948 A mathematical theory of communication *Bell Syst. Techn. J.* **27** 379–423 623–56

[4] Binh L N 2014 *Digital Processing: Optical Transmission and Coherent Receiving Techniques* (Boca Raton, FL: CRC Press)

Le Nguyen Binh

Appendix A

Technical data of single-mode optical fibers

A.1 Remarks

Optical fibers are the main transport medium of ultra-high speed and ultra-high capacity in data centers (DCs) for massive data interchanges and serving demand. Initially multimode fibers (MMFs) step index were employed but now there is graded index type, OM4, to reduce intermodal delay broadening, widening the bandwidth over a short distance. However, most transport is over single-mode fibers (SMF). Over the years there have been quite a number of types of SMF and eventually a standard SMF (SSMF) was employed, the G.652 fiber. Coring Inc. has produced this SSMF as SMF-28 and SMF-28e as well as large effective area fiber (LEAF). The difference between SMF-28e and SMF-28 is that the OH-absorption peak at 1400 nm is depressed, as shown in figure A.1, allowing channel carriers at the 1300–1500 nm spectral window possibly used. At 1300 nm the chromatic dispersion factor is almost zero and thus one can a transmit high-speed data sequence with many carriers in this spectral window, the O-band. The fiber attenuation in this O-band is higher than those in the C- (1525–1565 nm) and L-band (1565–1620 nm).

Emerging service networks, such as 5G, cloud computing, video retrieving, and ultra-HD video, continuously drive increasing service traffic, imposing higher requirements on the transmission rate and performance of transmission networks. The single-wavelength rate and performance improvement demand more fiber spectrum resources. Due to slow fiber deployment, the most effective way to cope with the ever-increasing network traffic is to improve the single-fiber capacity by expanding the fiber spectrum.

Widening the spectrum over which the attenuation is lowest is necessary for longer transmission distance. Presently the extended C+-band or super-C band are spectrally widened from the traditional C-band of 80 channels of 50 GHz spacing to 120 channels of 75 GHz spacing, as shown in figure A.2. With this extension of C-band and with appropriate gain of optical amplifiers of EYDFA (erbium

doi:10.1088/978-0-7503-5849-1ch8

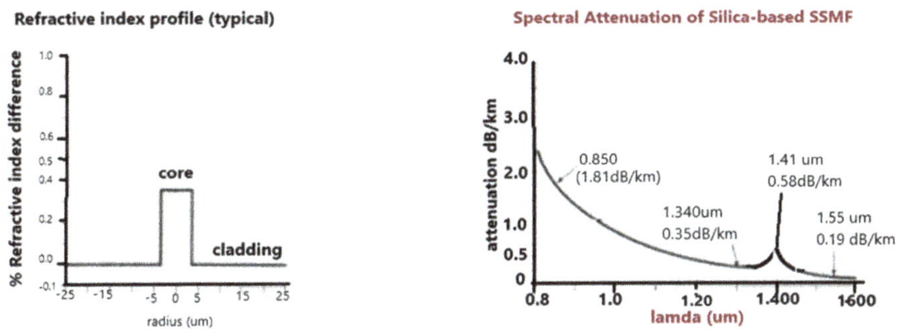

Figure A.1. SMF-28 Standard SMF, Fiber refractive index profile and spectral attenuation distribution. Note the absorption peaks at 1410 nm.

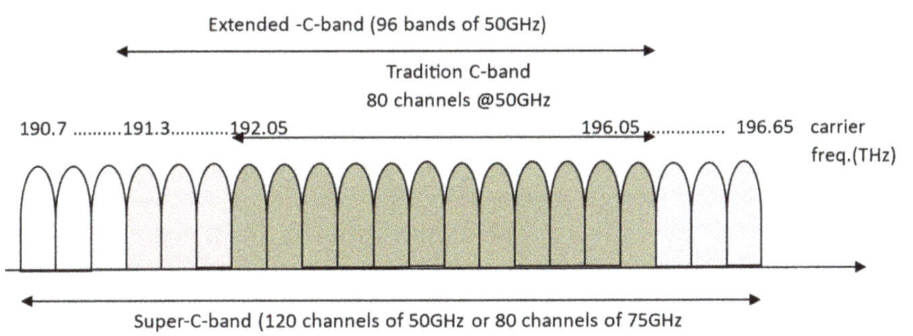

Figure A.2. Traditional C-band, extended C-band and super-C band over the 1550 nm spectral window.

ytterbium doped fiber amplifiers in combination with Raman amplifier types), the full super-C band can be fully employed and offers unique media for transporting of ultra-high capacity information systems, especially in DCN (DC networking) and DCI (inter DC networking).

A.2 Single-mode fibres

This appendix gives the main features of the fibers, the SSMF, G.652 standard type. The parameters on geometrical dimensions, the dispersion factors, attenuation as a function of wavelength, and the polarization mode dispersion. These data sets give users some picture on how much the signals transmitting though these fibers can suffer from distortion and degrading of their total average power. Figure A.2 shows the main technical parameters of SSMF with the refractive index profile, the geometrical parameters of the core, the cladding and outside coating for fiber strengthening. The attenuation as a function of wavelength is also shown to indicate the absorption peak at 1400 nm due to water OH contamination during the collapsing of the preform. Figure A.3 shows the technical parameters of the

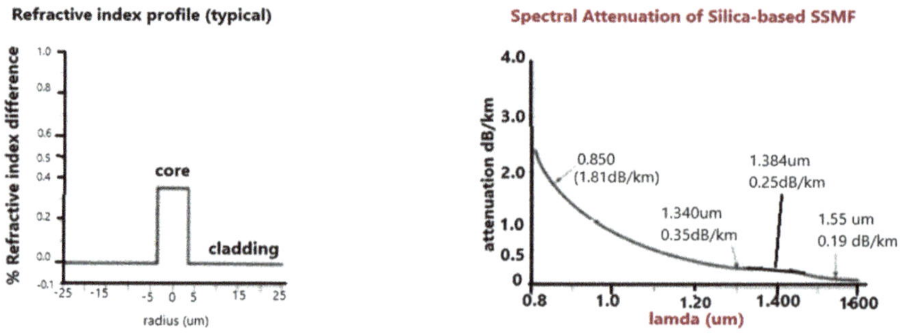

Figure A.3. SMF-28e Fiber refractive index profile and spectral attenuation distribution. Note: no absorption peak at 1410 nm—suppression of water absorption peaks. Refractive index profile and attenuation spectral property given in the inserted figures at the end.

enhanced SSMF where the water absorption peak is suppressed to allow low loss transmission over the O-band and C-band. The dispersion properties of the two SSMF types are quite similar. That is, the dispersion is near zero in the 1310 nm and the dispersion factor is about 17 ps nm^{-2} km^{-1}, that is, a pulse broadening of 17 ps for a 1.0 nm bandwidth wide (or 125 GHz) signals propagating over one kilometer of fiber. The dispersion due to group velocity different at the wavelength carriers over the signal bandwidth can be scaled appropriately for a signal of different bit rate or symbol rate, hence bandwidth, and the propagation distance.

Furthermore, the LEAF fiber or large effective area fibers are also given with their technical data. This large effective fiber allows higher power of all multiplexed channels to be supported with minimum distortion due to nonlinear effects, especially the self-phase modulation.

All parameters here are extracted from a number of fiber manufacturers, especially Corning Inc.

A.3 Standard single-mode optical fibers

Main parameters for standard SSMF are as follows.

A.3.1 mode field diameter (MFD) and mode spot size (r_o)

We refer to figure A.2 for geometrical profile and the attenuation spectra of SSMF. A summary of SSMF parameters is tabulated in table A.1.

MFD is defined as the diameter of the guided optical field distribution over the fiber cross-section, that is normally the Gaussian distribution, at 1/e intensity with respect to the maximum intensity of the guided mode at the center.

Similarly, the mode spot size is defined based on the radius of the guided mode intensity distribution. It depends on the first or second order method of moment to determine the 1/e value with respect to that at the center of the fiber (table A.2).

Table A.1. Mode field and spot size (micro-meter).

MFD (@1310 nm)	9.2 ± 0.5 um
MFD (@1550 nm)	10.5 ± 0.5 um
Dispersion (Units ps nm^{-2} km^{-1})	
D (1550 nm)	$\leqslant$ 18.0
D (1310 nm)	$\leqslant$ 22.0
Zero dispersion wavelength < λ_o	1304 < λ_o < 1324 nm
Zero dispersion slope (S_o)	$\leqslant$0.092 ps nm^{-2} km^{-1}

Table A.2. Polarization mode dispersion.

Polarization	
Mode Dispersion (PMD) unit—ps/(km)$^{1/2}$	
Optical link design	<0.04
Maximum PMD of fiber unit of link	<0.1

Table A.3. Attenuation.

Wavelength (nm)	Attenuation (dB km^{-1})	Notes
1310	$\leqslant$0.35	Bare fiber
1550	$\leqslant$0.25	Bare fiber—higher when cabled
Water peak (OH peak)	<2.1 absorption	At wavelength around 1383 nm
Under bending (diameter—mm)—attenuation (dB)		
32 @1550 nm	<0.5	
50 @1310 nm and @1550 nm	<0.05 and 0.1	

A.3.2 Attenuation

See table A.3 for the attenuation. Note the water absorption of water contamination in the 1340–1348 nm spectral window (see figure A.3). This leads to development of enhanced SMF type by suppression of this peak. The suppression of this absorption peak (as shown in figure A.2), allows the uses of more multiplexed channels, or coarse wavelength division multiplexing technique, to increase the total communication capacity for intra-data-center (DCI) transmission as the attenuation can be at an acceptable loss in the power budget.

A.3.3 cut-off

Cut-off wavelength: $\lambda_c \leqslant 1260$ nm

Cable cut-off wavelength: $\lambda_c < 1260$ nm. Note: cut-off wavelength is defined as the wavelength at which the second order mode starts appearing, that is, when the fiber can support more than one mode guiding in its core. A fiber can be single-mode at one wavelength but be no longer 'single' when the wavelength of the exciting laser into the fiber, is lower than its cut-off wavelength.

A.3.3.1 Dispersion and cut-off wavelength

The relationship between the fiber dispersion with the dispersion slope and cut-off wavelength can be given as

$$D(\lambda) = \frac{S_0}{4}\left[\lambda - \frac{\lambda_0^4}{\lambda^3}\right] \text{....In units ps nm}^{-2}\text{ km}^{-1}$$

Enhanced standard single-mode optical fibers

A.3.3.2 Large effective area fiber

LEAF fibers have a large effective area or the area over which the fundamental mode field diameter[1] (or twice the mode spot size) of near Gaussian distribution is covering. This large effective mode area allows the reduction of nonlinear effects, hence phase distortion, when many information channels of different carriers are propagating in parallel, the dense wavelength division multiplexing (DWDM) technique. The large effective area fibre (LEAF) allows the higher level of optical power of the guided mode of the optical channel without suffering any nonlinear effects. Further information can be found in LEAF fiber by Corning Inc. Furthermore, in order to avoid the nonlinear parametric conversion, hence the cross-talk noises in DWDM channels, NZ-DSF with large area fibers is commonly used. NZ-DSF is non-zero dispersion shifted fiber, which is a single mode fiber whose zero-dispersion wavelength is shifted away from the common C-band located at 1550 nm at which if zero dispersion value, would result in the nonlinear four-wave mixing.

[1] MFD (mode field diameter) or twice mode spot size is the 3 dB intensity points of the near Gaussian intensity distribution of the fundamental guided mode.

IOP Publishing

Le Nguyen Binh

Appendix B

A coherent balanced receiver and method for noise suppression

It can be shown that a balanced optical receiver can suppress the excess noise intensity generated from the local oscillator (LO) [1–3]. In a balanced receiver an optical coupler is employed to mix a weak optical signal with an LO field of average powers of E_S^2 and E_L^2, respectively. Figure B.1 shows the generic diagram of a balanced optical receiver under a coherent detection scheme in which two photo-detectors (PDs) and amplifiers are operating in a push–pull mode.

Usually the magnitude of the optical field of the LO, E_L, is much greater than that of E_S, the signal field. The fields at the output of a 2 ×2 coupler can thus be written as:

$$\begin{bmatrix} \tilde{E}_1 \\ \tilde{E}_2 \end{bmatrix} = \begin{bmatrix} \sqrt{1-k} & \sqrt{k}\, e^{j\pi/2} \\ \sqrt{k}\, e^{j\pi/2} & \sqrt{1-k} \end{bmatrix} \begin{bmatrix} \tilde{E}_S \\ \tilde{E}_L \end{bmatrix} \tag{B.1}$$

where E_S is the field of the received optical signal, E_L is the field of the LO (possibly a distributed feedback laser); k is the intensity coupling coefficient of the coupler. For a 3 dB coupler we have $k = 0.5$. E_S and E_L fields can then be written as:

$$\begin{aligned} E_S &= \sqrt{2}\,\tilde{E}_S\, e^{j(\omega_1 t+\phi_1)} \\ E_L &= \sqrt{2}\,\tilde{E}_L\, e^{j(\omega_2 t+\phi_2)} \end{aligned} \tag{B.2}$$

where $\tilde{E}_S$, $\tilde{E}_L$ are the magnitude of the optical fields of the signal and the LO laser. The electronic current generated from the photo-detectors (PDs) corresponding to the average optical power and the signals are given by:

$$i_1(t) = \frac{\Re}{2}\,|E_1|^2 = \frac{\Re_1}{2}\left\{ \begin{array}{l} \frac{1}{2}(E_S^2 + E_L^2) + \\ E_S E_L \cos\left[(\omega_1 + \omega_2)t + \phi_1 + \phi_2 - \frac{\pi}{2}\right] \end{array} \right\} + N_1(t) \tag{B.3}$$

doi:10.1088/978-0-7503-5849-1ch9

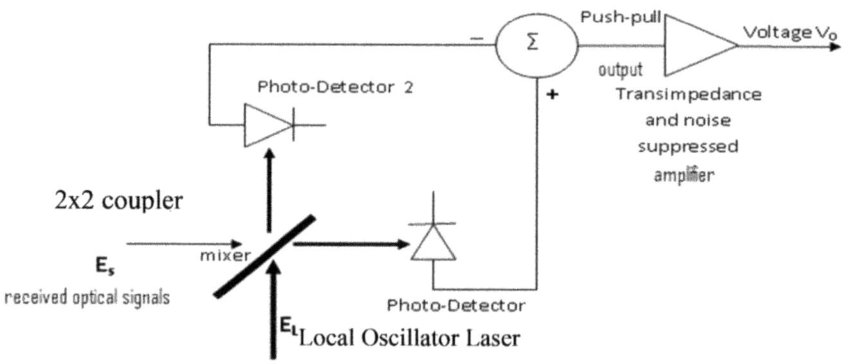

Figure B.1. Generic block diagram of a coherent balanced detector optical receiver. LO = local oscillator.

$$i_2(t) = \frac{\Re}{2} \mid E_2 \mid^2 = \frac{\Re_2}{2} \begin{Bmatrix} \frac{1}{2}(E_S^2 + E_L^2) \\ +E_S E_L \cos\begin{bmatrix} (\omega_1 + \omega_2)t \\ + \phi_1 + \phi_2 - \frac{\pi}{2} \end{bmatrix} \end{Bmatrix} + N_2(t) \tag{B.4}$$

where ϕ_1, ϕ_2 and ω_1, ω_2 are the phase and frequency of the signal and LO, respectively. $N_1(t)$ and $N_2(t)$ are the noises resulting from the photodetection process, $\Re_1$, $\Re_2$ are the responsivity of PDs 1 and 2, respectively. It is assumed that the two photodiodes have the same quantum efficiency, thence their responsivity. Thus we have the total current of a back-to-back PD pair referring to the input of the electronic amplified, as

$$i_{\text{Teq}}(t) = \begin{Bmatrix} I_{\text{dc}1} - I_{\text{dc}2} \\ +(\Re_1 + \Re_2) E_S E_L \cos\begin{bmatrix} (\omega_1 - \omega_2)t \\ +\phi_1 - \phi_2 - \frac{\pi}{2} \end{bmatrix} \end{Bmatrix} + N_1(t) + N_2(t) \tag{B.5}$$

where the first two terms $I_{\text{dc}1}$ and $I_{\text{dc}2}$ are the detector currents generated by the detector pair due to the reception of the power of the LO which is a continuous wave source, thus these terms appear as the DC constant currents that are normally termed as the shot noises due to the optical power of the LO. The third term is the beating between the LO and the signal carrier, the signal envelope. The noise currents $N_{1\text{eq}}$(t) and $N_{2\text{eq}}$(t)are seen as the equivalence at the input of the electronic amplifier from the noise processes which are generated by the PDs and the electronic amplifiers. Noise components are the shot noise due to bias currents, the quantum shot noise that is dependent on the strength of the signal and the LO, the thermal noise due to the input resistance of the amplifier and the equivalent noise current with reference to the input of the electronic amplifier contributed by all the noise sources at the output port of the electronic amplifier. We denote $N_{\text{lep}}(t)$ as the quantum shot noise generated due to the average current produced by the PD pair in reception of the average optical power of the signal sequence.

The difference of the produced electronic currents can be derived using the following techniques: (i) the generated electronic currents of the PDs can be coupled through a 180° microwave coupler and then fed to an electronic amplifier; (ii) the currents are fed to a differential electronic amplifier; or (iii) a balanced PD pair connected back to back and then fed to a small signal electronic amplifier [4]. The first two techniques require stringent components and are normally not preferred as contrasted by the high performance of the balanced receiver structure of (iii).

Regarding the electronic amplifier, the electronic preamplifier, a trans-impedance configuration is selected due to its wideband and high dynamics. However, it suffers high noise due to the equivalent input impedance of the shunt feedback impedance, normally around a few hundred ohms. In the next section, the theoretical analysis of noises of the optical balanced receiver is described.

B.1 Analytical noise expressions

In the electronic preamplifier, the selection of the transistor as the first stage amplifying device is very critical. Either a field effect transistor (FET) or bipolar junction transistor (BJT) could be used. However, the BJT is preferred for wideband application due to its robustness to noise. The disadvantages of using the BJT as compared with the FET are due to its small base resistance that leads to high thermal noise. However, for a shunt feedback amplifier the resistance of the Millers' equivalent resistance with reference to the input of the amplifier is much smaller and thus dominates over that of the base resistance. The advantage of the BJT over the FET is that its small signal gain follows an exponential trend with respect to the small variation of the driving current derived from the photodetection as compared with the parabolic for FET. The FET may also offer high input impedance between the gate to source of the input port, but may not offer much improvement in a feedback amplification configuration in terms of noise. This section focuses on the use and design of BJT multi-stage shunt feedback electronic amplifiers.

Noise of electronic amplifying devices can be represented by superimposing all the noise current generators to the small signal equivalent circuit. These noise generators represent the noise introduced into the circuit by physical sources/processes at different nodes. Each noise generator can be expressed by a noise spectral density current square or power, as shown in figure B.2. This figure gives a general model of a small signal equivalent circuit including noise current sources of any transistor, which can be represented by the transfer [y] matrix parameters. Indeed, the contribution of the noise sources of the first stage of the electronic preamplifier is dominant plus that of the shunt feedback resistance.

The current generator $\frac{di_n'^2}{df}$ is the total equivalent noise generator contributed by all the noise sources to node 1, including the thermal noise of all resistors connected to the node. Similarly, for the noise source $\frac{di_n''^2}{df}$ refers to the output node.

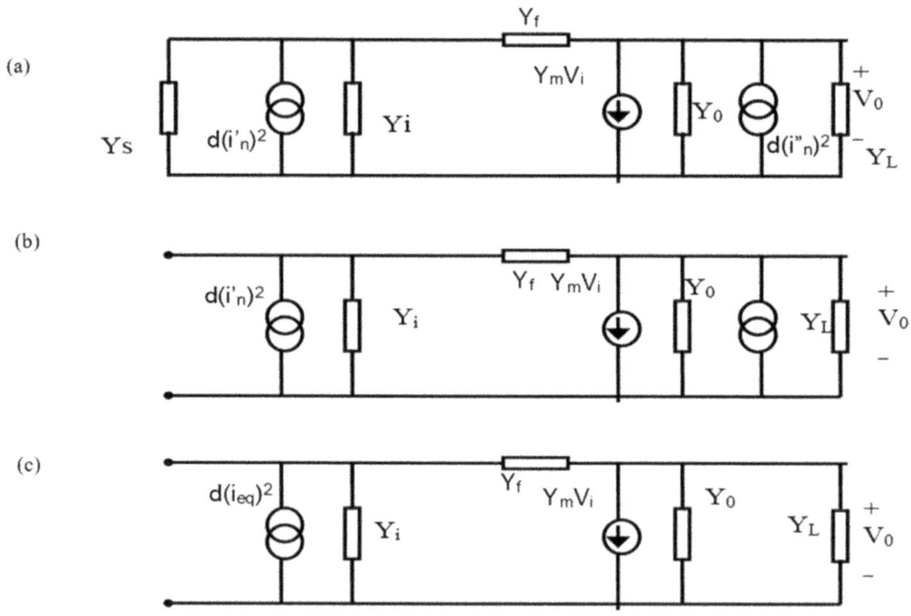

Figure B.2. Small equivalent circuits including noise sources (a) *Y*-parameter model representing the ideal current model and all current noise sources at the input and output ports, (b) with noise sources at input and output ports, and (c) with a total equivalent noise source at the input.

B.2 Noise generators

Electrical shot noise is generated by the random generation of streams of electrons (current). In optical detection, shot noise is generated by: (i) biasing currents in electronic devices, and (ii) photocurrents generated by the photodiode.

A biasing current I generates a shot noise power spectral density S_I given by:

$$S_I = \frac{d(i_I^2)}{df} = 2qI \text{ A}^2\,\text{Hz}^{-1} \tag{B.6}$$

where q is the electronic charge. The quantum shot noise $\langle i_s^2\rangle$ generated by the PD by an optical signal with an average optical power P_{in} is given by

$$S_Q = \frac{d\langle i_s^2\rangle}{df} = 2q\langle i_s^2\rangle\ A^2/Hz \tag{B.7}$$

If the PD is an avalanche photo-detector (APD) type then the noise spectral density is given by

$$S_Q = \frac{d\langle i_s^2\rangle}{df} = 2q\langle i_s^2\rangle\langle G_n^2\rangle\ A^2/Hz \tag{B.8}$$

It is noted here again that the dark currents generated by the PD must be included in the total equivalent noise current at the input after it is evaluated. These currents are generated even in the absence of the optical signal. These dark currents can be eliminated by cooling the PD to at least below the temperature of liquid nitrogen (77 K).

At a certain temperature, the conductivity of a conductor varies randomly. The random movement of electrons generates a fluctuating current even in the absence of an applied voltage. The thermal noise spectral density of a resistor R is given by:

$$S_R = \frac{d(i_R^2)}{df} = \frac{4k_B T}{R} \quad \text{in A}^2/\text{Hz} \tag{B.9}$$

where k_B is the Boltzmann's constant, T is the absolute temperature (in K) and R is the resistance in ohms, and i_R denotes the noise current due to resistor R.

B.3 Equivalent input noise current

Our goal is to obtain an analytical expression of the noise spectral density equivalent to a source looking into the electronic amplifier including the quantum shot noise of the PD. A general method for deriving the equivalent noise current at the input is given by representing the electronic device by a Y-equivalent linear network, as shown in figure B.3. The two current noise sources $di_n^{'2}$ and $di_n^{'2}$ represent the summation of all noise currents at the input and at the output of the Y-network. This can be transformed into a Y-network circuit with the noise current with reference to the input as follows.

The output voltages V_0 of figure B.3(a) can be written as:

$$V_0 = \frac{i_N'(Y_f - Y_m) + i_N'(Y_i + Y_f)}{Y_f(Y_m + Y_i + Y_o + Y_L) + Y_i(Y_o + Y_L)} \tag{B.10}$$

and for figure B.3(b)

$$V_0 = \frac{(i_N')_{eq}(Y_f - Y_m)}{Y_f(Y_m + Y_i + Y_o + Y_L) + Y_i(Y_o + Y_L)} \tag{B.11}$$

Thus, comparing the above two equations, we can deduce the equivalent noise current at the input of the detector as:

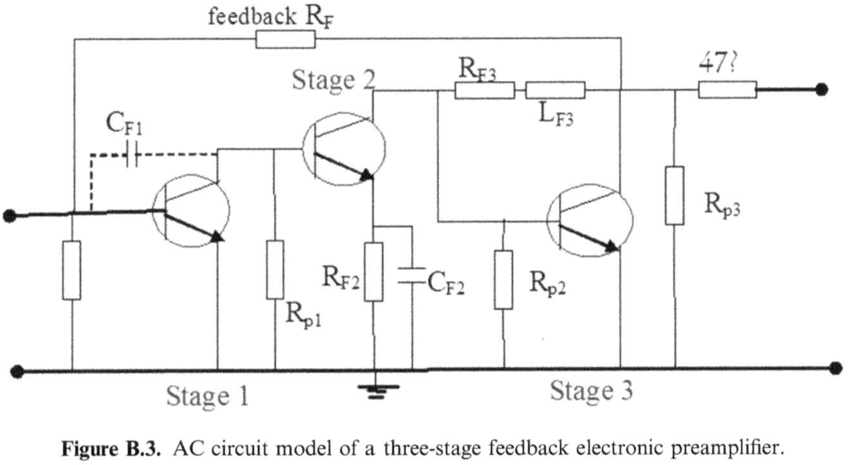

Figure B.3. AC circuit model of a three-stage feedback electronic preamplifier.

$$i_{\text{Neq}} = i_N' + i_N' \frac{Y_i + Y_f}{Y_f - Y_m} \tag{B.12}$$

Then reverting to mean square generators for a noise source we have

$$d(i_{\text{Neq}})^2 = d(i_N')^2 + d(i_N')^2 \left| \frac{Y_i + Y_f}{Y_f - Y_m} \right|^2 \tag{B.13}$$

It is therefore expected that if the Y-matrix of the front end section of the amplifier is known, the equivalent noise at the input of the amplifier can be obtained by using (B.13).

We propose a three-stage electronic preamplifier in AC configuration as shown in figure B.3. The details of the design of this amplifier are given in the next section. Small-signal and associated noise sources of this amplifier are given in section B.7. It can be seen that this general configuration is a forward path of shut, series and shut stages which reduces the interaction between stages due to the impedance levels [8]. Shunt–shunt feedback is placed around the forward path, hence stable transfer function is the transfer impedance, which is important for transferring the generated electronic photodetected current to the voltage output for further amplification and data recovery.

For a given source, the input noise current power of a BJT front end can be found by

$$i_{\text{Neq}}^2 = \int_0^B d(i_{\text{Neq}})^2 = a + \frac{b}{r_{\text{E}}} + c r_{\text{E}} \tag{B.14}$$

where B is the bandwidth of the electronic preamplifier, r_{E} is the emitter resistance of the front end transistor of the preamplifier. The parameters a, b, and c are dependent on the circuit elements and amplifier bandwidth. Hence an optimum value of r_{E} can be found and an optimum biasing current can be set for the collector current of the BJT such that i_{Neq}^2 is at its minimum, as

$$r_{\text{Eopt}} = \sqrt{\frac{b}{c}} \text{ ... hence ... } \rightarrow i_{\text{Neq}}^2 \Big|_{r_{\text{E}} = r_{\text{Eopt}}} = a + 2\sqrt{bc} \tag{B.15}$$

If two types of BJT are considered, Phillips BFR90A and BFT24 then a good approximation of the equivalent noise power can be found as:

$$\begin{aligned} a &= \frac{8\pi B^3}{3}\left\{ r_{\text{B}} C_{\text{s}}^2 + (C_{\text{s}} + C_{\text{f}} + C_{\text{tE}})\tau_{\text{T}} \right\}; \\ b &= \frac{B}{\beta_N}; \text{ and} \\ c &= \frac{4\pi B^3}{3}(C_{\text{s}} + C_{\text{f}} + C_{\text{tE}})^2 \end{aligned} \tag{B.16}$$

with B as the amplifier 3dB bandwidth. C_{s}, C_{f} and C_{tE} are the capacitances of the transistor and source and feedback path, respectively. The theoretical estimation of the parameters of the transistors can be derived, from the measured scattering parameters as given by the manufacturer [5] as

$$\begin{gathered}\text{For transistor type BFR90A}\\ r_{\mathrm{Eopt}} = 59\ \Omega\ \text{for} I_{\mathrm{Eopt}} = 0.44\ \mathrm{mA}\\ I_{\mathrm{eq}}^2 = 7.3 \times 10^{-16}\ \mathrm{A}^2\ \text{hence} \frac{I_{\mathrm{eq}}^2}{B} = 4.9 \times 10^{-24}\ \mathrm{A}^2\ \mathrm{Hz}^{-1}\\ I_{\mathrm{eq}} = 27\ \mathrm{nA}\ \text{thus} \frac{I_{\mathrm{eq}}}{\sqrt{B}} = 2.21\ \mathrm{pA}/\sqrt{\mathrm{Hz}}\end{gathered} \tag{B.17}$$

$$\begin{gathered}\text{For transistor type BFT24}\\ r_{\mathrm{Eopt}} = 104\ \Omega\ \text{for} I_{\mathrm{Eopt}} = 0.24\ \mathrm{mA}\\ I_{\mathrm{eq}}^2 = 79.2 \times 10^{-16}\ \mathrm{A}^2\ \text{hence} \frac{I_{\mathrm{eq}}^2}{B} = 6.1 \times 10^{-24}\ \mathrm{A}^2\ \mathrm{Hz}^{-1}\\ I_{\mathrm{eq}} = 30.2\ \mathrm{nA}\ \text{thus} \frac{I_{\mathrm{eq}}}{\sqrt{B}} = 2.47\ \mathrm{pA}/\sqrt{\mathrm{Hz}}\end{gathered} \tag{B.18}$$

Note that the equivalent noise current depends largely on some not well-defined values such as the capacitance, the transit times and the base spreading resistance and the short circuit current gain β_N. The term $\frac{I_{\mathrm{eq}}}{\sqrt{B}}$ is usually specified as the noise spectral density equivalent with reference to the input of the electronic preamplifier.

B.4 Pole-zero pattern and dynamics

An AC model of a three-stage electronic preamplifier is shown in figure B.3, and the design circuit is shown in figure B.4. As briefly mentioned above, there are three stages with feedback impedance from the output to the input. The subscripts of the resistors, capacitors and inductors indicate the order of the stages. The first stage is a special structure of shunt feedback amplification in which the shunt resistance is increased to infinity. The shunt resistance is in the order of hundreds of ohms for the required bandwidth, thus contributing to noises of the amplifier. This is not acceptable. The shunt resistance increases the pole of this stage and approaches the origin. The magnitude of this pole is reduced by the same amount as that of the forward path gain. Thus, the poles of the closed loop amplifier remain virtually unchanged. As R_{F} is increased the pole p_1 decreases but G_1 is increased. Hence $\frac{G_1 p_1}{s - p_1}$ remains constant. Thus, the position of the root locus is almost unchanged.

A compensating technique for reducing the bandwidth of the amplifier is to add capacitance across the base-collector of the first stage. This may be necessary if oscillation occurs due to the phase shift becoming unacceptable at $GH = 1$ with G being the open loop gain and H the feedback transfer function.

The second stage is a series feedback stage with feedback peaking. The capacitance C_{F2} is chosen such that at high frequencies it begins to bypass the feedback resistor. Thence the feedback admittance partially compensates for the normal high frequency drop in gain associated with the base and stray capacitances. Further, if the capacitance is chosen such that $R_{\mathrm{F2}}C_{\mathrm{F2}} = \tau_{\mathrm{r}}$ the transfer admittance and input impedance become single-pole, which would be desirable [6]. The first and

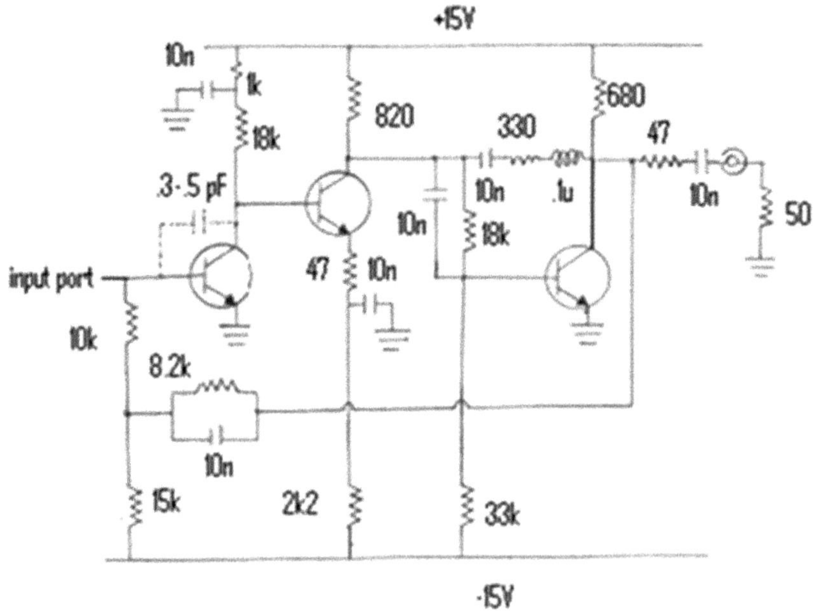

Figure B.4. Design circuit of the electronic preamplifier for the balanced optical amplifier.

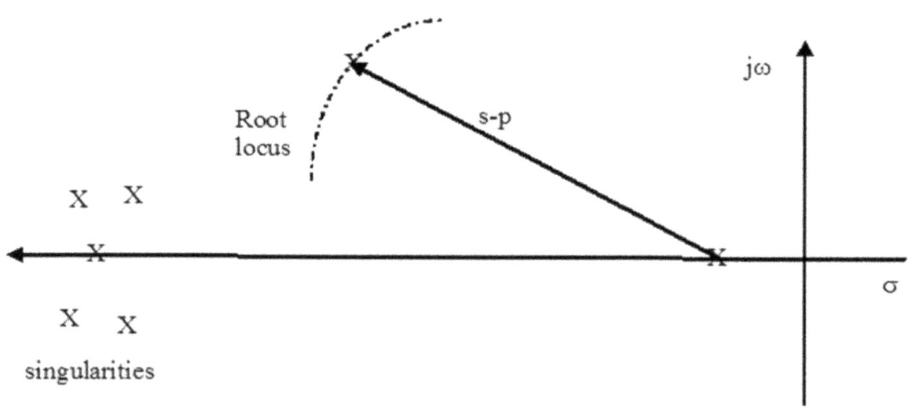

Figure B.5. Effect of first stage pole on root locus of a shunt feedback electronic amplifier. The root locus is given by $1 = (GH)_{\text{mid-band}}\sum\frac{s-z_i}{z_i}\sum\frac{p_i}{s-p_i}$.

second stages are direct-coupled (figure B.5). Figure B.5 displays the pole-zero pattern in the Laplace plane that includes the real and complex parts of the denominator and numerator of the amplifier frequency response. The root locus of such pole-zero patten determines the stability of the amplification without reaching oscillation, that is, that the dominant pole due to the feedback circuit would never reach the imaginary axis jw.

The third stage uses inductive peaking technique. For a shunt stage with a resistive load, the forward path gain has only one pole and hence there is only one

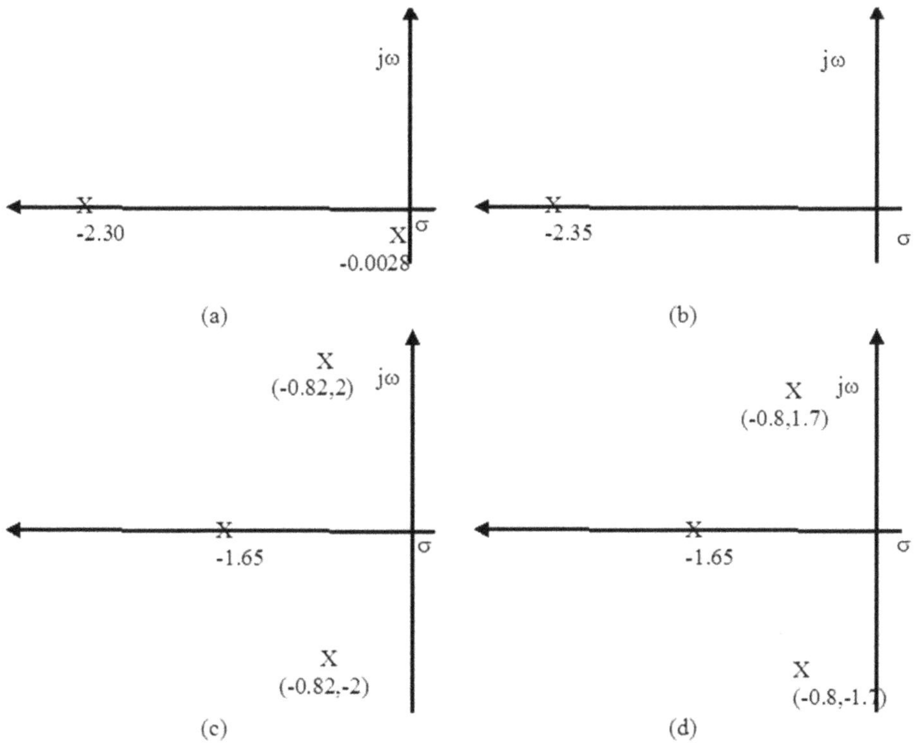

Figure B.6. High-frequency root-locus diagram (a) singularity pattern for first stage, (b) second stage (c) third stage (d) root locus for closed loop pattern for $GH = 220$.

real pole in the closed loop response. A complex pole pair can be obtained by placing a zero in Z_{F2} and hence a pole in the feedback transfer function $H(\omega)$. Figure B.6(a)–(c) shows the high frequency singularity pattern of individual stages. Figure B.6(d) shows the root locus diagram. It can be calculated that the poles take up the positions in figure B.6(e) when the loop gain GH is 220.

In the low frequency region, the loop gain has five poles and three zeroes on the negative axis, and two zeroes at the origin. The largest low frequency pole is set at approximately the input coupling capacitor and the input resistance leading to a low frequency cut-off of around tens of kHz.

The singularities of individual stages can be approximated by:

Stage 1: The dominant pole at

$$p = -\frac{1}{\beta_N R_{L1} C_{F1}} \tag{B.19}$$

and the other pole located at

$$p = -\frac{C_{F1}}{C_{F1} + C_{L1}} \frac{1}{\tau_T + C_{F1}/g_m} \tag{B.20}$$

and the mid-band gain $= \beta_N R_{L1}$ with the load of stage 1 of R_{L1}.

Stage 2: The dominant pole at

$$p = -\frac{1}{r_B G_{T2} \tau_T}; \text{ and the mid-band gain} = G_{T2} = 1/R_{F2} \qquad \text{(B.21)}$$

Stage 3: Complex pole pair at

$$|p|^2 = -\frac{R_L}{\tau_T^2 R_{F3}}; \sigma = -\frac{(R_L + R_{F3})}{2\tau_F R_{F3}}; \tau_F = \frac{L_{F3}}{R_{F3}}; \text{ and a zero at } -\frac{1}{\tau_F}; \text{ with a gain} = -R_{F3} \quad \text{(B.22)}$$

The feedback configuration of the circuit of the trans-impedance electronic preamplifier has a 10 K or 15 K resistor whose noises would contribute to the total noise current of the amplifier. The first and second stages are direct-coupled, hence eliminating the level shifter and a significant amount of stray capacitance. The peaking capacitor required for the second stage is in the order of 0.5 pF, hence no discrete component can be used. This may not require any component as the stray capacitance may suffice.

The equivalent noise current at the input of the trans-impedance amplifier is approximately proportional to the square of the capacitance at the base of the transistor. Therefore, minimization of the stray capacitance at this point must be conducted by shortening the connection lead as much as possible. As a rough guide, critical points on the circuit where stray capacitance must be minimized, are not at signal ground, and have impedance and small capacitance to ground. For instance, at the base of the output stage, the capacitance to the ground is more tolerable at this point and therefore at the collector to the ground of the second transistor, than say the collector of the last stage, which should ideally have no capacitance to ground.

An acceptable step response can be achieved by manipulating the values of changing C_{F1}, R_{F3}, L_{F3}. Since C_{F1} contributes to the capacitance at the base of the first stage, C_{F1} is minimized. The final value of C_{F1} is 0.5 pF (about half-twist of two wires), R_{F1} is 330 ohms and L_{F1} is about 0.1 μF (2 cm of wire). The amplifier is sensitive to parasitic capacitance between the feedback resistor and the first two stages. Thus, a grounded shield is placed between the 10 K resistor and the first and second stages.

The expressions for the singularities for the first two stages are similar as described above. The base-collector capacitance of the third stage affected the position of the poles considerably. The poles of the singularities of this stage are as follows: the mid-band trans-resistance is $-R_{F3}$, with a zero@$z = \frac{1}{\tau_F}$; and a complex pole pair given by

$$|p_p|^2 = \frac{1}{L_F C_F + R_F C_F \tau_T + \tau_T \tau_F \frac{R_F}{R_L}}; \sigma = \frac{1}{2}|p_p|^2\left(R_F C_F + \tau_T\left(1 + \frac{R_F}{R_L}\right)\right) \qquad \text{(B.23)}$$

In addition, there is a large pole given approximately at:

$$p = -\frac{1}{L_F C_F \tau_T |p_p|^2} \qquad \text{(B.24)}$$

Shown in figure B.7 are the open loop singularities in (a), the root locus diagram in (b) and the close loop in (c). The two large poles can be ignored without any

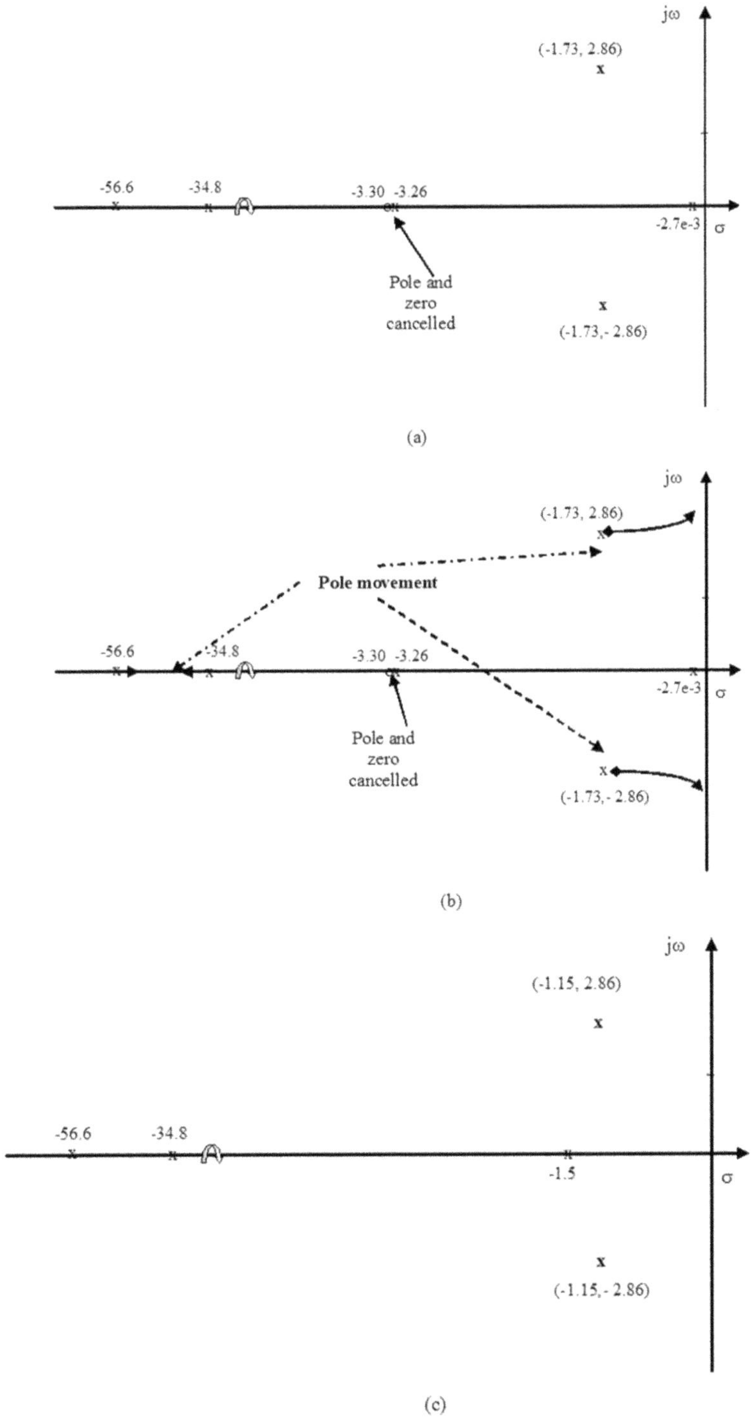

Figure B.7. Pole-zero patterns in the s-plane and their dynamics of (a) open loop; (b) root locus; and (c) closed loop.

significant difference. Similarly, for the root-locus diagram, the movement of the large poles is negligible, the pole and zero pair can be ignored and the movement of the remaining three poles can be calculated by considering just these three poles as the open loop singularities.

B.5 Responses and noises measurements

B.5.1 Rise-time and 3 dB bandwidth

Based on the pole-zero patterns given in the previous section, the step responses can be estimated and contrasted with the measured curve, as shown in figure B.8. The experimental set-up for the rise time measurement is shown in figure B.9 in the electrical domain without using a PD. An artificial current source is implemented using a series resistor with minimum stray capacitance. This testing in the electrical domain is preferred over the optical technique as the rise time of an electrical time domain reflectometer (TDR) is sufficiently short so that it does not influence the measurement of the rise time. If the optical method is used the principal problem is that we must be able to modulate the intensity of the lightwave with a very sharp step function. This is not possible when the Mach–Zehnder intensity modulator (MZIM) is used as its transfer characteristics follow a square of a cosine function. A very short pulse sequence laser such as mode-locked fiber laser can be used. However, this is not employed in this work as the bandwidth of our amplifier is not very wide and the TDR measurement is sufficient to give us reasonable value of the rise time.

The passband of the trans-impedance amplifier is also confirmed with the measurement of the frequency response by using a scattering parameter test set HP 85046A, the scattering parameter S_{21} frequency response is measured and the 3 dB passband is confirmed at 190 MHz.

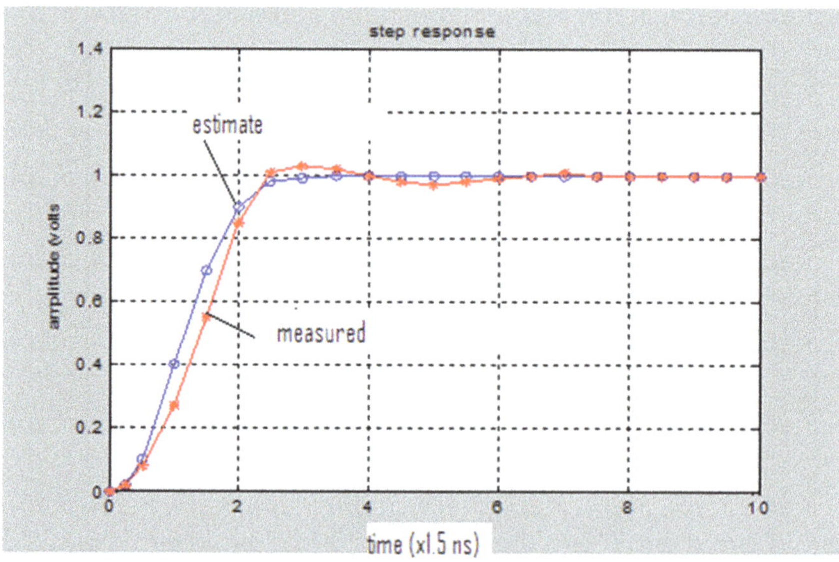

Figure B.8. Step response of the amplifier: o—estimated *—measured.

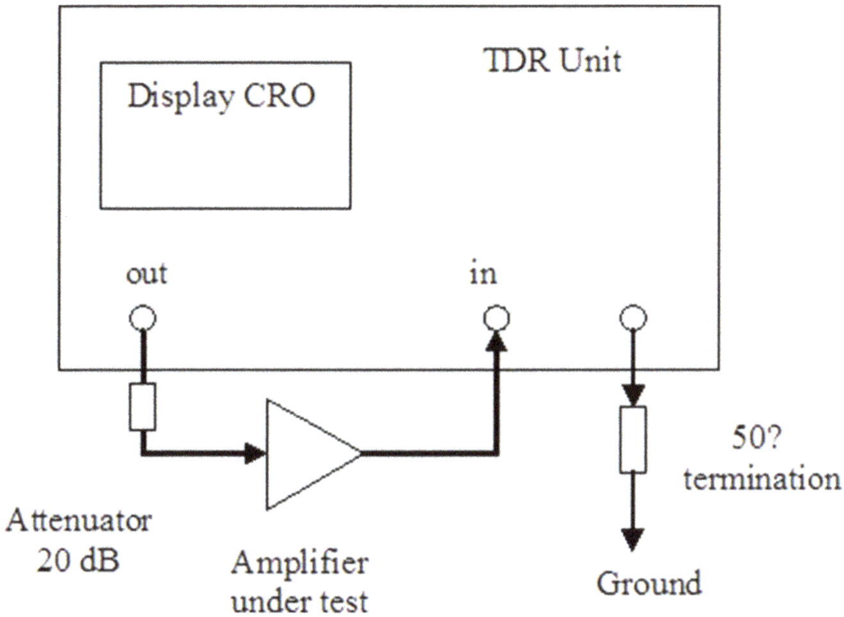

Figure B.9. Experimental set-up for measurement of the rise time in the electrical domain using a TDR.

B.5.2 Noise measurement and suppression

Two pig-tailed PDs are used and mounted with back-to-back configuration at the input of the transistor of Stage 1. A spectrum analyzer is used to measure the noise of the amplifier, as shown in figure B.10. The background noise of the analyzer was measured to be −88 dBm. The expected noise with reference to the input is 9.87×10^{-24} A^2 Hz^{-1}. When the amplifier is connected to the spectrum analyzer this noise level is increased to −85 dBm, which indicates the noise at the output of the amplifier is around −88 dBm.

Since this power is measured as 50 Ω, using a spectral bandwidth of 300 kHz, the input current noise can be estimated as:

$$\int di_{\mathrm{Neq}}^2 = \frac{v_N^2}{R_T^2} = \frac{10^{-11} \times 50}{(5k)^2} = 2 \times 10^{-17.5}\,\mathrm{A}^2 \rightarrow \frac{di_{\mathrm{Neq}}^2}{df} = \frac{2e^{-17.5}\,\mathrm{A}^2}{3 \times 10^5} = 1.06 \times 10^{-23}\,\mathrm{A}^2\,\mathrm{Hz}^{-1} \quad \text{(B.25)}$$

Thus, the measured noise is very close to the expected value.

B.5.3 Requirement for quantum limit

The noise required for near quantum limit operation can be estimated. The total shot noise with reference to the input is

$$i_{\mathrm{T-shot}}^2 = (\Re_1 + \Re_2)|E_L|^2 \quad \text{(B.26)}$$

Where $\Re_{1,2}$ is the responsivity of PDs 1 and 2, respectively. The total excess noise referred to the input is

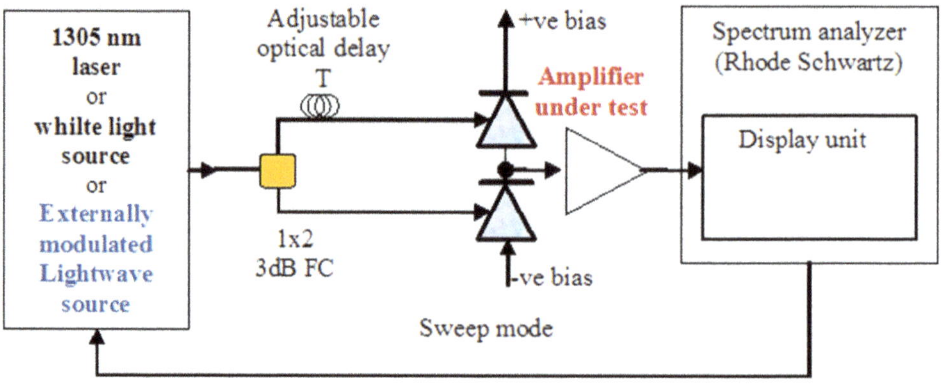

Figure B.10. Experimental set-up to measure excess noise impact and cancellation of a balanced receiver. FC = fiber coupler.

$$i^2_{\text{NT-shot}} = (\Re_1 + \Re_2)|E_L|^2; \quad i^2_{\text{N-excess}} = 2q\gamma(\omega)(\Re_1 + \Re_2)^2|E_L|^4 \tag{B.27}$$

The excess noise from each detector is correlated so γ is a function of frequency, typically between 10^4 and 10^{10} A^{-1}. A receiver will operate within 1 dB of the quantum limit if the shot noise is about 6 dB above the excess noise and the amplifier noise. The amplifier noise power spectral density is 6.09×10^{-14} A^2 Hz^{-1}. Assuming that a complete cancellation of the excess noise can be achieved, the LO required power is given as

$$i^2_{\text{NT-shot}} > 4i^2_{\text{NTeq}} \rightarrow |E_L|^2 > 0.24 \text{ mW} \approx -6.2 \text{ dBm} \tag{B.28}$$

Since there is bound to be excess noise due imperfection of the balancing of the two detectors, the power of the LO is expected to be slightly greater than this value.

The output voltage of the amplifier can be monitored so noise contributed by different processes can be measured. When the detectors are not illuminated we have

$$v_0^2 = Z_T^2 i^2_{\text{Neq}} \tag{B.29}$$

where Z_T is the trans-impedance and v_o is the output voltage of the amplifier. It is known that incoherent white light is free from excess intensity noise, so when the detectors are illuminated with white light sources we have

$$v_0^2 = Z_T^2\left(i^2_{\text{Neq}} + i^2_{\text{N-shot}}\right) \tag{B.30}$$

When the LO is turned on illuminating the detector pair then

$$v_0^2 = Z_T^2\left(i^2_{\text{Neq}} + i^2_{\text{N-shot}} + i^2_{\text{N-excess}}\right) \tag{B.31}$$

B.5.4 Excess noise cancellation technique

The presence of un-cancelled excess noise can be shown by observing the output voltage of the electronic preamplifier using a spectrum analyzer. First, the detectors are illuminated with incoherent white light and then the output noise is measured.

Then the detectors are illuminated with the laser LO at a power level equal to that of the white light source, thus the shot noise will be at the same magnitude. Thus, the increase in the output noise will be due to the un-cancelled intensity noise. This, however, does not demonstrate the cancellation of the LO intensity noise.

A well-known method for this cancellation is to bias the LO just above its threshold [7]. This causes the relaxation resonance frequency to move over the passband. Thus, when the outputs of the single or balanced receivers are compared, the resonance peak is observed for the former case and not in the latter case in which the peak is suppressed. Since the current is only in the μA scale in each detector for an LO operating in the region near the threshold, the dominant noise is that of the electronic amplifier. This method is thus not suitable here. However, the relaxation resonance frequency is typically between 1 and 10 GHz. A better method can be developed by modulating the LO with a sinusoidal source and measuring the difference between single and dual photodetection receivers. The modulation can be implemented at either one of the frequencies to observe the cancellation of the noise peak power or by sweeping the LO over the bandwidth of the amplifier so that the cancellation of the amplifier can be measured. For fair comparison, the LO power will be twice of that of the single detector case so the same received power of both cases can be almost identical and hence the signal-to-noise-ratio (SNR) at the output of the amplifier can be derived and compared.

B.5.5 Excess noise measurement

An indication of excess noise and its cancellation within 0–200 MHz frequency range by the experimental set-up can be observed as shown in figure B.10. A laser of 1305 nm wavelength is used with an output power of −20 dBm employed as the LO so that the dominant noise is that of the electronic trans-impedance preamplifier as explained above. The LO is intensity modulated by the sweep generator from the spectrum analyzer of 3 GHz frequency range (Rhode Schwartz model). Noise suppression pattern of the same trend can be observed with suppression about every 28 MHz interval over 200 MHz, the entire range of the trans-impedance amplifier.

Figure B.11 shows the noise spectral density for the case of single and dual detectors. Note that the optical paths have different lengths. This path difference is used for cancellation of the excess noise as described in references [8, 9]. It can be shown that the optical delay line leads to a cancellation of noise following a relationship of $\{1 - \cos(2\pi fT)\}$ with T as the delay time of the optical path of one of the fibers from the fiber coupler to the detector. Hence the maxima of the cancellation occur at $f = m/T$ where m is an integer. From figure B.10 we can deduce the optical path length as $f = 30$ MHz or $T = 31.7$ ns, thus $d = cT/n_{\text{eff}} = 6.8$ m with n_{eff} (~1.51 @1550 nm) being the effective refractive index of the guided mode in the fiber. The total length of the fiber delay is equivalent to about 6.95 m in which the effective refractive index of the propagation mode is estimated to be 1.482 at the measured wavelength. Thus, there is a small discrepancy, which can be accounted for by the uncertainty of the exact value of the effective refractive index of the fiber.

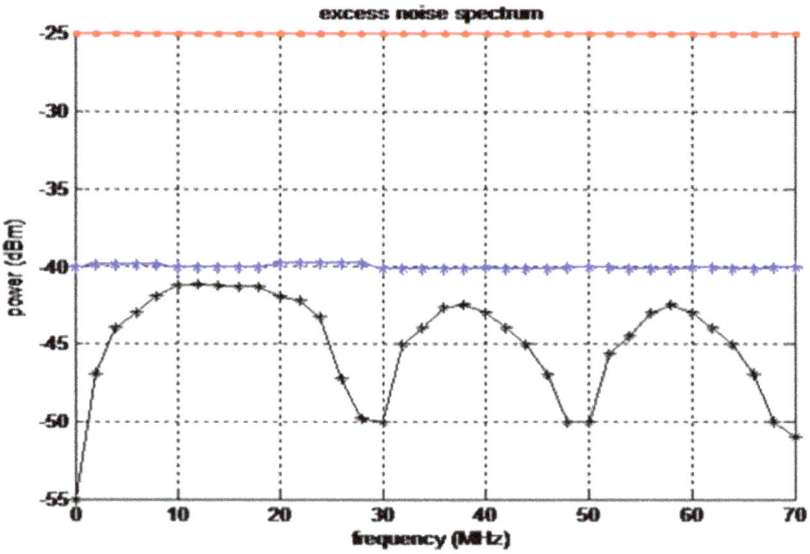

Figure B.11. Excess noise for single detector and dual detectors with noise cancellation using delay and filter at the input port of the preamplifier.

Using this method, Abbas *et al* [1] achieved a cancellation of 3.5 dB, while 8 dB is obtained in our set-up. This discrepancy can be accounted for by a combination of (i) the photodiode pair is to be indentical and (ii) the signal test paths must be the same so as to minimize the cable impedance effects added to the system transfer function.

In our initial measurements we have observed some discrepancies which are due to: (i) unmatched properties of the photodetection of the balanced detector pair; (ii) the delay path having three extra optical connections and therefore more loss, so even when there is no delay the excess noise would not be completely cancelled. Thus, at some stage the longer path was implemented with a tunable optical delay path so as to match the delay path to the null frequency of the RF wave to achieve cancellation of the excess noise. This noise cancellation can lead to an improvement of 16 dB in the optical SNR (OSNR) due to the dual-detector configuration and thus an improvement of near 100 km length of standard single-mode fiber.

The noise cancellation is periodic following a relationship with respect to the delay time T imposed by the optical path of the interferometer T as $(1 - \cos 2\pi fT)$. The optical delay path difference is estimated by adjusting the optical delay line, which is implemented by a fiber path coupled out of a fiber end to open air path and back to another fiber path via a pair of Selfoc lenses.

It is noted that the LO light source can be directly modulated (DD) with a sweep sinusoidal signal derived from the electrical spectrum analyzer (ESA). This light source has also been externally modulated using an MZIM of bandwidth much larger than that of the electronic amplifier (about 25 GHz). The excess noise characteristics obtained using both types of sources are almost identical.

Furthermore, with the excess noise reduction of 8 dB and the signal power for the balanced receiver operating in a push–pull mode then the received electrical current would be double that of a single detector receiver. That means a gain of 3 dB in the SNR by this balanced receiver.

B.6 Remarks

In this appendix, we consider the design, implementation of a balanced receiver using a dual detector configuration and a BJT front end preamplifier. In view of recent growing interest in coherent optical transmission systems with electronic digital signal processing [9–11] a noise-suppressed balanced receiver operating in the multi-GHz bandwidth range would be essential. Noise cancellation using optical delay line in one of the detection paths leads to suppression of excess noise in the receiver.

In this section we demonstrate, as an example, by design analysis and implementation, a discrete wideband optical amplifier with a bandwidth of around 190 MHz and a total input noise equivalent spectral density of 10^{-23} A^2 Hz^{-1}. This agrees well with the predicted value using a noise model analysis. It is shown that with sensible construction of the discrete amplifier, minimizing the effects of stray capacitances and appropriate application of compensation techniques, a bandwidth of 190 MHz can be easily obtained for transistors with a few GHz transition frequencies. Further, by using an optimum emitter current for the first stage and minimizing the biasing and feedback resistance and capacitance at the input, a low noise amplifier with an equivalent noise current of about 1.0 $pA/\sqrt{Hz}$ is achieved. Furthermore, the excess noise cancellation property of the receiver is found to give maximum SNR of 8 dB with a matching of the two PDs.

Although the bandwidth of the electronic amplifier reported here is only 190 MHz which is about 1% of the 100 Gb s^{-1} target bit rate for modern optical Ethernet, the bandwidth achieved by our amplifier has reached around the maximum region that can be achieved with discrete transistor stages whose transition frequency is only 5 GHz. The amplifier configuration reported here can be scaled up to multi-GHz region for 100 Gb s^{-1} receiver using an integrated electronic amplification device without much difficulty. We thus believe that the design procedures for determining the pole and zero patterns on the s-plane and their dynamics for stability consideration are essential so that any readers who are interested in its implementation using monolithic microwave integrated circuit technology can use them. Should the amplifier be implemented in the multi-GHz range then the microwave design technique employing noise figure method may be most useful when incorporating the design methodology reported in this section.

B.7 Noise equations

Referring to the small signal and noise model given in figures B.12 and B.13, the spectral density of noise due to collector bias current I_C and small signal transconductance g_m and base conductance g_B generated in a BJT can be expressed as:

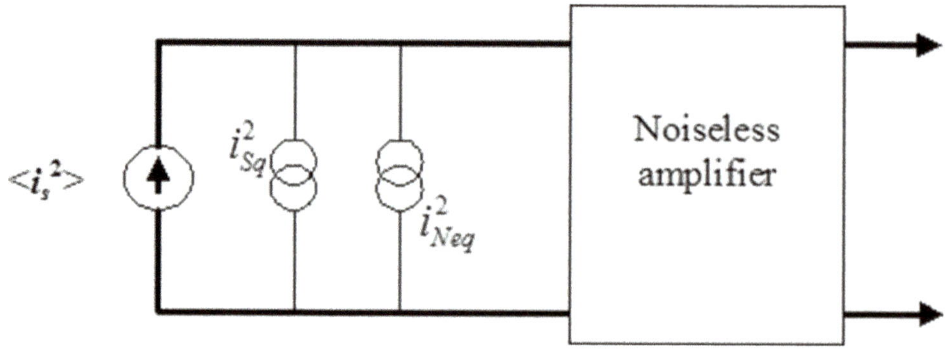

Figure B.12. Equivalent noise current at the input and noiseless amplifier model. i_{Sq}^2 is the quantum shot noise, which is signal dependent; i_{Neq}^2 is the total equivalent noise current with reference to the input of the electronic amplifier.

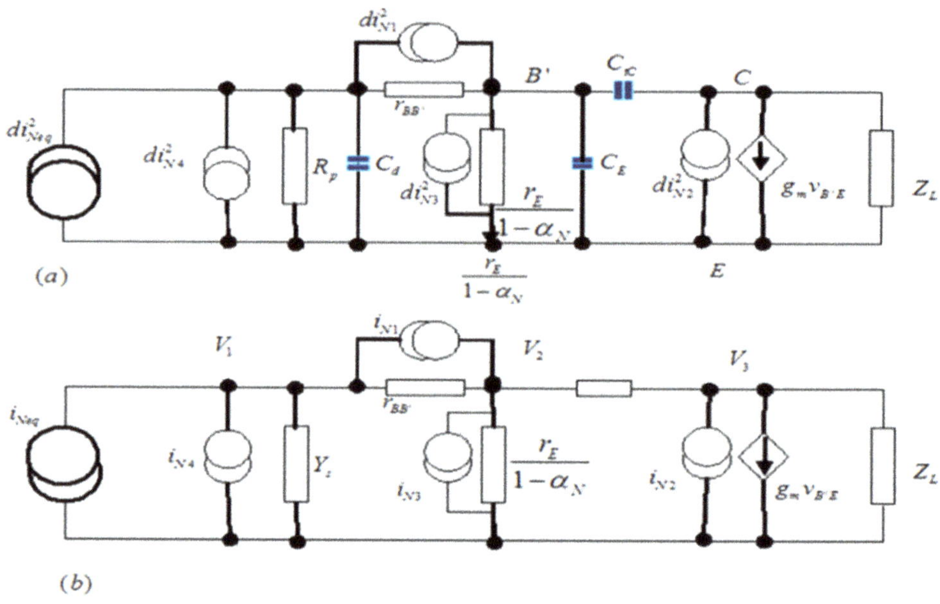

Figure B.13. (a) Approximated noise equivalent and small signal model of a BJT front end; (b) generalized noise and small signal model circuit. Note that $r_B = r_{sd} + r_{BB'}$ and $C_d = C_p + C_i$ with $r_{BB'}$ being the base resistance, r_{sd} the diode resistance, C_d the photodiode capacitance and C_i the input capacitance.

$$\begin{aligned} di_1^2 &= 4k_B T g_B df;\; di_2^2 = 2qI_c df + 2k_B T g_m df \\ di_3^2 &= 2qI_B df + 2k_B T(1-\alpha_N) g_m df;\text{ and} \\ di_4^2 &= \sum \text{shot noise of diodes and thermal noise of bias in } g_B \text{ resistors} \end{aligned} \tag{B.32}$$

From nodal analysis of the small signal equivalent circuit given in figure B.13 we can obtain the relationship

$$\begin{bmatrix} Y_s + g_B & -g_B & 0 \\ -g_B & g_B + y_1 + y_f & -y_f \\ 0 & g_m - y_f & y_f + y_2 \end{bmatrix} \begin{bmatrix} V_1 \\ V_2 \\ V_3 \end{bmatrix} = \begin{bmatrix} i_{eq} - i_{N1} \\ i_{N1} + i_{N3} \\ i_{N2} \end{bmatrix} \tag{B.33}$$

Hence, V_3, V_2, V_1 can be found by using Euler's rule for the matrix relationship

$$V_3 = \frac{\Delta_{13}}{\Delta}\left(i_{eq} - i_{N1}\right) + \frac{\Delta_{23}}{\Delta}\left(i_{N1} + i_{N3}\right) + \frac{\Delta_{33}}{\Delta}(i_{N2}) \tag{B.34}$$

Hence the noise currents with reference to the input are

$$d(i_1'^2) = \left|\frac{Y_s}{g_B}\right|^2 dI_1^2 = \omega_2 C_s^2 r_B 4kT df \tag{B.35}$$

$$\begin{aligned} d(i_2'^2) &= \left|(Y + y_1 + y_f) + \frac{Y_s}{g_B}(y_1 + y_f)\right|^2 \left|\frac{1}{y_f - g_m}\right|^2 = \left|\frac{1}{y_f - g_m}\right|^2 dI_2^2 \\ &= \left(\frac{1}{\beta_N r_E^2} + \omega^2\left(C_0^2 - \frac{2C_B r_B}{\beta_N r_E}\right) + \omega^4 2C_s r_B C_B^2\right) 2kTr_E df \end{aligned} \tag{B.36}$$

and

$$d(i_3'^2) = \left|\frac{Y_s}{g_B} + 1\right|^2 dI_3^2 = \left(\omega C_s^2 r_B^2 + 1\right)\frac{2kT}{\beta r_E} df \tag{B.37}$$

in which we have assumed that

$$\begin{aligned} &\omega C_f \ll g_m \\ &C_0 = C_s + C_E + C_f + \frac{C_s r_s}{\beta r_E} \approx C_s + C_{tE} + C_f + \frac{\tau_t}{r_E} = C_x + \frac{\tau_t}{r_E} \\ &\text{with } C_x = C_E + C_f;\; C_E = C_{tE} + \frac{\tau_t}{r_E} \end{aligned} \tag{B.38}$$

The total noise spectral density with reference to the input of the amplifier is shown in figures B.12 and B.13 and is thus given as

$$\frac{di_{Neq}^2}{df^2} = \frac{di_{1eq}^2}{df^2} + \frac{di_{2eq}^2}{df^2} + \frac{di_{3eq}^2}{df^2} = 2k_B T\left[\frac{1}{\beta r_E}\frac{\beta+1}{\beta} + \omega^2\left(\frac{2}{r_B} + \frac{1}{\beta r_E}\right) + C_0 r_E - \frac{2C_s r_B C_x}{\beta} + \omega^2 C_s^2 r_B^2 C_x^2 r_E\right] \tag{B.39}$$

Thence the total noise power with reference to the input is given as:

$$\begin{aligned}
i_{\mathrm{Neq}}^2 &= \int_0^B di_{\mathrm{Neq}}^2 = a + \frac{b}{r_{\mathrm{E}}} + cr_{\mathrm{E}};\ \text{with} \\
a &\approx 2k_{\mathrm{B}}T\left[\frac{8\pi^2B^3}{3}\left(C_{\mathrm{s}}r_{\mathrm{B}} + C_x\tau_t + \frac{C_{\mathrm{s}}^2r_{\mathrm{B}}^2}{\beta}\right) + \frac{(2\pi)^4}{5}B^5C_{\mathrm{s}}^2r_{\mathrm{B}}^22C_y\tau_t\right] \\
b &\approx 2k_{\mathrm{B}}T\left[\frac{B}{\beta} + \frac{4\pi^2B^3}{3}\left(\frac{C_{\mathrm{s}}r_{\mathrm{B}}^2}{\beta} + \tau_{\mathrm{t}}^2\right) + \frac{16\pi^4B^5}{5}C_{\mathrm{s}}^2r_{\mathrm{B}}^2\tau_t^2\right] \\
c &\approx 2k_{\mathrm{B}}T\left[\frac{4\pi^2B^3C_x^2}{3} + \frac{16\pi^4B^5}{5}C_{\mathrm{s}}^2r_{\mathrm{B}}^2C_y^2\right]\text{with } C_y = C_{\mathrm{tE}} + \frac{\tau_{\mathrm{t}}}{r_{\mathrm{E}}}
\end{aligned} \tag{B.40}$$

where the first term in each coefficient a, b, and c is dominant for a bandwidth of 190 MHz of the two transistor types Phillip BFR90A and BFT24 under considerations.

If the dependence of C_{E} on I_{E} is ignored then (B.40) becomes

$$i_{\mathrm{Neq}}^2 = 2k_{\mathrm{B}}T\left[\frac{1}{r_{\mathrm{E}}}\frac{B}{\beta} + r_{\mathrm{E}}\left(\frac{4\pi^2B^3}{3}C_0^2\right) + \frac{16\pi^4B^5}{5}C_{\mathrm{s}}^2r_{\mathrm{B}}^2C_x^2 + \frac{8\pi^2B^3C_{\mathrm{s}}^2r_{\mathrm{B}}}{3}\right] \tag{B.41}$$

For the transistor BFR90A the optimum emitter resistance is $r_{\mathrm{Eopt}} = 49\ \Omega$ and the total equivalent noise power with reference to the input is $4.3 \times 10^{-16}\ \mathrm{A}^2$, which is moderately high.

Clearly, the noise power is a cubic dependence on the bandwidth of the amplifier. This is thus very critical for an ultra-wideband optical receiver. Therefore, it is very important to suppress the excess noise of the optical receiver due to the quantum shot noise of the LO.

References

[1] Abbas G L, Chan V and Yee T K 1985 A dual-detector optical heterodyne receiver for local oscillator noise suppression *IEEE J. Lightwave Technol.* **LT-3** 1110–22

[2] Kasper B, Burns C, Talman J and Hall K 1986 Balanced dual detector receiver for optical heterodyne communication at Gb/s rate *Electron. Lett.* **22** 413–4

[3] Alexander S 1987 Design of wide-band optical heterodyne balanced mixer receivers *IEEE J. Lightwave Technol.* **LT-5** 523–37

[4] Binh L N 2008 *Digital Optical Communications* (Boca Raton, FL: CRC Press) ch 4

[5] Philips Semiconductors Data handbook part 10 1981, Wideband Transistors and Wideband Hybrid IC Modules by Philips Electronic Components and materials, 1981, open source, Internet Archive, Identifier: s-10-1981-wideband-trs-n-hybrid-modules, ark:/13960/t8ch2gw8

[6] Cherry E M and Hooper D E 1968 *Amplifying and Low Pass Amplifier Design* (New York: Wiley) chs 4 and 8

[7] Alexander S 1987 Design of wideband optical heterodyne balanced mixer *IEEE J. Lightwave Technol.* **LT-5** 523–37

[8] Sterlin R, Battiig R, Henchoz P and Weber H 1986 Excess noise suppression in a fiber optic balanced heterodyne detection system *Opt. Quantum Electron.* **18** 445–54
[9] Kahn J and Ip E 2009 Principles of digital coherent receivers for optical communications *Paper OTuG5, Proc. OFC 2009 (San Diego, CA)* March 2009
[10] Zhang C, Mori Y, Igarashi K, Katoh K and Kikuchi K 2009 Demodulation of 1.28-Tbit/s polarization-multiplexed 16-QAM signals on a single carrier with digital coherent receiver *Paper OTuG3, Proc. OFC 2009 (San Diego USA)* March 2009
[11] Binh L N 2008 *Digital Optical Communications* (Boca Raton, FL: CRC Press/Taylor & Francis Group) ch 11

IOP Publishing

Transmission, Processing, and All-Optical Routing for Ultra-High Capacity Data Center Networking (Second Edition)

Le Nguyen Binh

Appendix C

Power budget

C.1 Overview

Power budget is the counterpart of the rise time budget to ensure that the signals entering the reception sub-system have sufficient energy for detection that satisfies a certain criterion for error-free reception. The two scenarios of distinct reception sensitivities of a communication receiver are when forward error coding (FEC) is employed or not. If no FEC is employed then for a Gaussian pdf (probability density function), the bit error rate (BER) is required to reach 10^{-9} or the amplitude ratio of the average levels between the '1' and '0' must be at least six times the total standard deviation of the noise levels of the '1' and '0'. With FEC then this BER can be lowered to about 10^{-3} or thereabouts depending on the error coding method.

Thus, the following steps must be employed to determine the power budget:

- Identify the modulation format of the signals and hence the Euclidean distance between the states. Thence the required ratio between the energy level and the noise power for a specified BER can be determined.
- Identify the probability density function of the transmission and detection processes, hence the noises. Then identify the noise sources contributed to the modulated and transmitted signals or signal levels.
- Obtain the attenuation factor, e.g. dB km^{-1} in fiber and any noise sources superimposing on the signals during transmission, e.g. the amplification spontaneous emission (ASE) noise of inline optical amplifiers.
- Then determine the required receiver sensitivity of the reception systems, which is defined as the minimum power level required of the signals to be available at the input of the receiver at a specified BER.

doi:10.1088/978-0-7503-5849-1ch10

Work back, using a linear log scale diagram to estimate the required signal power to be launched at the output of the transmitter in order to achieve this reception-receiving sensitivity.

The above steps can be done in our head without resorting to computing devices, or even a simple calculator. This can be done by converting all power and attenuation in log scale or in dB level, then all estimations are just additions and subtractions.

C.2 Power budget estimation example

As an example, consider the transmission system given in figure C.1. The modulation is assumed to be binary on–off keying (OOK), the transmission link can be a multi-span standard single-mode fiber (SSMF) optically amplified type and the reception system is based on direct detection technique with a photodetector (PD) and followed by a trans-impedance amplifier (TIA) and automatic gain control (AGC) if required.

Under a binary OOK modulation scheme we can estimate the required signal-to-noise ratio (SNR) to be 6 for a BER of 1e−9. With some uncertainty in the transmission system, then a BER =1e−12 can be used leading to an SNR of 7 (the *Q*-factor in linear scale) is required. Figure C.2 depicts the required SNR for BER values over 10 decades. That is, the *Q*-factor increases by one unit required for a decade of BER for a Gaussian pdf detection process. Thus corresponding to *Q* of 6 and 7 the optical SNR (OSNR) can be determined to be $10\log_{10}6 \sim 15$ dB.

So for an optical receiver, how do we determine the noise power at its input port, i.e. at the front of the PD. That means that we have to estimate all noise equivalent power to the input port of the receiver, or alternatively we find all equivalent noise currents with reference to the input port of the receiver, as shown in figure C.3, in which we can identify: (a) $\langle i_s^2 \rangle$ represents average input current as converted by the photodetection process of the input signal; (b) total equivalent noise current i_{Neq}^2 as

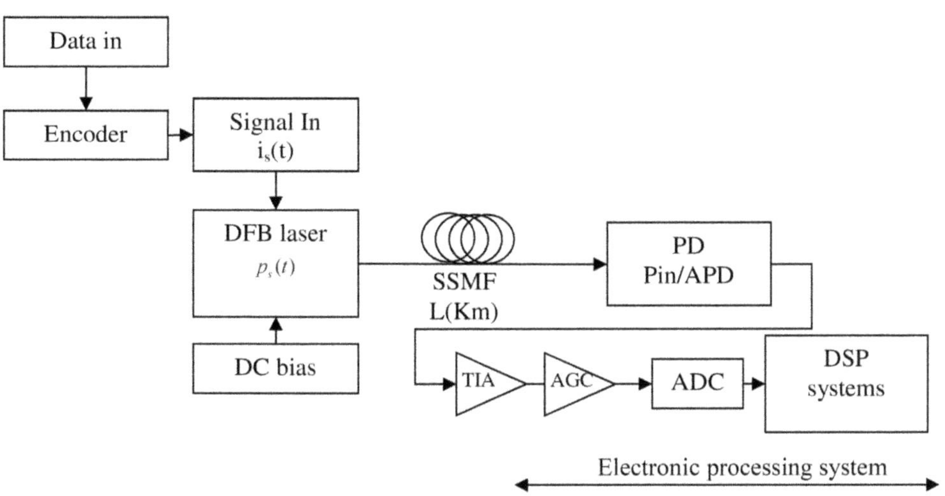

Figure C.1. Intensity modulation direct detection transmission scheme.

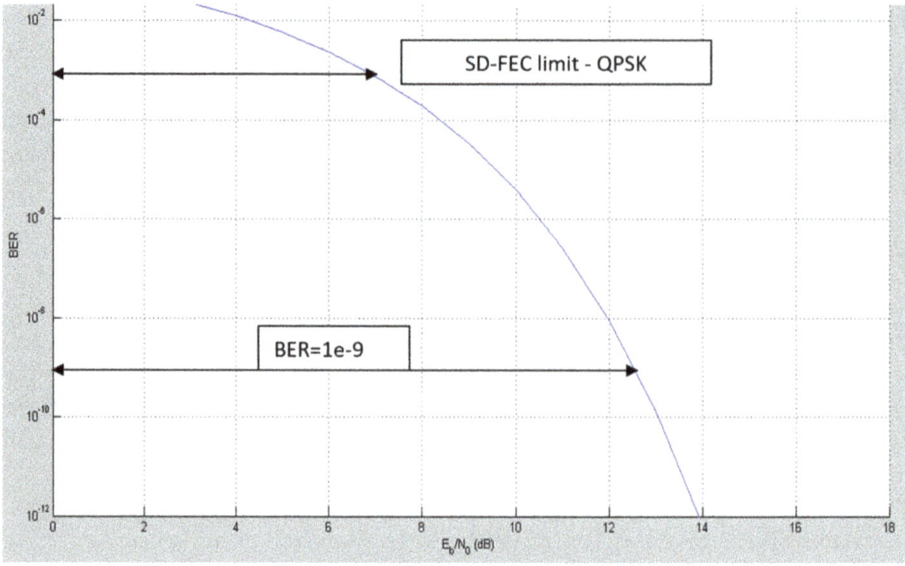

Figure C.2. Theoretical estimation for BER versus SNR for binary OOK of Gaussian noisy system. SD-FEC = software defined forward error coding; BER = bit error rate.

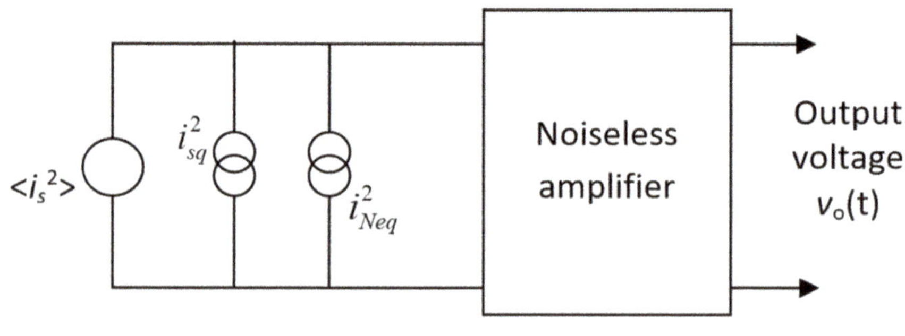

Figure C.3. Equivalent current model with reference to the input of the optical receiver, average signal current and equivalent noise current of the electronic preamplifier as seen from its input port.

referred to the input of the electronic preamplifier followed by the PD[1]; and (c) total noise equivalent, i_{sq}^2, generated during the opto-electronic conversion such as the quantum shot noise by the signal input and the dark current noise of the PD.

[1] Commercial amplifiers normally specify the noise characteristics as the total equivalent noise spectral density with reference to the input in units of A/sqrt(Hz). This parameter is normally measured by connecting the output of the amplifier to a spectrum analyzer with no signals input. The output is the integrated over the bandwidth of the amplifier to obtain the total noise voltage at the output. In the case of a TIA, this output voltage noise is then transferred back to the input by a simple division. Hence we can get the total equivalent noise current with reference to the input. This noise current with reference to the input can be converted to equivalent optical power by using the responsivity of the PD. From this value and the OSNR we can justify whether the transmission link would operate to give the required BER.

It is noted that under direct detection the amplifier noise current normally dominates over the quantum shot noises, while this is opposite in the case of coherent detection in which the quantum shot noise is due to the high average power of the coherent sources.

Now reverting back to the optical fiber transmission link, we have for direct modulation and biased at current $I_b = 5I_{th}$ with a magnitude of $4.5I_{th}$ then an average optical power at the output of the distributed feedback (DFB) laser can reach 10 dBm. But for the '0' we would have to drive with a magnitude of about $1.5I_{th}$. So we expect noisy signals for signals at the base level, a 'high' level, and we expect a distinction ratio of more than 20 dB. Under this driving condition let assume that the average power of 0 dBm can be obtained at the output of the DFB laser.

The attenuation and power requirements for each sub-system of the link can now be estimated as:

- The fiber loss for 80 and 100 km SSMF would be about 20–25 dB.
- Using a p–i–n or avalanche photodetector (APD) (avalanched gain of 10) and a differential TIA (Inphi model 3250A) with a noise spectral density with reference to the input of $(\frac{di_N^2}{df})^{1/2} = 20 - 40 \ \text{pA}/\sqrt{\text{Hz}}$, we would have −7 dBm and −17 dBm, respectively, for a required OSNR of 15 dB and BER of 1e−9 to 1e−12 (δ or Q-factor of 6 or 7 under Gaussian assumption of pdf).
- This indicates that an APD is preferred as the PD so that we can avoid the use of an optical amplifier. The projection variation of transmission distance with respect to the launched power at the output of the transmitter is shown in figure C.4.
- With the p–i–n PD, we expect to transmit 25G non-return-to-zero (NRZ) channel over only 40 km with a launched power of 4 dBm. However, with the APD of a gain of 10, then the transmission distance limited by the power budget can reach over 80 km on condition that the pulse sequence is equalized at the digital signal processor (DSP)-based receiver or a single sideband (SSB) modulation scheme is employed such as the duo-binary modulation format.

C.3 SNR and optical SNR

We refer to figure C.3, representing the total noise with reference to the input and the generated signal current from the input optical signals. It must be noted that superimposing on the signals there is also the noise commonly known as the quantum shot noise, which is signal-dependent and is dependent on the signal average current whose spectral density is given by:

$$S_{NQ} = 2q\langle i_S \rangle \text{ in A}^2/\text{Hz} \tag{C.1}$$

Therefore, by integrating this noise spectral density given in equation (C.1), over the 3 dB bandwidth of the system, thence taking the square root to give the noise current i_{NQ}. Furthermore, there are noise currents coming from the dark current of the photodetector i_{Nd}, that is, the electronic current in the semiconductor junction at

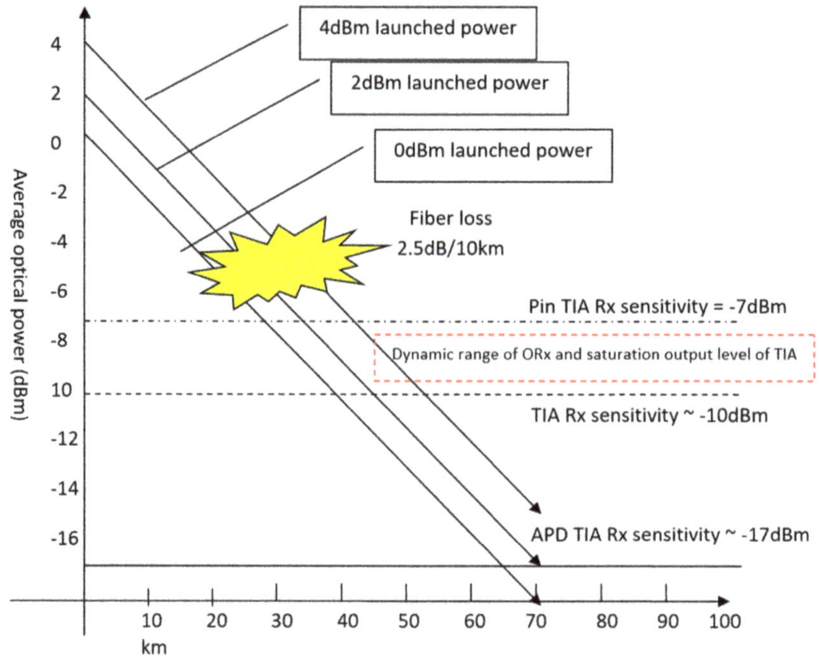

Figure C.4. Power budget with an average launched power of 0, 2 and 4 dBm with receiver sensitivity of −7 dBm and −17 dBm and fiber loss of 2.5 dB/10 km.

temperature above the absolute reference temperature. So let the total noise current with reference to the input of an optical receiver be written as:

$$i_{\text{NT}} = i_{\text{NA}} + i_{\text{NQ}} + i_{\text{Nd}} \tag{C.2}$$

Thus the SNR, in terms of power, can be formed as

$$\text{SNR} = \frac{\langle i_S^2 \rangle}{i_{\text{NT}}^2} \tag{C.3}$$

The BER can be estimated for amplitude shift keying (ASK) binary signals under Gaussian noise assumptions as

$$\text{BER} = \frac{1}{2}\text{ercf}\left(\frac{Q}{\sqrt{2}}\right) \text{ where } Q = \frac{I_1 - I_0}{\sigma_1 - \sigma_0} \tag{C.4}$$

with I_1, I_0 as the average current detected by the PD for the '1' and '0', respectively, and σ_1, σ_0 the standard deviation of the noise levels due to the '1' and '0', respectively. If the modulation format is quadrature amplitude modulation (QAM) then the errors of detection can be considered as the summation of the two orthogonal 'binary' signals, the complex and real components. The BER can be estimated for both directions, as given in (C.4).

Thus, we could see that by estimating the total current noise with reference to the input of the electronic preamplifier, normally a TIA, and then the dark current and the signal-dependent signal current noise with a typical average optical power at the input, we can obtain a rough estimate of the SNR and the optical OSNR. Note that the estimation of the noise should be considered for the dominating noise term.

Naturally the electronic current at the output of the preamplifier of the receiver would normally not be sufficient for the ADC and signal processor, so a voltage gain amplifier with or without an AGC stage, see figure C.1, would commonly be employed to boost the signal level. The noise with reference to the input of this voltage gain stage with reference to the input of the optical receiver is normally much smaller than that of the preamplifier, so it can be ignored.

Other modulation can be derived using the principles of superposition of probability of errors contributed by the considered constellation point by neighboring points.

C.4 TIA: differential and non-differential types

The TIA is the critical amplifying device which is to be connected in cascade with a high-speed PD to provide the preamplification of the current produced in the PD to the voltage output whose level would be appropriate for an analog-to-digital conversion so that digitized signals can then be processed in the digital domain by a DSP. The total equivalent noise with reference to the input port of this amplifier is very critical on the impact of the SNR of the optical receiver. The method for calculation of noise processes and noise currents as well as the total equivalent noise with reference to the input are given in appendix B.

Traditionally single-input or non-differential input TIA has been extensively exploited over the years [1]. However, currently differential TIAs (DTIAs) offer wider bandwidth and high differential trans-impedance gain of around 3000–5000 Ω over 30–40 GHz bandwidth by Si–Ge technology whose transition frequency f_T can reach 280 GHz[2] [2, 3]. However, these DTIAs [4] are limited in their dynamic range and normally an AGC stage would be required. A typical set-up of PD pair and a DTIA is shown in figure C.5. Typical circuit diagrams of the DTIA employing Si–Ge and InP are shown in figures C.6(a) and (b), respectively, in which there are the main differential long-tail pair followed by two common collector stages connected to the differential outputs of this long-tail pair. The outputs of these stages are then fed back by a shunt feedback to obtain the highest transfer-impedance and wideband properties. The outputs are fed back to the differential inputs. These differential ports are then fed into a further differential voltage gain stage (without any feedback stage).

The DTIA can offer a transfer impedance of 4000–5000 Ω. Thus, with an optical signal average power input of 0 dBm (or 1 mW) and a responsivity of the PD of 0.9–1.0 A W^{-1} a current of 1 mA is produced and fed into the DTIA. This in turn produces a differential voltage at the output of 4–5 V signals. Unfortunately, this level will saturate the outputs of the DTIA. So normally an automatic gain control stage would be employed to increase the dynamic range of the optical receiver. The

[2] The transition frequency f_T is defined as the frequency at which the short circuit current gain of a transistor equals unity.

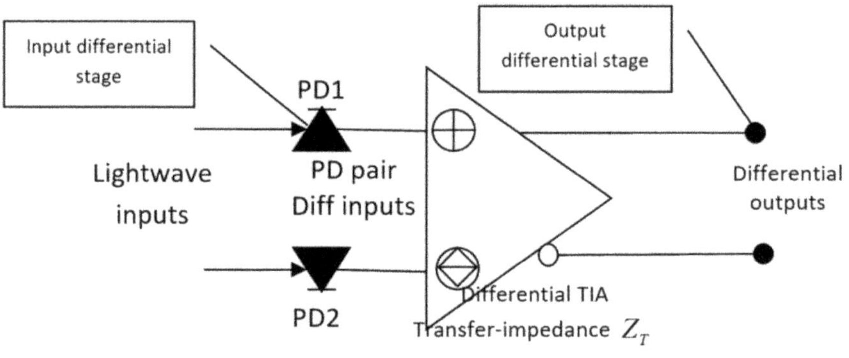

Figure C.5. DTIA and PD pair.

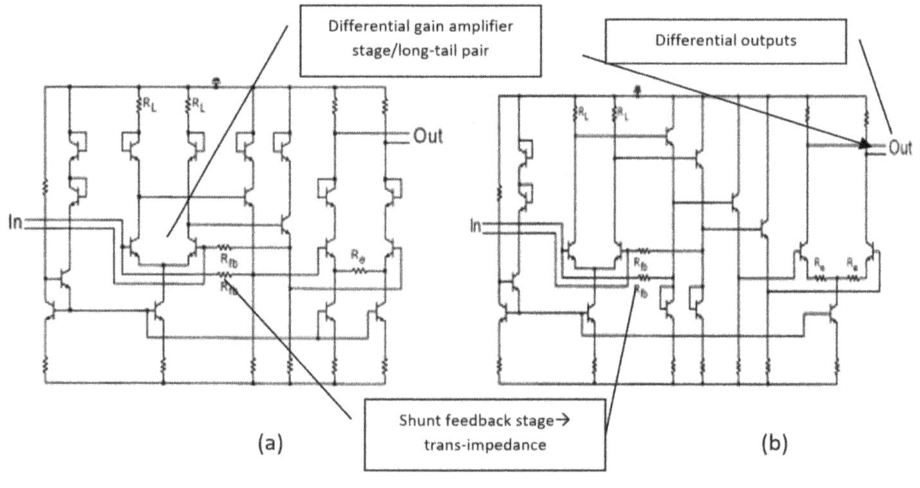

Figure C.6. Typical circuit topologies of DTIA by Si–Ge and InP technology.

parameters of diff TIA can be with the input-referred current noise of the InP TIA is approximately $25\text{pA}/\sqrt{\text{Hz}}$ on average to a bandwidth of 40 GHz.

References

[1] van Muoi T 1984 Receiver design for high speed optical fiber systems *IEEE J. Lightwave Technol.* **2** 243–66

[2] Kempf P and Racanelli M 2002 Silicon germanium BiCMOS technology *IEEE GaAs IC Symp. Dig.* Oct. 2002 pp 3–6

[3] Jagannathan B *et al* 2002 Self-aligned site NPN transistors with 285 GHz_{fand} 207 GHz f_T in a manufacturable technology, *IEEE Electron Device Lett.* **23** 258–60

[4] Weiner J S *et al* 2003 SiGe differential trans-impedance amplifier with 50-GHz bandwidth *IEEE J. Solid-State Circuits* **38** 1512

www.ingramcontent.com/pod-product-compliance
Ingram Content Group UK Ltd.
Pitfield, Milton Keynes, MK11 3LW, UK
UKHW051904100926
8556IPUK00005B/761

9 780750 358477